MOLECULAR APPROACHES TO EUCARYOTIC GENETIC SYSTEMS

ACADEMIC PRESS RAPID MANUSCRIPT REPRODUCTION

ICN–UCLA Symposia on Molecular and Cellular Biology
Vol. VIII, 1977

MOLECULAR APPROACHES TO EUCARYOTIC GENETIC SYSTEMS

edited by

GARY WILCOX
Department of Bacteriology
and Molecular Biology Institute
University of California, Los Angeles
Los Angeles, California

JOHN ABELSON
Department of Chemistry
University of California, San Diego
La Jolla, California

C. FRED FOX
Department of Bacteriology
and Molecular Biology Institute
University of California, Los Angeles
Los Angeles, California

ACADEMIC PRESS INC. New York San Francisco London 1977
A Subsidiary of Harcourt Brace Jovanovich, Publishers

ACADEMIC PRESS, INC.
111 Fifth Avenue, New York, New York 10003

United Kingdom Edition published by
ACADEMIC PRESS, INC. (LONDON) LTD.
24/28 Oval Road, London NW1

Library of Congress Cataloging in Publication Data
Main entry under title:

Molecular approaches to eucaryotic genetic systems.

(ICN-UCLA symposia on molecular and cellular biology; v. 8)
Proceedings of a symposium held in Park City, Utah, Feb. 27–Mar. 4, 1977.
1. Genetics—Congresses. I. Wilcox, Gary, Date
QH426.E9 575.1 77-10829
ISBN 0-12-751550-X

PRINTED IN THE UNITED STATES OF AMERICA

Contents

Preface xi

I. RAPID DNA SEQUENCING TECHNIQUES

1. Nucleotide Sequence of Bacteriophage ϕX174 DNA 1
John C. Fiddes

2. A Promotor Region for Yeast 5S RNA 15
Walter Gilbert, Allan M. Maxam, Richard Tizard, and Konstantin G. Skryabin

II. RECOMBINATION DNA TECHNOLOGY: METHODS AND RESULTS

3. The Mechanism of Phase Variation 25
Janine Zieg, Michael Silverman, Marcia Hilmen, and Melvin Simon

4. Workshop Summary: Plasmid Vectors ,37
H. Boyer

5. Workshop Summary: Phage Vector Systems 39
Dan S. Ray

6. Model Recombinants for the Development and Manipulation of EK2 Phage Vector Systems 41
Daniel J. Donoghue and Phillip A. Sharp

7. Bacteriophage fl as a Vector for Constructing Recombinant DNA Molecules 55
Gerald F. Vovis, Mariko Ohsumi, and Norton D. Zinder

8. Cloning of Satellite DNA Sequences with the λ Vector λgt-araB 63
Peter Philippsen, Chong S. Lee, and Ronald W. Davis

9. Workshop Summary: Screening for Recombinant DNAs 73
J. E. Dahlberg

10. Screening for Recombinant DNAs with *In Situ* Immunoassays 75
A. Skalka and L. Shapiro

11. Molecular Cloning of *λh80dara* Restriction Fragments with Noncomplementary Ends 85
Donald A. Kaplan, Lawrence Greenfield, and Gary Wilcox

12. Horizontal Slab Gel Electrophoresis of DNA 103
Donald A. Kaplan and Gary Wilcox

13. Workshop Summary: Vehicles for Molecular Cloning in Mammalian Host Cells 111
George C. Fareed

14. SV40 Vectors for Molecular Cloning in Mammalian Cells 113
George C. Fareed

15. Construction of SV40 Vectors and Expression of Inserted Sequences 129
Stephen P. Goff and Paul Berg

16. Episomal States of an SV40-*Escherichia Coli* Recombinant Genome in Different Mammalian Cell Lines 141
P. Upcroft, J. A. Upcroft, H. Skolnik, and G. C. Fareed

17. Mapping of Sequences with Twofold Symmetry on the SV40 Genome 153
Che-Kun James Shen and John E. Hearst

18. Studies on an SV40 DNA Carrier State in Monkey Cells in Culture 167
John M. Jordan

III. GENETICS AND MOLECULAR BIOLOGY OF YEAST

19. The *HIS4* Fungal Gene Cluster Is Not Polycistronic 179
Ramunas Bigelis, Joseph Keesey, and G. R. Fink

20. Workshop Summary: Yeast 189
E. P. Geiduschek

21. Mating Type Interconversion in Yeast and Its Relationship to Development in Higher Eucaryotes 193
Ira Herskowitz, Jeffrey N. Strathern, James B. Hicks, and Jasper Rine

22. Co-ordinate Regulation of the Synthesis of Yeast Ribosomal Proteins 203
Charles Gorenstein and Jonathan R. Warner

23. Isolation and Characterization of *Escherichia Coli* Clones Containing Genes for the Stable Yeast RNA Species 213
Jacques S. Beckmann, Peter F. Johnson, John Abelson, and Shella A. Fuhrman

24. Organization of Ribosomal DNA in Yeast 227
Jane Haris Cramer, Frances W. Farrelly, Joy T. Barnitz, and Robert H. Rownd

25. Organization of Yeast Ribosomal DNA 239
Thomas D. Petes, Lynna M. Hereford, and David Botstein

IV. GENETICS AND MOLECULAR BIOLOGY OF NEUROSPORA AND SLIME MOLDS

26. Tricks for RNA Labeling and Partial Restriction Digests 247
Nancy Maizels

27. Genetic Control of Phosphorus Metabolism in Neurospora 253
Robert L. Metzenberg and Robert E. Nelson

28. Regulation of Quinate Catabolism in Neurospora: The *QA* Gene Cluster 269
J. W. Jacobson, J. A. Hautala, M. C. Lucas, W. R. Reinert, P. Strøman, J. L. Barea, V. B. Patel, M. E. Case, and N. H. Giles

29. Workshop Summary: Neurospora 285
Richard Weiss

30. Workshop Summary: Slime Molds 287
Richard A. Firtel and William F. Loomis

31. Poly(A) Metabolism in *Dictyostelium Discoideum* 291
Allan Jacobson, Carl Mathew Palatnik, Cheryl T. Mabie, and Carol Wilkins

32. Analysis of Recombinant Plasmids Carrying Gene Sequences from *Dictyostelium* 301
Karen L. Kindle and Richard A. Firtel

33. Organization of Ribosomal and 5S RNA Coding Regions in *Dictyostelium Discoideum* 309
William C. Taylor, Andrew F. Cockburn, Gary A. Frankel, Mary Jane Newkirk, and Richard A. Firtel

V. CONTROL MECHANISMS AND GENE ORGANIZATION IN DROSOPHILA

34. A New Approach for Identifying and Mapping Structural Genes in *Drosophila Melanogaster* 315
Michael W. Young and David S. Hogness

35. Identification of a Cis-Acting Control Element in *Drosophila Melanogaster* 333
Arthur Chovnick, Margaret McCarron, and William Gelbart

VI. GENETICS AND DEVELOPMENTAL MECHANISMS IN NEMATODES

36. Expression of Genes Essential for Early Development in the Nematode, *C. Elegans* 347
David Hirsh, William B. Wood, Ralph Hecht, Stephen Carr, and Rebecca Vanderslice

37. Workshop Summary: Nematodes 357
W. B. Wood

38. Mutants of Acetylcholine Metabolism in the Nematode *Caenorhabditis Elegans* 359
Richard L. Russell, Carl D. Johnson, James B. Rand, Stewart Scherer, and Maurice S. Zwass

39. Studies on Two Body-Wall Myosins in Wild Type and Mutant Nematodes 373
Frederick H. Schachat, Harriet E. Harris, Robert L. Garcea, Janice W. LaPointe, and Henry F. Epstein

VII. MAMMALIAN GENETIC SYSTEMS

40. The Isolation of a Suppressible Nonsense Mutant in Mammalian Cells 381
M. R. Capecchi, R. A. Vonder Haar, N. E. Capecchi, and M. M. Sveda

41. Cyclic AMP, Microtubules, Microfibrils, and Cancer 399
Theodore T. Puck

42. Gene Transfer in Somatic Cell Populations 413
Frank H. Ruddle

43. Workshop Summary: Mammalian Genetic Systems 419
Immo Scheffler

44. Genetic Control of Argininosuccinate Synthetase in Human Lymphoblasts 421
Joseph D. Irr and Lee B. Jacoby

45. Small Stable RNA Molecules in the Nucleus: Possible Mediators in Gene Expression 431
David Apirion, Imre Berek, Bikram S. Gill, and Uma S. Podder

Author Index 441
Subject Index 443

Preface

The purpose of this ICN–UCLA symposium on "Molecular Approaches to Eucaryotic Genetic Systems" was to bring molecular biologists who have pioneered the new techniques of recombinant DNA research and rapid DNA sequencing together with the geneticists who have established promising eucaryotic genetic systems. The meeting was held in Park City, Utah from February 27 to March 4, 1977. The highlight of the meeting was the special lecture on "The Structure of Chromatin" by Nobel laureate Francis Crick. In closing his lecture, Dr. Crick gave a rather concise summary of the meeting:

> My impression of this meeting, if we look around and see what has been presented, is that we all know the sorts of problems. Of course, some people are looking at more ambitious problems, they are interested in problems of cell-cell interaction, and the actual details of differentiation. Before we do all that, we really want to know the nature of the eucaryotic genome, and how it is controlled, never mind what happens when you get the products of all that. If you look at what people have done here, most of them have been stating that they have systems which look promising, or they have methods which look exciting, and we have some results, but I think it's fair to say we don't have any deep results yet. We're setting up the problem at this stage. At least that's my impression of the thing. On the other hand, I must tell you that I can't help feeling that we're in an era of very rapid progress. It will be surprising if what I've said is true in two or three years time, maybe even one. In other words, we have so many methods to attack some of these things, and things can go so rapidly, it would be surprising if we didn't begin to see daylight. Once we begin to see daylight, I think we can design our experiments better. At the moment, we're really groping, and I think we should recognize that. When we begin to see what it's all about, then I think things will go even faster, and I think many of the things that we've heard today, in different parts of the meeting, will then begin to knit together. So it seems to me one of the great advantages of this meeting is, it's made people aware of other people's approaches and methods though I don't think it has produced solutions. That's just my impression. Thank you.

We are indebted to the National Institutes of Allergy and Infectious Diseases for their support which helped defray speakers' travel expenses, and to ICN Pharmaceuticals for their sponsorship of the symposia in general. We would also like to thank Fran Stusser and Robert Williams of the ICN–UCLA staff for their excellent help with the organization of this meeting.

MOLECULAR APPROACHES TO EUCARYOTIC GENETIC SYSTEMS

NUCLEOTIDE SEQUENCE OF BACTERIOPHAGE ØX174 DNA

John C. Fiddes

MRC Laboratory of Molecular Biology
Hills Road, Cambridge CB2 2QH, England

ABSTRACT. The plus and minus technique has been used by Sanger et al. (1) to obtain an almost complete nucleotide sequence for the DNA of the single stranded bacteriophage ØX174. The locations of the initiation and termination sites for the nine known genes and the extent of the intercistronic regions were determined. Two pairs of genes have been demonstrated to overlap, with the nucleotide sequence being read in different phases. Gene B is thus entirely contained within gene A, and likewise gene E is within gene D. In this article some features of the ØX174 nucleotide sequence are discussed.

INTRODUCTION

The plus and minus technique (2) for DNA nucleotide sequence analysis is based on primed synthesis of DNA by DNA polymerase I. This is usually accomplished with single stranded DNA as template annealed to a restriction enzyme fragment as specific primer. The details of the method will not be discussed here.

The technique is ideally suited for studying single stranded DNA bacteriophages such as ØX174 since the viral DNA provides a naturally occurring single stranded DNA template. The ØX174 complementary DNA strand can also be used since it is easily separable from the viral DNA strand when ØX174 double stranded DNA is centrifuged in alkaline caesium chloride gradients. Detailed cleavage maps of the ØX174 replicative form (RF) DNA are known for a range of restriction enzymes (3-9) and these have been correlated with the ØX174 genetic map by the marker rescue method (10-12).

Another reason for sequencing ØX174 DNA is that the biology of the phage has been studied extensively (see review by Denhardt, ref. 13). ØX174 has one of the smallest known DNA genomes containing only about 5400 nucleotides. Nine genes have been identified by recombination and complementation analysis and protein products have been related to them. The combined molecular weights of these proteins more than account for the coding capacity of the phage genome. The locations of the three promoters and possible sites for the termination of transcription (14-16) and the position of the origin of viral strand DNA synthesis are known (17,18). The nucleotide sequence of ØX174 would thus be expected to provide information about the overall organisation of the phage genome and of the nature of these control sites.

Largely through use of the rapid plus and minus technique an almost complete nucleotide sequence has now been obtained by Sanger et al. (1) for the 5375 nucleotides of ØX174. In this paper it is intended to discuss some features of this sequence.

RESULTS AND DISCUSSION

ØX174 NUCLEOTIDE SEQUENCE

Coding sequences. Genes D, J, F, G and H of ØX174 are transcribed from a promoter preceding gene D (see Fig. 1).

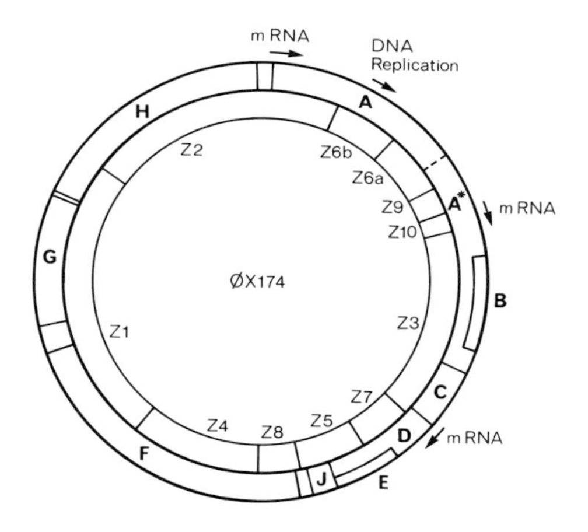

Fig. 1. Genetic map of bacteriophage ØX174, based on the data of Sanger et al. (1). The ØX174 genes A, A*, B, C, D, E, F, G, H and J are indicated along with the size of the intercistronic spaces, the locations of the transcriptional starts and the origin of viral strand DNA synthesis. The inner circle represents the fragments produced by the enzyme Hae III (Z).

This half of the genome contains the structural genes. The F protein is the major component of the viral capsid, proteins G and H form the spikes at the vertices of the icosahedral capsid, while the small J protein is a minor capsid component. The D protein does not form part of the virion but it is produced in large amounts during infection like the structural proteins.

Due to the ready availability of these proteins they were chosen for amino acid sequence analysis. Complete amino acid sequences were not derived by amino acid sequencing methods but sufficient data was collected to complement the tentative DNA sequencing results obtained from the plus and minus technique and from the older techniques of cleavage by bacteriophage T_4 endonuclease IV and transcription into RNA. This combination of approaches has enabled the complete primary structure to be determined for genes D, J and G and for extensive regions of gene F (1,19-21).

In common with the filamentous DNA phage the viral strand of ØX174 DNA is rich in thymidine. For ØX174 the value is 31.2% (1). Examination of the distribution of this nucleotide in the ØX174 structural genes shows that it is accommodated largely in the third, degenerate, position of the codons. Thus the percentages of codons in genes F and G terminating in T are 55.9 and 54.3 respectively (20,21). The significance of this is not clear. It was initially thought that the high T content was required for a particular conformation adopted by the DNA when packaged in the virion. However, this now seems less likely since the closely related single strand DNA containing isometric phage G4 does not show this feature (G.N. Godson, unpublished results). Alternatively there may be some relationship between this effect and the presence of the nucleotide A adjacent to the anticodon in many tRNA species.

Use has been made of this high frequency of third position T to confirm the reading frame in regions of ØX174 where amino acid sequence data had not been derived independently (1). Thus, only a few regions of the H protein have been sequenced (including the N and C termini) but the correct reading phase is that in which the largest number of codons terminate in T. This was used to confirm uncertainties in the DNA base sequence.

With genes A (double stranded DNA replication), B and C (single strand DNA synthesis) and E (the lysis function) amino acid sequence data is not available, since the proteins are synthesised in smaller amounts than those of the structural genes. However, despite this lack of information there is sufficient circumstantial evidence to define the boundaries of these coding regions and the correct translational phases.

The evidence for the identification of the start and reading frame of the A gene is as follows. Sinsheimer _et al._

(22,23) have determined the 5' terminal sequences of the major in vitro species of RNA initiated from ØX174 RF DNA. These sequences have been related by hybridisation to the restriction enzyme fragment map of ØX174 (14). One RNA species with the initiation sequence 5' pppA-A-A-U-C-U-U-G-G 3' was identified in Hind II fragment 4 which is in the correct position for the RNA to represent the start of the A gene transcript.

The corresponding DNA nucleotide sequence 5' A-A-A-T-C-T-T-G-G 3' is only found once in ØX174 and has been located in Hind II fragment 4 (1). Twenty nucleotides from the 5' end of this mRNA start site is the nucleotide sequence 5' A-T-G 3' which is the proposed fMet initiation codon for the A protein (Fig. 2).

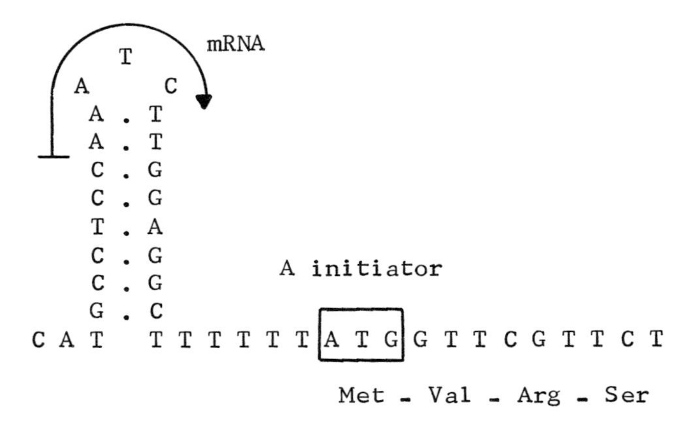

Fig. 2. ØX174 DNA sequence at the initiation of gene A showing the proposed secondary structure, based on Sanger et al. (1). The A-T-G initiation codon of gene A is indicated along with the sequence of the first four amino acids of the protein. The position of the mRNA synthesised is also shown.

This prediction is supported by the fact that the preceding nucleotide sequence has the characteristics that Shine and Dalgarno proposed for a ribosome binding site (24). From an examination of known ribosome binding site nucleotide sequences they observed that the region preceding the initiation codon was characteristically complementary to the 3' terminal sequence of the E. coli 16S rRNA. This complementarity ranges from between three to eight nucleotides and, in the case of the bacteriophage R17 gene A site, a complex between the two RNA species has been demonstrated (25). In each of the ØX174 genes for which the N terminal amino acid sequence is known definitely (D, J, F, G and H) this homology is observed. In the case of the ØX174 A gene the nucleotide sequence 5' G-G-A-G-G 3', the 3' end of which is seven nucleotides to the 5' side of the initiator codon, is complementary to part of the 16S rRNA terminal sequence 5' G-A-U-C-A-C-C-U-C-C-U-U-A_{OH} 3'.

Following this proposed initiation codon the nucleotide sequence can be read in one phase for 1536 nucleotides before encountering a termination codon. This would produce an A protein of 512 amino acids which is in reasonable agreement with the accepted molecular weight of about 55,000-67,000 daltons (13). Further confirmation of the validity of this reading frame comes from the high level of T found in the third codon position in the 5' terminal region of gene A and from the identification of the base changes of two gene A amber mutants, *am*33 and *am*86, which map close to the 5' terminus of gene A (1).

Overlapping coding regions. The exact location and boundaries of the genes D, J, F, G, H and A in the genome have thus been determined. The total coding requirement for these genes is 4890 nucleotides, which leaves 265 nucleotides for the remaining genes B, C and E. This estimate takes into account the 220 nucleotides of untranslated space between the start of gene D and the end of gene A. The combined molecular weights of these proteins, using the lowest estimates available (13), would require a total of about 900 nucleotides coding capacity. This discrepancy is resolved by the fact that the gene B is contained entirely within gene A (26,27) and likewise gene E is within gene D (28). In both cases translation of the pairs of overlapping genes is in different phases.

The evidence substantiating the A-B overlap comes mainly from the nucleotide sequence analysis of gene A and B mutants by Brown and Smith (27) and Smith *et al*. (26). The mutants involved are *am*16 in gene B, *am*18 and *am*35 in gene A and *ts*116 which is a revertant of the A mutant *am*18 yet shows temperature sensitive properties in gene B.

The first indication of a situation of overlapping genes is that the B mutant *am*16 maps closer to the 5' terminus of gene A than does the A mutants *am*18 and *am*35. Nucleotide sequence analysis of this mutant shows it to be a G → T transversion which introduces a chain termination (amber) codon in a phase other than that already designated the A phase. This mutant defines the reading phase for gene B. In the A phase the change converts a leucine codon into a phenylalanine codon but this does not appear to have any phenotypic effect on the A protein.

Sequence analysis of the independently isolated A mutants *am*18 and *am*35 show them both to have the same C → T change producing an amber codon in the A phase. The change in the amino acid sequence in the proposed B phase is an alanine to a valine which is not phenotypically recognised. Mutant *ts*116 which is a B revertant of the gene A mutant *am*18 causes a G → C transversion which converts the A amber codon of *am*18 into a tyrosine codon, thus eliminating the chain termination of A. In the B phase, however, a glutamic acid changes to a glutamine which, in addition to the alanine to valine change still present from the *am*18 mutant, causes the gene B product to be temperature sensitive. Figs. 3 and 4 summarise these changes.

B protein wt	Glu - Ile - Glu - Ala - Gly
DNA wt	G-A-G-A-T-T-G-A-G-G-C-T-G-G-G
	↓
DNA *am*16 (B)	G-A-G-A-T-T-T-A-G-G-C-T-G-G-G
A protein wt	Arg - Leu - Arg - Leu - Gly
	↓
A protein *am*16	Arg - Phe - Arg - Leu - Gly

Fig. 3. Summary of the nucleotide and amino acid sequence changes caused by the B mutant *am*16 (ref. 27).

A protein *ts*116	Met - Thr - Tyr - Lys - Leu
	↑
A protein wt	Met - Thr - Gln - Lys - Leu
DNA wt	A-T-G-A-C-G-C-A-G-A-A-G-T-T-A
	↓
DNA *am*18, *am*35 (A)	A-T-G-A-C-G-T-A-G-A-A-G-T-T-A
	↓
DNA *ts*116 (B)	A-T-G-A-C-G-T-A-C-A-A-G-T-T-A
B protein wt	Asp - Ala - Glu - Val
	↓
B protein *am*18, *am*35	Asp - Val - Glu - Val
	↓
B protein *ts*116 (B)	Asp - Val - Gln - Val

Fig. 4. Summary of the nucleotide and amino acid sequence changes caused by the A mutants *am*18 and *am*35 and the B mutant *ts*116 (ref. 26).

The precise location of the initiation codon for the B gene has been determined. Ravetch *et al.* (29) isolated a ribosome binding site from ØX174 mRNA. Sequence analysis of this RNA showed that the site corresponded to a region in the ØX174 DNA sequence which is in the appropriate position, and has an A-T-G in the correct phase, to code for the start of the B protein.

Good evidence has also been obtained by Barrell *et al.* (28) to substantiate the D-E overlap. The genetically determined map of ØX174 (30,31) gave the following gene order: D-E-J-F. However, amino acid and nucleotide sequence analysis has shown that genes D and J are contiguous, thus leaving no space for gene E.

Marker rescue experiments (28,32) showed that this anomaly is explained by gene E being contained within gene D. In the marker rescue technique *E. coli* spheroplasts are infected with mutant viral DNA to which is annealed a restriction fragment made from wild type RF DNA. If the fragment covers the region of DNA containing the mutation wild type phage are produced. These experiments showed that gene E mutants (*am*3, *am*27, *am*34 and N11) could be rescued by fragments which were known from sequence analysis to be contained entirely within gene D.

The gene E reading phase was determined by nucleotide sequence analysis of the E amber mutants mentioned above and the E initiation codon was identified as the only A-T-G or G-T-G in the appropriate phase which was preceded by a sequence showing the 16S rRNA complementarity.

Features of overlapping genes. Some indication as to the order of evolution of the overlapping genes in ØX174 can be gained from a study of the distribution of codons ending in T (27). In the A-B overlap a high level of third position T is found in the B phase and a low level in the A phase. This was interpreted to mean that ØX174 originally had distinct genes A and B and that the present form of A arose as a translational read through into B. In the D-E overlap the high third position T is a feature of the D phase, suggesting in this case that gene E evolved at a later stage. The gene E product, the phage's lysis function, is rich in hydrophobic amino acids such as leucine and phenylalanine which is the result of the high third position T of gene D being transferred to the first position of the codons in the E phase.

The significance to other systems of the overlapping genes of ØX174 is not clear. Possibly the single stranded DNA phage represent a unique case due to the constraints imposed on their size by the requirement to package the DNA in a capsid of fixed size. Obviously the major problem with eukaryotic systems is one of an excess rather than an inadequate amount of DNA though the existence of overlapping genes in such systems cannot be ruled out.

The nucleotide sequence of ØX174 is now stored on a computer and this has facilitated the searches for particular sequences (R. Staden and B.G. Barrell, personal communication). One interesting feature which has emerged from this study is that ØX174 appears to have the potential for other systems of overlapping genes. Thus a total of 15 A-T-G codons, which are preceded by sequences with characteristics of ribosome binding sites, have been identified. Proteins, ranging in size from 22 amino acids long to one example in gene H which would be 191 or 201 amino acids long, could conceivably be synthesised from these initiation codons.

This observation probably indicates that the full nucleotide sequence requirements for ribosome binding are not understood and that these extra sites are not used. There are, however, indications that ØX174 has extra genes. SDS polyacrylamide gels of ØX174 infected cells show minor reproducible bands which have not been identified (13). Also there are several viral induced functions which have not yet been correlated with the known viral proteins. For example, on infection ØX174 induces a methylase which introduces a single 5-methyl cytosine into the viral DNA (33) and a phage function is responsible for the shut off of host cell DNA synthesis, though this may be attributed to A*, the translational, in phase, restart of gene A (34). Also the proteins which are responsible for sizing unit length viral DNA and ligating it have not been unequivocally determined.

Intercistronic sequences. The three genes of the RNA phage such as MS2 and Qβ are separated by two distinct intercistronic regions which range in size from about 25 to 40 nucleotides (35). With ØX174 the situation is not so straightforward due to the existence of overlapping initiation and termination codons. Between the 5' termini of genes A and J in ØX174, a region which encompasses 42% of the phage genome, there is no untranslated region since the initiators and terminators of genes A and C, C and D, and D and J overlap (see Fig. 1). The extent of the C coding region has recently been established (B.G. Barrell, unpublished results) from the sequence analysis of a C mutant, och6.

These three examples of overlapping initiators and terminators in ØX174 are shown in Fig. 5. Two types of overlapping codons are found -

A-T-G-A and T-A-A-T-G

though it may be significant that both result in the two genes having different reading phases. A similar example has been observed between the B and A genes of the *E. coli trp* operon (36).

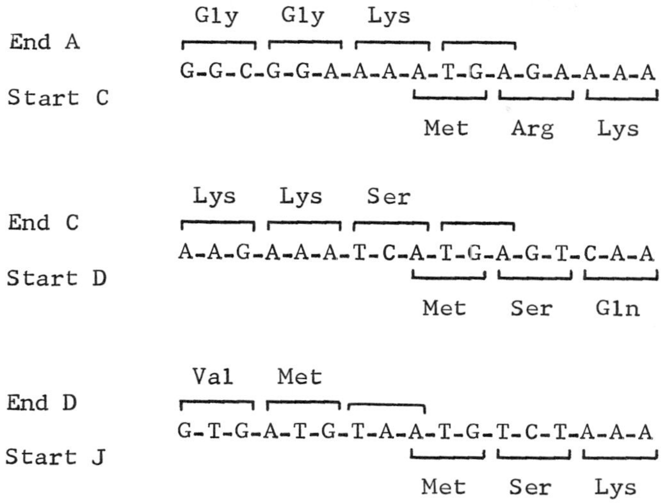

```
               Gly   Gly   Lys
End A        ┌───┐ ┌───┐ ┌───┐ ┌───┐
             G-G-C-G-G-A-A-A-A-T-G-A-G-A-A-A-A
Start C                        └───┘ └───┘ └───┘
                                Met   Arg   Lys

               Lys   Lys   Ser
End C        ┌───┐ ┌───┐ ┌───┐ ┌───┐
             A-A-G-A-A-A-T-C-A-T-G-A-G-T-C-A-A
Start D                        └───┘ └───┘ └───┘
                                Met   Ser   Gln

               Val   Met
End D        ┌───┐ ┌───┐ ┌───┐
             G-T-G-A-T-G-T-A-A-T-G-T-C-T-A-A-A
Start J                        └───┘ └───┘ └───┘
                                Met   Ser   Lys
```

Fig. 5. The three examples in ØX174 of overlapping initiation and termination codons, based on Sanger *et al.* (1) and Barrell *et al.* (28).

Intercistronic regions of 39, 111, 11 and 66 nucleotides respectively are found between genes J and F, F and G, G and H and H and A. The secondary structures proposed for these spaces are shown in Fig. 2 (H-A) and Fig. 6 (J-F and F-G).

The A-H region has been mentioned already in connection with the identification of the start of the A coding region. The sequence, T-T-T-C-A-T-G, which precedes the mRNA start by five nucleotides, is similar to the T-A-T-Pu-A-T-Pu sequence proposed by Pribnow (37) and Schaller *et al.* (38) to be characteristic of promoters recognised by *E. coli* RNA polymerase. However, following the mRNA initiator site and immediately preceding the proposed initiation codon for gene A is the sequence T-T-T-T-T-T-A which bears a strong resemblance to the sequence $U_{5-6}A$ found at the 3' end of certain naturally occurring mRNA species. These include the 4S (OOP) and 6S RNAs of bacteriophage λ and the bacteriophage Ø80 induced RNA species.

There is, however, evidence that the H-A junction in ØX174 is the site of the major transcriptional termination in the phage (15,16). This comes from an alignment of the RNA species with the genetic and restriction fragment map. If the

T_6A does represent this terminator then its position immediately after the promoter would appear strange since it would presumably result in premature termination of transcription.

However, it has been proposed that part of the recognition signal for transcriptional termination is a hairpin loop structure in the mRNA (Rosenberg et al., unpublished, and 39). RNA species synthesised from positions before the A promoter would have such a secondary structure (Fig. 2) and might therefore be expected to terminate at the sequence T_6A. However, mRNA which was initiated at the A promoter would not have this structure and would therefore not be subject to termination.

The secondary structure between genes J and F (Fig. 6) may also be involved in the termination of transcription since this site has been associated with one of the weak transcriptional terminators of ØX174. There is, however, no sequence of the form $T_{5-6}A$. There is no evidence for a promoter in this region.

The largest untranslated space in ØX174 is the 111 nucleotide gap between genes G and F (40). There is evidence to substantiate the existence in ØX174 viral DNA of both the hairpin loops in this region (Fig. 6). The depurination products obtained from the viral DNA fragment resistant to the action of the single strand specific nucleases of N. crassa (41) match those predicted from the nucleotide sequence of the major loop. The depurination products from a similar fragment isolated through its resistance to the enzyme S_1 (H. Schaller, personal communication) indicates that both of the loops shown in Fig. 6 are found.

The function of this region is not known. The loop structures may represent the weak transcriptional termination site observed between genes F and G or they may be involved in the binding of a specific viral protein in the processes of assembly or penetration. Alternatively the function of this region may be in the double stranded form of the DNA since the symmetrical region formed by the loop has an unusual base distribution with a central run of 17 adjacent A-T base pairs flanked by regions in which seven out of nine and 10 out of 12 base pairs are G-C.

Sequences which have both a control and a coding function. In several regions of the ØX174 genome the DNA sequence has both a coding and a control function. The most striking example of this economy of DNA usage is shown by the sequence which includes the 3' termini of genes D and E and the 5' terminus of gene J (see Fig. 7).

In this region the nucleotide sequence 5' A-A-G-G-A-G-T-G-A-T 3' has three functions. With one nucleotide looped out

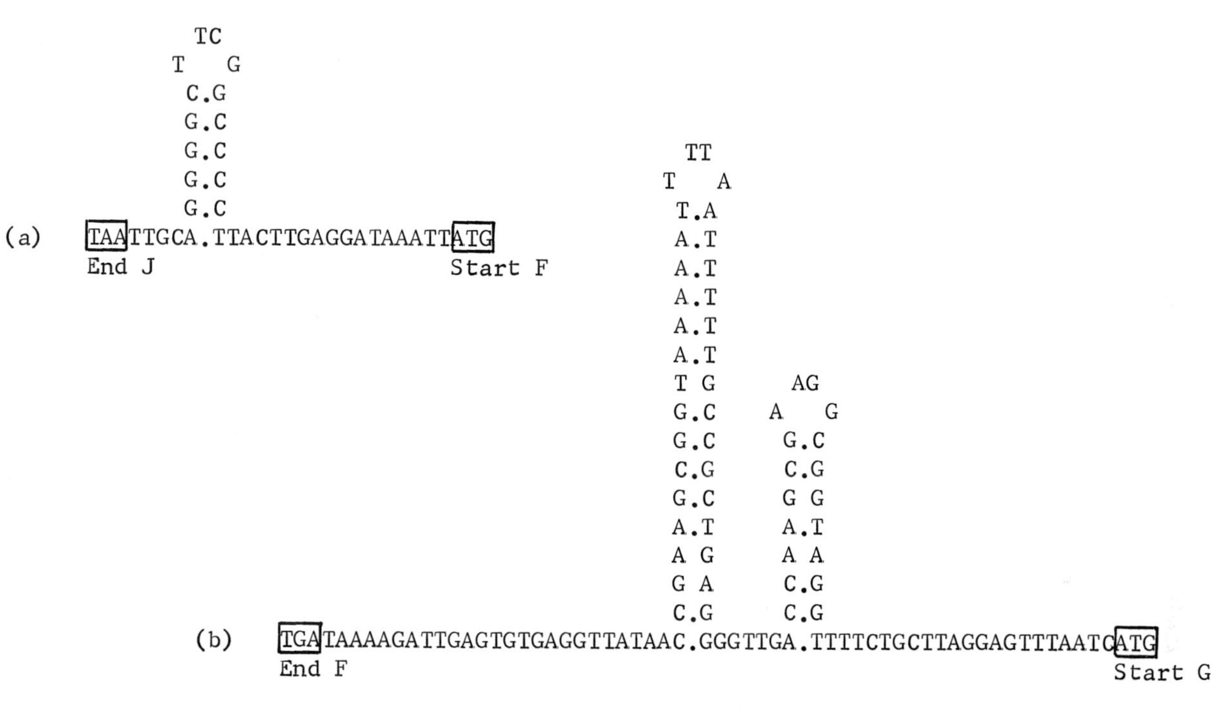

Fig. 6. (a) Intercistronic sequence between genes J and F (1). (b) Intercistronic sequence between genes F and G (40). In both cases the proposed secondary structure is shown.

```
Gene D   Ala   Glu   Gly   Val   Met  terminator

       G-C-G-G-A-A-G-G-A-G-T-G-A-T-G-T-A-A-T-G-T-C-T

Gene E   Arg   Lys   Gln  terminator   fMet  Ser   Gene J
```

Fig. 7. Nucleotide and amino acid sequences at the 3' (C) termini of genes D and E and the 5' (N) terminus of gene J, based on Barrell et al. (28).

it is complementary to the 3' terminal sequence of the 16S rRNA and thus serves as the ribosome binding recognition signal for gene J. However, this sequence also codes, in two different phases, for the E and D proteins. Other examples of ribosome binding sites which also code for the preceding protein are found with genes D, C and B and with A*, the translational restart of gene A. In the case of the B gene ribosome binding to the ØX174 mRNA has been demonstrated (29).

Two of the three ØX174 promoters are found in coding regions. The A promoter, which was identified as correspon-ding to one of the mRNA initiation sequences of Grohmann et al. (23) has been discussed already and is found in the untrans-lated region between genes H and A. However, the other two mRNA initiation sequences have been located in coding regions. Thus the site for transcriptional initiation which precedes gene D is located in gene C and the one preceding gene B is in gene A (see Fig. 1). The leader sequence from the 5' end of the mRNA to the A-U-G initiation codon is 32 nucleotides long in the case of gene D and 176 in the case of gene B. Despite both promoters being in coding regions they still bear a certain resemblance to the characteristic T-A-T-Pu-A-T-Pu (37, 38). Thus the sequences T-A-C-T-A-T-C and T-A-C-A-G-T-A are found in the appropriate positions in the D and B promoters respectively.

The origin of the viral strand DNA synthesis in ØX174 appears to be in the A coding region. The precise location of this has not yet been assigned. Replicating double stranded molecules (RF II) which have a specific gap in the plus strand have been isolated (17). It is proposed that this gap is related to the initiation of DNA synthesis.

The in vitro repair of these gapped RF II molecules was used to locate the gap in restriction enzyme fragments which correspond to gene A (18). In an extension of this approach pyrimidine tract analysis of the repaired material showed the characteristic product C_6,T (42). The sequence 5' C-T-C-C-C-C-C 3' has been found in the appropriate position in ØX174 DNA (2). The surrounding sequence does not show any unusual features such as symmetry, though there is a region rich in A-T base pairs flanked by regions rich in G-C base pairs.

REFERENCES

1. Sanger, F., Air, G.M., Barrell, B.G., Brown, N.L., Coulson, A.R., Fiddes, J.C., Hutchison, C.A. III, Slocombe, P.M. and Smith, M. (1977) Nature, in press.
2. Sanger, F. and Coulson, A.R. (1975). J. Mol. Biol. 94, 441.
3. Lee, A.S. and Sinsheimer, R.L. (1974) Proc. Nat. Acad. Sci. USA 71, 2882.
4. Hayashi, M.N. and Hayashi, M. (1974) J. Virol. 14, 1142.
5. Vereijken, J.M., van Mansfeld, A.D.M., Baas, P.D. and Jansz, H.S. (1975) Virology 68, 221.
6. Jeppesen, P.G.N., Sanders, L. and Slocombe, P.M. (1976) Nuc. Acids Res. 3, 1323.
7. Brown, N.L., Hutchison, C.A. III and Smith, M. (1977) J. Mol. Biol., in press.
8. Sato, S., Hutchison, C.A. III and Harris, J.I., Proc. Nat. Acad. Sci. USA, submitted.
9. Baas, P.D., van Heusden, G.P.H., Vereijken, J.M., Weisbeek, P.J. and Jansz, H.S. (1976) Nuc. Acids Res. 3, 1947.
10. Edgell, M.H., Hutchison, C.A. III and Sclair, M. (1972) J. Virol. 8, 574.
11. Middleton, J.H., Edgell, M.H. and Hutchison, C.A. III (1972) J. Virol. 10, 42.
12. Weisbeek, P.J., Vereijken, J.M., Baas, P.D., Jansz, H.S. and van Arkel, G.A. (1976) Virology 72, 61.
13. Denhardt, D.T., CRC Critical Reviews in Microbiology (1975) 4, 161.
14. Smith, L.H. and Sinsheimer, R.L. (1976) J. Mol. Biol. 103, 699.
15. Axelrod, N. (1976) J. Mol. Biol. 108, 753.
16. Hayashi, M., Fujmura, F.K. and Hayashi, M. (1976) Proc. Nat. Acad. Sci. USA 73, 3519.
17. Johnson, P.H. and Sinsheimer, R.L. (1974) J. Mol. Biol. 83, 47.
18. Baas, P.D., Jansz, H.S. and Sinsheimer, R.L. (1976) J. Mol. Biol. 102, 633.
19. Air, G.M., Blackburn, E.H., Sanger, F. and Coulson, A.R. (1975) J. Mol. Biol. 96, 703.
20. Air, G.M., Blackburn, E.H., Coulson, A.R., Galibert, F., Sanger, F., Sedat, J.W. and Ziff, E.B. (1976) J. Mol. Biol. 107, 445.
21. Air, G.M., Sanger, F. and Coulson, A.R. (1976). J. Mol. Biol. 108, 519.

22. Smith, L.H., Grohmann, K. and Sinsheimer, R.L. (1974) Nuc. Acids Res. 1, 1521.
23. Grohmann, K., Smith, L.H. and Sinsheimer, R.L. (1975) Biochemistry 14, 1951.
24. Shine, J. and Dalgarno, L. (1974) Proc. Nat. Acad. Sci. USA 71, 1342.
25. Steitz, J.A. and Jakes, K. (1975) Proc. Nat. Acad. Sci. USA 72, 4734.
26. Smith, M., Brown, N.L., Air, G.M., Barrell, B.G., Coulson, A.R., Hutchison, C.A. III and Sanger, F. (1977) Nature, in press.
27. Brown, N.L. and Smith, M. (1977) J. Mol. Biol. submitted.
28. Barrell, B.G., Air, G.M. and Hutchison, C.A. III (1976) Nature 264, 34.
29. Ravetch, J.V., Model, P. and Robertson, H.D. (1977) Nature, in press.
30. Benbow, R.M., Hutchison, C.A. III, Fabricant, J.D. and Sinsheimer, R.L. (1971) J. Virol. 7, 549.
31. Benbow, R.M., Zuccarelli, A.J., Davis, G.C. and Sinsheimer, R.L. (1974) J. Virol. 13, 898.
32. Hutchison, C.A. III and Edgell, M.H. (1971) J. Virol. 8, 181.
33. Razin, A. (1973) Proc. Nat. Acad. Sci. USA 70, 3773.
34. Martin, D.G. and Godson, G.N. (1975) Biochem. Biophys. Res. Commun. 65, 323.
35. Fiers, W., Contreras, R. Duerinck, F., Haegeman, G., Iserentant, D., Merregaert, J., Min Jou, W., Molemans, F., Raeymakers, A., van den Berghe, A., Volckaert, G. and Ysebaert, M. (1976) Nature 260, 500.
36. Platt, T. and Yanofsky, C. (1975) Proc. Nat. Acad. Sci. USA 72, 2399.
37. Pribnow, D. (1975) Proc. Nat. Acad. Sci. USA 72, 784.
38. Schaller, H., Gray, C. and Herrman, K. (1975) Proc. Nat. Acad. Sci. USA 72, 737.
39. Sugimoto, K., Sugisaki, A., Okamoto, T. and Takanami, M. (1977) J. Mol. Biol., in press.
40. Fiddes, J.C. (1976) J. Mol. Biol. 107, 1.
41. Bartok, K., Harbers, B. and Denhardt, D.T. (1975) J. Mol. Biol. 99, 93.
42. Eisenberg, S., Harbers, B., Hours, C. and Denhardt, D.T. (1975) J. Mol. Biol. 99, 107.

A PROMOTOR REGION FOR YEAST 5S RNA

Walter Gilbert, Allan M. Maxam,
Richard Tizard, and Konstantin G. Skryabin*

Department of Biochemistry and Molecular Biology
Harvard University
Cambridge, Massachusetts 02138

ABSTRACT. We have sequenced the DNA region immediately preceding the structural gene for the 5S ribosomal RNA of yeast (*Saccharomyces cerevisiae*). Since the 5S RNA, a component of the larger ribosomal subunit, has a 5' triphosphate, this molecule is likely to be the direct product of the gene, unmodified by maturation at the 5' end. Thus the DNA sequence immediately before this region must contain the recognition sequences where the RNA polymerase binds to begin synthesis, i.e. this sequence must represent a eukaryotic promoter.

INTRODUCTION

How is the mystery of the functioning of the genes of higher organisms to be unravelled? To understand the detailed organization of the genetic information of higher organisms and to discover the signals used in the control of gene expression, we must analyze many elements of the genetic structure, including DNA sequences. The technical problem has three components: How are we to obtain large amounts of a specfic DNA fragment from a higher organism? How are we to work out its sequence? And how are we to identify those parts of the sequence which are interesting and relevant?

Obtaining DNA segments of interest from higher organisms can be done by recombinant DNA techniques. Once a fragment of DNA has been inserted in a plasmid that can replicate in a bacterium, a clone of such cells will produce adequate copies of that DNA. Furthermore, recent advances in DNA sequencing provide simple ways of sequencing any DNA molecule that can be obtained in picomole

* Present address: Institute of Molecular Biology
USSR Academy of Sciences
32 Vavilov Street
Moscow 117312, USSR

amounts. Thus any molecule from a bacterial plasmid, a bacterial virus, or an animal virus, can be obtained in large enough amounts for sequencing. A fragment of such eukaryotic DNA, carried in a bacterial plasmid, can be mapped with restriction enzymes; the availability of restriction enzymes is now such that a map, breaking such a DNA region into pieces of the order of 100 to 200 base pairs long, can easily be achieved. Our chemical sequencing method (I), beginning with these restriction fragments, inserts a 5' or 3' terminal label and works out their sequence by analyzing partial breakage patterns. Our method works directly with the restriction fragments, either as double-stranded DNA or, after denaturation, as single-stranded material. The power of the method is such that one can by simple manipulations develop a sequence extending over 100 nucleotides from each point of end labeling. Thus the first two problems, obtaining enough DNA and working out the sequence, are solved in principle. However, how shall we identify the sequence of a relevant, interesting region and understand how it functions as a signal? This is much harder to arrange. We lack mutants that would characterize control regions. We have large gaps in our understanding of the connection between the organization of the higher cell genome and the final product of a structural gene. A direct approach would be to work out the sequence of regions at which the RNA polymerase and control proteins must act, segments immediately before the actual point of initiation of RNA synthesis (in bacterial cells other regions are also involved in transcriptional control: attenuators (2) and controlled termination signals (3) lie between the point of initiation of RNA synthesis and the structural region). However, in order to work out such regions for any structural gene, one needs to know the relationship between the messenger RNA, finally expressed in the cytoplasm, and its precursers, going back ultimately to the identification of the point of initiation of the transcription unit. We must know where the initiating nucleoside triphosphate paired with the DNA, in order to make sense of DNA sequences in this relevant region. No doubt this will be done for a number of structural genes in the next few years. In this paper we sketch such an approach for one simple system where the final product has an unmodified 5' end.

RESULTS AND DISCUSSION

Yeast 5S RNA provides our example. Hindley and Page worked out the sequence of this RNA (4), a component of the larger ribosomal subunit, and showed that the mature species, 121 nucleotides long, has a triphosphate at its 5' end. Thus this molecule is a direct product, at least in so far as the 5' sequence is concerned, of a transcription unit on the DNA. Furthermore, the 5S gene, like the other ribosomal RNA genes of yeast, is repeated, and therefore it is easy to obtain plasmids carrying this region. For these reasons, we sought to identify the 5S region among plasmids carrying portions of the yeast ribosomal RNA region, with a view to sequencing the DNA region immediately before that coding for the 5S RNA.

Petes, Wensink and Botstein (unpublished) isolated a series of plasmids containing yeast DNA. Hybridization to 5S, 18S and 28S ribosomal RNA's of yeast identified rDNA clones. Digestion of these plasmids with Eco Rl, coupled with hybridization of the Rl fragments to the ribosomal RNA's generated a map of the ribosomal area (Petes, Hereford and Skryabin, to be published). The map was derived by noting the associations between Rl fragments in different clones. In *Saccharomyces cerevisiae*, there is a repeating structure for the ribosomal region:

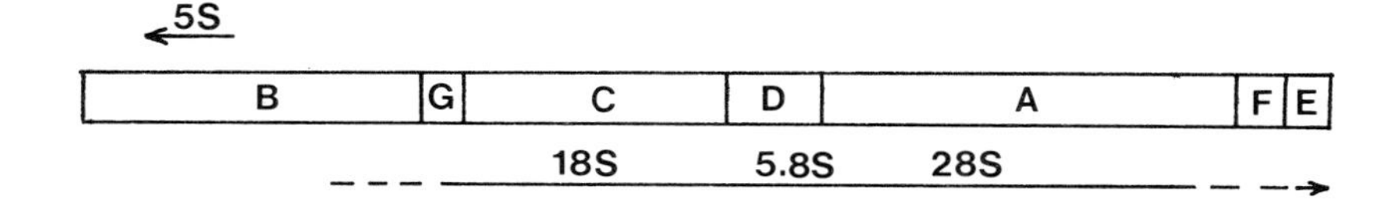

Eco Rl map of the yeast ribosomal region

Fragment B, the second largest Rl fragment, 2600 base pairs long, hybridizes to the yeast 5S sequence. Hindley and Page's sequence of yeast 5S RNA (4) (although from *S. carlsbergensis*) suggested several restriction cuts in the structural gene, the one of primary interest being a Hae III cut (*Hemophilus aegyptis*) eight nucleotides from the 5' end of the RNA. To construct a restriction map of the B fragment, we end-labeled the fragment with polynucleotide kinase, cut the fragment close to either one or the other end, and then examined partial digests of the resulting long, uniquely terminally labeled fragments with a series of restriction enzymes (5).

```
-100                -80                 -60                 -40                 -20
AAATATAATAAAAATTGTCCTCCACCCATAACACCTCTCACTCCCACCTACTGAACATGTCTGGACCCTGCCCTCATATCACCTGCGTTTCCG
TTTATATTATTTTTAACAGGAGGTGGGTATTGTGGAGAGTGAGGGTGGATGACTTGTACAGACCTGGGACGGGAGTATAGTGGACGCAAAGGC

      pppGGUUGCGGCCAUAUCUACCAGAAAGCACCGUUUCCCGUCCGAUCAACUGUAGUUAAGCUGGUAAGAGCCUGACCGAGUAGUGUAGUGG

                     20                  40                  60                  80
TTAAACTATCGGTTGCGGCCATATCTACCAGAAAGCACCGTTTCCCGTCCGATCAACTGTAGTTAAGCTGGTAAGAGCCTGACCGAGTAGTGTAGTGG
AATTTGATAGCCAACGCCGGTATAGATGGTCTTTCGTGGCAAAGGGCAGGCTAGTTGACATCAATTCGACCATTCTCGGACTGGCTCATCACATCACC
               HaeIII

GUGACCAUACGCGAAACUCAGGUGCUGCAAUCU(OH)

     100                 120                 140                 160                 180
GTGACCATACGCGAAACTCAGGTGCTGCAATCTTTATTTCTTTTTTTTTTTTTTTTTTTTTTTTTTTTCTAGTTTCTTGGCTTCCTATGCT
CACTGGTATGCGCTTTGAGTCCACGACGTTAGAAATAAAGAAAAAAAAAAAAAAAAAAAAAAAAAAAAGATCAAAGAACCGAAGGATACGA
```

Fig. 1

Nucleotide sequence before, within, and beyond the *S. cerevisiae* gene for 5S ribosomal RNA

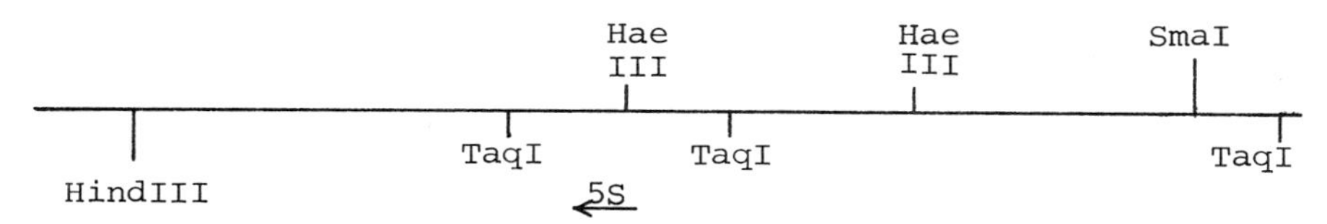

Restriction map of the B fragment

We sequenced the ends of all of the Hae III fragments in order to identify which fragment junction lay within the 5S ribosomal RNA gene. From this junction, we then sequenced out 100 bases before the gene, and on the other side, the gene itself. Fig. 1 shows the DNA sequences and the corresponding 5S RNA.

The structure we obtain for 5S rDNA confirms most of the RNA sequences worked out by Miyazaki (6) for S. cerevisae and by Hindley and Page (4) for S. carlsbergensis. Fig. 2 compares our sequence with that derived from three yeast strains (4,6,7). Some of the differences may be strain differences. However, except for two base changes in Torulopsis utilis, the differences are most likely to be minor errors in RNA sequencing. Fig. 2 shows the RNase T_1 fragments that we predict to be different; the variations look like the typical difficulties found in RNA sequencing: the hardest problem is to order a series of pyrimidines within a single T_1 fragment.

The sequencing determines the direction of transcription of the 5S gene to be from right to left on the B fragment, as shown, and thus to be opposite in direction to the major 18S and 28S rRNA synthesis.

The region immediately to the left of the initial triphosphate in Fig. 1 must be the site for a yeast RNA polymerase. The 5S ribosomal DNA of animal cells is trancribed by RNA polymerase III (8); thus it is likely that this sequence contains the promoter for yeast polymerase III. Several elements in this yeast promoter sequence are curious. There is a purine/pyrimidine bias in the 90 bases before the triphosphate. 60 out of 90 purines are on the lower strand. There is no particular A-T/G-C bias. There is a triple partial repeat of an 18-base-pair long sequence; 14 bases are repeated in the promoter and 11 bases early in the gene. This may be a remnant of the process that created this DNA region. Fig. 3 shows these properties of this yeast promoter.

The promoter and the gene are flanked on the left and right with dA-dT rich regions almost serving to indentify the gene itself. Before the promoter there is a region of 16 AT pairs; after the gene there is an oligo dT stretch pppGGUUGCGGCCAUAUCUACCAGAAAGCACCGUUUCCCGUCCGAUCAACUGUAGUUAAGCUGGUAAGAGCCUGACCGAGUAGUGUAGUGGGU

GACCAUACGCGAAACUCAGGUGCUGCAAUCU$_{OH}$ *S. cerevisiae* 5S rRNA, inferred from DNA

UUCUCCG ψAG

pppGGUUGCGGCCAUAUCUACCAGAAAGCACCGUUCUCCGUCCGAUCAACUGψAGUUAAGCUGGUAAGAGCCUGACCGAGUAGUGUAGUGGGU

GACCAUACGCGAAACUCAGGUGCUGCAAUCU$_{OH}$ *S. cerevisiae* 5S rRNA (6).

CCAUACCAUCUAG UUCUCCG AUAACCUG

pppGGUUGCGGCCAUACCAUCUAGAAAGCACCGUUCUCCGUCCGAUAACCUGUAGUUAAGCUGGUAAGAGCCUGACCGAGUAGUGUAGUGGGU

AAACCUAG

GACCAUACGCGAAACCUAGGUGCUGCAAUCU$_{OH}$ *S. carlsbergensis* 5S rRNA (4).

CCAUAUCUAGCAG UUCUCCG CUAAG AUCG

pppGGUUGCGGCCAUAUCUAGCAGAAAGCACCGUUCUCCGUCCGAUCAACUGUAGUUAAGCUGCUAAGAGCCUGAUCGAGUAGUGUAGUGGGU

GACCAUACGCGAAACUCAGGUGCUGCAAUCU$_{OH}$ *T. utilis* 5S rRNA (7).

Figure 2. Nucleotide sequences of yeast 5S ribosomal RNAs. RNase T_1 oligonucleotides which differ with those predicted by 5S rDNA have been extracted from each sequence.

```
                                                                                    pppG
CCTCCACCCATAACACCTCTCACTCCCACCTACTGAACATGTCTGGACCCTGCCCTCATATCACCTGCGTTTCCGTTAAACTATCG
GGAGGTGGGTATTGTGGAGAGTGAGGGTGGATGACTTGTACAGACCTGGGACGGGAGTATAGTGGACGCAAAGGCAATTTGATAGC

CCT C C CATA CACCT                                CCT C C CATA CACCT
GGA G G GTAT GTGGA                                GGA G G GTAT GTGGA

  T  A   ATAA A  T T A T   A  TA T AA AT T T  A   T   T ATAT A  T   TTT   TTAAA TAT
  A  T   TATT T  A A T A   T  AT A TT TA A A  T   A   A TATA T  A   AAA   AATTT ATA

CC CC CCC    C CC C C C CCC CC  C G  C  G C GG CCC GCCC C   C CC GCG   CCG    C   CG
GG GG GGG    G GG G G G GGG GG  G C  G  C G CC GGG CGGG G   G GG CGC   GGC    G   GC

     A   A AA A      A     A   A  GAA A G   GGA   G    A A  A   G G    G  AAA  A  G
GGAGG GGG A  G GGAGAG GAGGG GGA GA   G A AGA   GGGA GGGAG A AG GGA G AAAGG AA   GA AG

CCTCC CCC T  C CCTCTC CTCCC CCT CT   C T TCT   CCCT CCCTC T TC CCT C TTTCC TT   CT TC
     T   T TT T      T     T   T  CTT T C   CCT   C    T T  T   C C    C  TTT  T  C
```

Fig. 3. The promoter region for S. cerevisiae 5S ribosomal RNA. The guanosine triphosphate which initiates the RNA chain is shown above the DNA sequence at the extreme right. A 14-nucleotide repeating sequence and AT/GC and purine/pyrimidine distributions in this region are shown below.

29 bases long. The sequence of T's at the end of the gene is reminiscent of a number of bacterial terminations (reviewed in (9)), but the poly U sequence, found in bacteria, would have to have been removed.

If we compare this yeast promoter with the bacterial promoters known, we see no homologies, nor might we expect to. In bacteria, the RNA polymerase interacts with a region of the DNA extending out to some 40 bases before the beginning of the messenger (reviewed in (9)). The polymerase protein also shields the first 18-20 bases that will be transcribed against attack with pancreatic DNase. The bacterial promoter's most defined region is the "Pribnow box", a 7-base-pair long approximate homology to the sequence TATPuATG centered about one turn (8-11 bases) before the initiation of the RNA chain. Another region defined by the prokaryotic work is centered about 35 bases before the beginning of the messenger. Here there are some homologies, utilizing a GTTG sequence, and mutations in both the lac and the lambda promoters show that this region is important. Furthermore, in bacteria the area around 60 nucleotides before the beginning of the messenger can be a site for positive control proteins. The CAP factor (the cyclic AMP binding protein that stimulates promoters such as lac, gal or ara) or the lambda repressor (in its *alter ego* of a positive control factor for its own synthesis (10)) bind there, often to palindromic sites.

The yeast sequence does not resemble any of the common elements in the prokaryotic promoters; further explorations will be necessary to identify the critical nucleotides and regions with which the yeast polymerase interacts.

ACKNOWLEDGEMENTS

This work is supported by the NIH, grant GM 09541 to W.G. W.G. is an American Cancer Society Professor of Molecular Biology. K.G.S. was supported by a postdoctoral fellowship from the USSR Academy of Sciences.

REFERENCES

(1) Maxam, A.M. and Gilbert, W. (1977) Proc. Nat. Acad. Sci. USA, in press.
(2) Bertrand, K., Korn, L., Lee, F., Platt, T., Squires, C.L., Squires, C., and Yanofsky, C. (1975) Science 189, 22.
(3) Roberts, J.W. (1976) in RNA Polymerase, Cold Spring Harbor Laboratory, pp. 247-271.
(4) Hindley, J. and Page, S.M. (1972) FEBS Lett. 26, 157.

(5) Smith, H.O. and Birnstiel, M.L. (1976) Nucleic Acids Res 3, 2387.
(6) Miyazaki, M. (1974) J. Biochem. (Tokyo) 75, 1407.
(7) Nishikawa, K. and Takemura, S. (1974) FEBS Lett. 40, 106.
(8) Weinmann, R., and Roeder, R.G. (1974) Proc. Nat. Acad. Sci. USA 71, 1790.
(9) Gilbert, W. (1976) in RNA Polymerase, Cold Spring Harbor Laboratory, pp. 193-205.
(10) Ptashne, M., Backman, K., Humayun, M.Z., Jeffrey, A., Maurer, R., Meyer, B., and Sauer, R.T. (1976) Science 195, 156.

THE MECHANISM OF PHASE VARIATION

Janine Zieg, Michael Silverman,
Marcia Hilmen and Melvin Simon

Department of Biology (B-022)
University of California, San Diego
La Jolla, California 92093

ABSTRACT. Previous studies of the expression of flagellar antigens in *Salmonella* have shown that they are specified by two different genes, H1 and H2. The expression of these genes is regulated such that only one gene activity, or phase, is expressed at any given time. In contrast to other known regulatory systems, the switch in expression appears to be a result of a change at the DNA level. It has been suggested that this is an example of an alternate mechanism involving site specific recombination that is required for gene expression. In order to determine the mechanism involved, molecular cloning techniques were used to isolate the segments of *Salmonella* DNA that contained the H1 and H2 loci. Heteroduplex analysis of this DNA revealed an anomaly, i.e., an apparent inversion, which was shown to map adjacent to the gene controlling the structure of the H2 flagellar antigen. A correlation was demonstrated between the state of the H2 gene and the sequence of the adjacent segment, i.e., when the adjacent segment was in one orientation H2 was expressed; in the other orientation, H2 was not expressed. We propose that an inversion of this region is the phase determining event in flagellar gene expression in *Salmonella*.

INTRODUCTION

Bacterial systems have provided models for the molecular mechanisms involved in the regulation of gene expression. These generally involve transient interaction of a specific binding protein with a site on the DNA adjacent to the regulated gene. There are, however, examples of regulatory phenomena in prokaryotic cells which imply a completely different mechanism, one that involves a metastable change in the DNA structure (18,22,23). Phase variation in *Salmonella*, an example of such a phenomenon, was discovered by Andrewes in 1922 (1). He found that *Salmonella typhimurium* cultures contained two different flagellar serotypes. In 1949, Stocker showed that the variation was the result of the ability of *Salmonella* strains to switch from one phase (flagella antigen) to another with frequencies ranging from 10^{-5} to 10^{-3}

per bacterium per generation (21). These changes occurred at relatively high frequencies, and they consisted of an alteration between two specific antigens rather than the formation of new antigenic alleles. Therefore, they could not be explained as mutations. Lederberg and Iino studied the genetic basis of the phase transition (11). They concluded that there were two genes, H1 and H2, that controlled the alternative flagellar filament structural proteins. Furthermore, the ability to switch from the expression of H1 to H2 was controlled by a genetic element closely linked to the H2 gene. Finally, they found that the state of expression of the H2 gene could be transduced, and they concluded, therefore, that the regulation must take place at the DNA level. Iino and his co-workers subsequently extended the genetic characterization of the system in *Salmonella*. They demonstrated the presence of promotor-like mutations of the H1 and H2 genes ($aH1^-$ and $aH2^-$), and other mutations that affected the frequency of switching ($vH2^-$). $vH2^-$ was a genetic element that mapped near H2 and which decreased the frequency of phase transition several orders of magnitude. It was found to be effective only in a cis configuration with the H2 gene. Another element, rh1, functioned as a repressor of H1 activity and was coordinately expressed with H2. Thus, when H2 was expressed, i.e., H2-on, both the H2 and rh1 gene products were produced and rh1 repressed the synthesis of the H1 product. However, when H2 was "off", rh1 was not produced and the H1 gene was expressed. Very little is known about the molecular mechanism for phase switching. It has been suggested that this mechanism might be involved in an insertion of an heterologous sequence, i.e., an IS element, whose presence would inactivate H2 and rh1 (9,16). An alternative hypothesis may involve an inversion of a region adjacent to H2. Or the methylation of a regulatory sequence, e.g., the specific methylation of a promotor region in the H2 gene, could also explain the phenomenon (8).

METHODS AND RESULTS

These hypotheses can be tested by examining preparations of specific clones of DNA containing the H2 region. If an anomaly in this region were involved, it could be detected directly using the cloned DNA. It has been shown that the phase variation system in *Salmonella* will function after transfer to *E. coli* (4,12). Thus, *E. coli* strain C600 rk^-mk^- Hag^- was used as recipient for cloning the H2 gene. *Salmonella* DNA (SL4213) (4) and colcinogenic factor E1 were digested with Eco R1 endonucleases, ligated, and transformed into the recipient *E. coli* strain (5). The transformants were streaked onto motility agar containing colicin, and the clones that emerged as swarms from the streaks were isolated. Single colonies were picked and tested for their reactivity with specific anti-flagellar antibody. Non-cross reacting serum specific for H1-b antigen and H2-enx antigen was used. The clones tested were either immobilized by anti-b antiserum and not by anti-enx antiserum, or *vice versa*. DNA was isolated from a clone expressing H2-enx and another expressing the H1-b antigen. The DNA was treated with Eco R1 and examined by electron microscopy (3). No obvious anomalies were observed. Each DNA sample was then denatured, allowed to reanneal, and the products were examined. There were no anomalous features to the DNA derived from the H1-b clone. However, out of the DNA derived from the clone that expressed H2-enx flagella, about 10% of the restricted fragments contained a bubble-like structure (Fig. 1). The DNA was further characterized using restriction endonucleases. Eco R1 endonuclease treatment separated the *Salmonella* DNA fragment I from the Col factor DNA. The *Salmonella* DNA measured 4.96 $\pm$ .04 microns which corresponds to a molecular weight of 9.9×10^6 daltons. The bubble region was 0.27 $\pm$ .03 microns in length and corresponded to between 750 and 1000 base pairs. The shorter arm adjacent to the bubble was .49 $\pm$ .04 microns, approximately 1450 base pairs. The position of three Sal I endonuclease sites as determined by electron microscopy and agarose gel electrophoresis with respect to the R1 sites are shown in lower Fig. 1. Deletions in fragment I were obtained by isolating partial Sal I digest fragments from agarose gels. (The Col E1 factor does not have a site for Sal I endonuclease). The fragments (II and III) were then used to retransform C600 Hag^-. It was found that even Col E1-fragment III hybrids gave transformation and motile enx transformants. The DNA isolated from these transformants was shown to carry only fragment III. The arms of fragment III are each approximately .5 microns, sufficient to code for a 50,000 dalton protein which is the approximate size of the H2 gene product (10). Therefore, the DNA that codes for the H2 gene product must be adjacent to or included in the bubble.

The DNA that codes for H2 was also cloned by an alternative method. In this technique, *Salmonella* DNA carrying H2-enx was sheared to pieces ranging in size from 4 to 10 megadaltons. The sheared DNA pieces were then treated with terminal transferase in the presence of deoxy-CTP to introduce C-tails on the DNA (15). The plasmid, pA01, which carries kanamycin resistance (a derivative of Col E1-Kan) has only a single pst site and was used as the vehicle. It was treated with terminal transferase, and polydeoxy GTP tails were formed on the pst ends of the plasmid.

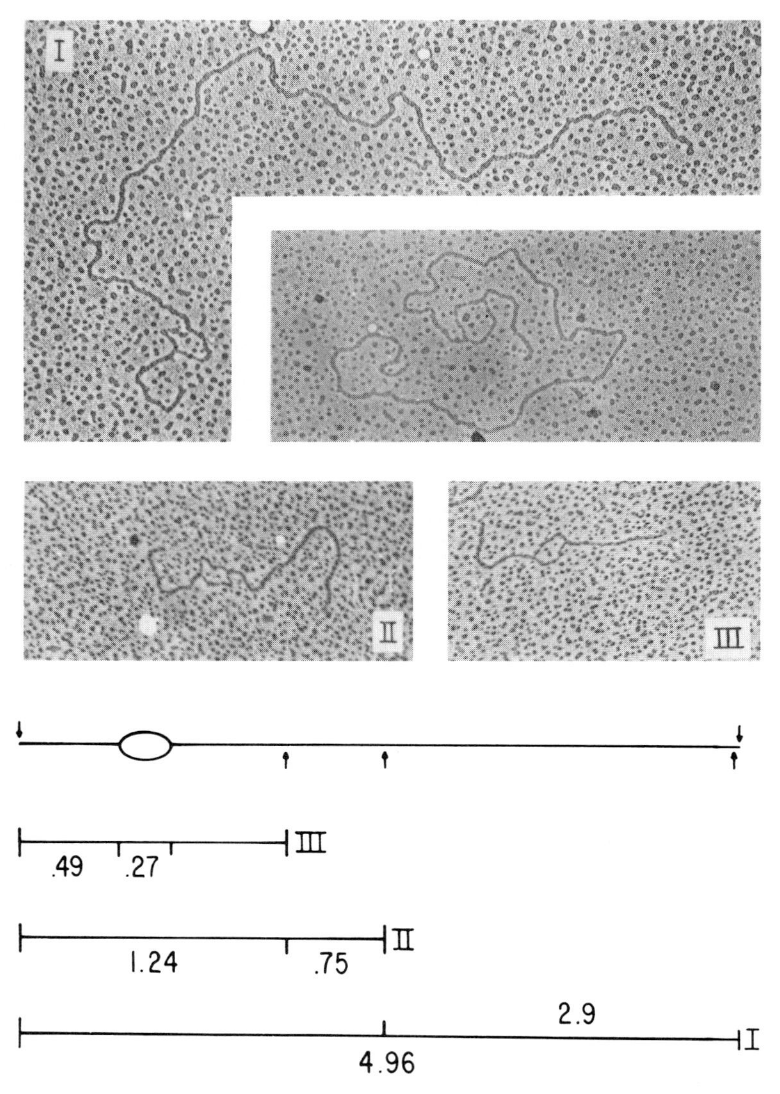

FIG. 1. Electron micrographs of heteroduplexed Salmonella H2 DNA fragments I, II, and III showing the bubble-like anomaly. The lower Figure indicates positions of Eco R1 sites (arrows above live) and Sal I sites (arrows below line). Covalently closed Col E1 and single strand ØX174 circles were used as standards in the measurements. Dimensions are shown in microns.

The plasmids and random fragments of DNA were then annealed and used for transformation into the E. coli C600 Hag$^-$ recipient strain. This technique has been shown to result in the insertion of DNA into the single pst site, and the poly GC tail regions serve to regenerate the pst site so that the inserted fragment may be removed by pst treatment. The transformants were spread on agar containing kanamycin and individual clones were picked. Each clone was then tested for motility, and clones that were found to be motile were tested in the presence of antiserum. One clone (pJZ100) was found that carried H2-enx flagella and also expressed kanamycin resistance. DNA was extracted and characterized by both electron microscopy and by restriction endonuclease digestion and agarose gel electrophoresis. Fig. 2 shows the results of these experiments. The cloned fragment contained two pst sites, both to the left of the bubble (as shown in Fig. 2). When the fragment III was treated with pst, the same sites were observed. The shearing point on the right hand side of the fragment was approximately .1 micron from the end of the bubble. Since this fragment contained the genetic information necessary to code for the H2 gene product, but did not carry sufficient DNA to the right of the bubble for the 50,000 molecular weight protein, the H2 gene must map on the left hand side of the bubble. To further confirm this location, fragment III was treated with pst, then religated and used to retransform. Its ability to express H2-enx was lost. Thus, removing DNA from the left-hand side of the bubble resulted in the loss in the ability to express H2. We conclude, therefore, that the gene that codes for H2-enx expression maps on the left side of the bubble (as drawn). Fig. 2 shows an agarose gel of endonuclease digestions of pJZ100 and the resulting map. Pst I digestion results in four fragments, Sal I digestion in two large fragments and Eco R1 digestion in four fragments.

The next question is, does the event that generates the bubble correspond with phase transition? In order to be able to follow the frequency of phase transition, fragment I DNA was recloned onto a lambda phage vehicle. Fig. 3 shows an electron micrograph of the lambda hybrid derived from Charon I phage (this volume, 2) carrying fragment I. DNA was denatured, renatured and the bubble was clearly visible. This also indicates that if there were a correlation between the bubble and phase variation, we would expect that the lambda phage should be demonstrating phase variation. The lambda phages were used as transducing particles to study the expression of H2. Single plaques were picked and the phage were tested for their ability to transduce H2-enx to E. coli that lacked the flagellin gene (MS5014) (20). High level transducers and low level transducers were found in the lambda

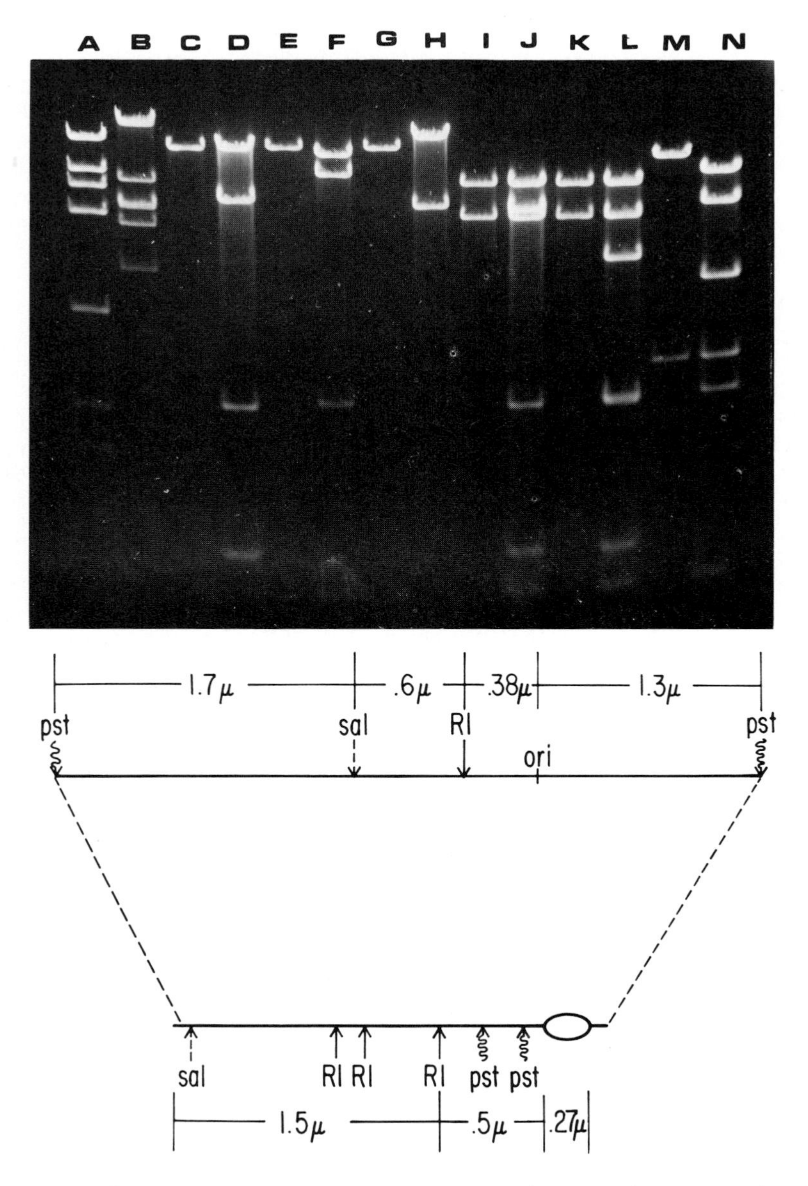

FIG. 2. Agarose gel electrophoresis showing restriction endonuclease digests of plasmid pJZ100. Fragment sizes are determined by comparison with molecular weight standards: (A) Hpa I digested R6K (8.5, 5.0, 4.2, 3.0, 1.5, 1.0, 0.8, and 0.2 megadaltons); (B) Eco Rl digested lambda phage (13.7, 4.7, 3.7, 3.6, 3.0 and 2.1 megadaltons); (C) pAOl/PST I; (D) pJZ100/Pst I; (E) pAOl/Eco Rl; (F) pJZ100/Eco Rl; (G) pAOl/

FIG. 2. continued.
SalI; (H) pJZ100/Sal I; (I) pA01/Pst I + Sal I; (J) pJZ100/Pst I + Sal I; (K) pA01/Pst I + Eco R1; (L) pJZ100/Pst I + Eco R1; (M) pA01/Sal I + Eco R1; (N) pJZ100/Sal I + Eco R1.

The resulting map is shown above. Measurements are in microns. Restriction endonuclease Eco R1 was obtained from R. Kolter; Sal I and Pst I were purchased from New England Bio-Labs, Beverly, Massachusetts.

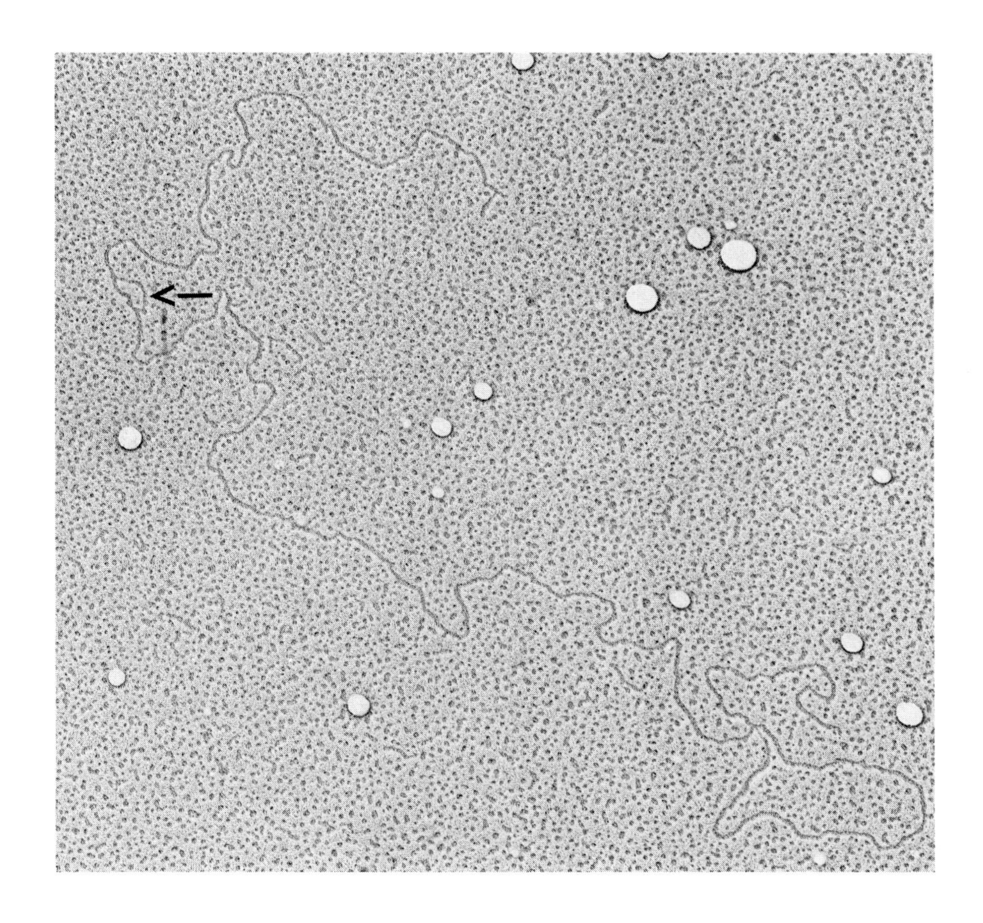

FIG. 3. Heteroduplex of lambda phage Charon I carrying fragment I. The position of the bubble region is indicated by the arrow.

population. When single plaques from a population of low level transducers were tested, 1-2% were found to give high levels. When plaques from phages that gave high levels of complementation were tested, 80% of them gave high levels and 20% low levels. This analysis could be repeated, and each time when single plaques were picked, they were found to be mixed. In general, those that showed high levels of transduction contained 5-20% phage that transduced at low levels, while phage that had predominately low levels of transduction also contained 1-5% high level transducing phage. We can assume that these two types represent the behavior of phage carrying phase 2-on and phase 2-off. There is an apparent asymmetry in the frequency of phase transition. This is similar to the initial observation of Stocker and also of Makela (12,21). They found that the phase transition proceeded from 4-10 times faster in the direction from phase 2-on to phase 2-off than in the reverse direction.

In order to determine if the bubble correlated with the expression of different states of H2, DNA from both lambda populations, i.e., those containing mostly the high level transducing phages and populations which contained mostly low-level transducing phages, was prepared. The DNA was renatured and reannealed and the frequency of inversions was measured. 300 molecules of lambda H2-off DNA were scored, and 2.5% had the bubble. 150 molecules of DNA carrying H2-on were scored, and 12% were found to carry the bubble. When equal amounts of these preparations were mixed, denatured and reannealed, 34% of 150 molecules were found to have the bubble. The large increase in the frequency obtained when the two types of lambda DNA were mixed indicated that it was a heteroduplex between phase 2-on and phase 2-off regions that form the bubble.

DISCUSSION

We suggest, therefore, that phase transition results from an inversion of a specific region of the DNA, i.e., in one orientation the H2 gene can be transcribed, while in the opposite orientation it is not transcribed. This hypothesis is shown in Fig. 4. Further evidence to support this model has come from examinations of fragments of the cloned DNA that were denatured, renatured and spread from a 30% formamide hypophase for electron microscopy. If the bubble is the result of an inversion, we would expect that after denaturation, the DNA in the bubble region could pair while the ends of the DNA, now in the opposite orientations, would not be homologous and would not be able to renature. We would, therefore, expect to see "H" like structures. These kinds of structures were in fact observed in preparations of pst treated pJZ100. Thus on the basis of these experiments, we

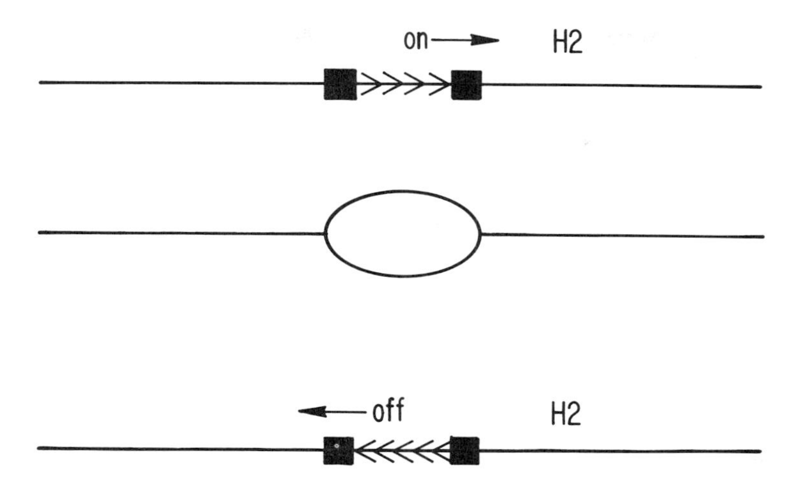

FIG. 4. A model for the mechanism of phase variation in Salmonella. The orientation of the region between the blocks could control transcription of the H2 gene.

conclude that phase variation is an example of a recombinational switch which controls gene expression. Presumably some site specific recombination enzyme is able to catalyze the inversion of the fragment of DNA which contains a promotor-like region and, thus, turns on and off expression of the H2 gene. Two interesting questions arise concerning these kinds of enzyme systems. First, if phase transition involves a simple recombinational event, why is the transition asymmetric? The ordinary recombination systems involving recA and recBC may not control phase transition, but rather, a completely different enzyme system may be involved. It is clear that there are a number of elements which reside as normal tenants of E. coli which are able to insert themselves in a variety of places on the chromosome and also may be removed and reinserted elsewhere (17). The enzymes which govern these processes must recognize specific regions in the insertion sequences. At least 3 different insertion sequences have been clearly defined: IS 1, IS 2, and IS 3. Thus, there may be a large class of site specific recombination enzymes which can be involved both in the insertion of

IS elements and in the kinds of phenomenon observed in phase transition. The second interesting question is the nature of the vH2⁻ state in Salmonella, in which phase transition is fixed. On the basis of the model proposed, the vH2⁻ could be an alteration in the sequence recognized by the enzyme that promotes recombination and, therefore, the frequency of inversion would be decreased. A mutation in this site no longer recognizable by the enzyme can also be used to select for modifications of the enzyme that can recognize the altered structure. These would suppress the effects of vH2⁻ mutations.

Thus, we have in prokaryotes an example of a regulatory mechanism which involves a change in the structure of the DNA. There are many observations in classical eukaryotic genetics which suggest that these kinds of changes also occur. It was initially pointed out in the early analysis of phase variation that the effects observed were very similar to those observed by McClintock in the genetic system of maize (13,14). Similar effects have now been found in yeast and in Drosophila genetics (this volume, 6,7,19). Regulatory events involved in development and differentiation could also involve such changes. These results with the phase system suggest that it may be proper to rehabilitate the notion of site specific recombinational events as playing important roles in development and differentiation.

ACKNOWLEDGEMENTS

This research was supported by a grant from the National Science Foundation (PCM 76-17197) and from the National Institutes of Health (USPHS Al 13008-01). J. Zieg was supported by National Institutes of Health (GM 07240-02).

We wish to thank W. Roewekamp and R. Firtel for their valuable assistance with the poly GC tailing technique, and A. Otsuka for the use of plasmid pAO1. We also thank D. Figurski for providing Col E1 DNA, and R. Kolter for the use of plasmid R6K.

REFERENCES

1. Andrewes, F.W. (1922). J. Pathol. Bacteriol. 25,515.
2. Blattner, F. and Williams, B. (1975). Workshop of Safer Prokaryotic Vehicles, La Jolla, California.
3. Davis, R.W., Simon, M., and Davidson, N. (1971). Methods of Enzymology XXI Part D, 413.
4. Enomoto, M., and Stocker, B.A.D. (1975). Genetics 81, 595.

5. Hershfield, V., Moyer, M.W., Yanofsky, C., Lovett, M., and Helinski, D. (1975). Proc. Nat. Acad. Sci. U.S.A. 71, 3455.
6. Hicks, J.B., and Herskowitz, I. (1977). Genetics, in press.
7. Hicks, J.B., Strathern, J.N., and Herskowitz, I., in DNA Insertion Elements, Plasmids and Episomes.
8. Holliday, R., and Pugh, J.E. (1975). Science, 187,226.
9. Iino, T. (1969) Bacteriol. Rev. 33.454.
10. Kondoh, H., and Hotani, H. (1974). Biochem. Biophys. Acta 336,117.
11. Lederberg, J., and Iino, T. (1956). Genetics 41,744.
12. Makela, P.H. (1964). J. Gen. Microbiol. 35,503.
13. McClintock, B. (1956). Cold Spring Harbor Symp. Quant. Biol. 21,197.
14. McClintock, B. (1967). 26th Symp. Soc. Develop. Biol. 184.
15. Otsuka, A. (1977). (University of California, San Diego) in press.
16. Pearce, U.B., and Stocker, B.A.D. (1967). J. Gen. Microbiol. 49,335.
17. Saedler, H., and Heiss, B. (1973). Molec. Gen. Genet. 122,267.
18. Saedler, H., Reif, H.J., Hu, S., and Davidson, N. (1974). Molec. Gen. Genet. 132,265.
19. Shapiro, A.J., and Adhya, S. (1977). Cold Spring Harbor Laboratory, Cold Spring Harbor, New York, in press.
20. Silverman, M., Matsumura, P., and Simon, M. (1976). Proc. Nat. Acad. Sci. U.S.A. 73,3126.
21. Stocker, B.A.D. (1949). J. Hyg., Camb. 47,398.
22. Tonegawa, S., Hozumi, N., Matthyssens, G., and Schuller, R. (1976). Cold Spring Harbor Symp. Quant. Biol., in press.
23. Toussaint, A. (1976). Virol. 70,17.

Summary of Plasmid Vectors Workshop
Convener H. Boyer
Participants: J. Crosa (University of Washington), D. Figurski (University of California, San Diego), F. Bolivar (University of California, San Francisco)

Dr. J. Crosa discussed various aspects of the biology of *E. coli* K12, the most widely used host bacterial strain for plasmid transformations. It was noted that *E. coli* K12 has no O or K surface antigens (as do most *E. coli* strains isolated from various organisms) and consequently does not colonize the human or other animal guts. The addition of two plasmids *ent* and K88 to *E. coli* K12 does not significantly add to the ability of this organism to colonize animal guts. In other coliforms these plasmids confer the property of intensive invasion of the small bowel. Dr. Crosa itemized some of the different types of plasmids found in bacteria and the transient and peripatetic nature of these molecules. The persistence and distribution of plasmids with various genetic make-ups is influenced greatly by the strong selective pressures applied by man, namely, the use of antibiotics in animal feeds and the inappropriate use of antibiotics by parts of the medical profession. The nature of plasmid transposons was also discussed. It was pointed out that the TnA transposon, carrying the gene for ampicillin resistance normally found in enteric strains, has now been found in pathogenic *Haemophilus* strains. In *Haemophilus* infections the antibiotic treatment of choice has been penicillin or one of its derivatives. Dr. Crosa also reviewed his work on the nature of the bivalent replication mechanism in the plasmid R6K.

Dr. D. Figurski discussed the replication properties of four plasmids: F (or mini-F) and RK2 which are "stringently" replicating molecules and two plasmids with relaxed replication modes, *col*EL and R6K. The general approach used for these experiments is the reduction of the plasmid in question to the minimum essential elements (origin of replication and possible structural genes) necessary for replication. This is done by restriction endonuclease fragmentation of the plasmid and the covalent joining of the fragments to a piece of DNA containing a selective marker such as Kan^r or Tc^r. Figurski and also Crosa reported that the R6K plasmid has two origins of replication and one termination site. The termination site is not absolutely necessary for replication and in each molecule one of the replication origins is picked for a given round of replication. The frequency of initiation is not equally divided between the two origins. Dr. Figurski described the construction of a hybrid molecule containing the RK2 origin and the colEL origin of replication. In this hybrid the RK2 origin of replication and copy number phenotype dominates the molecule's replication.

Dr. Bolivar discussed the construction of several multipurpose cloning vehicles, notably pBR313 and pBR322, and described their utility in cloning experiments. Another plasmid, pBH20, was described which contains the lac operon controlling signals with a single EcoRI site located about 24 nucleotides into the B-galactosidase structural gene. This plasmid can be used as a receptor for DNA fragments with structural information which can be put under the genetic control of the lac operon. Another plasmid, pBR345, was described which contains about 1000 base pairs and is capable of replication. The present information about the position of the origin of replication and the nucleotide sequence in that region of the molecule was presented. The possibility of a structural gene being present on the plasmid was suggested by the failure to clone DNA at a single HaeII restriction site mapping several hundred nucleotides away from the putative origin of replication.

Conclusion:

It is clear that aside from the various phenotypes exerted by different plasmids, the interesting features of these molecules deal with the replication of the DNA molecule. How is copy number and incompatibility controlled? Are these features of the origins of replication and the types of enzymes and proteins utilized by different plasmids? What role do membrane proteins play in these mechanisms? Do some plasmids only utilize host enzymes and proteins or are there structural genes contained on the plasmid which regulate initiation and/or membrane attachment? What determines the host range of a plasmid? What is the nature of the terminus of replication? Does it modulate the expression of certain regions of the plasmid genome? These are some of the questions to be examined in this field.

WORKSHOP SUMMARY: Phage Vector Systems, presented by Dan S. Ray, Molecular Biology Institute, University of California at Los Angeles, California.

Bacteriophage systems for cloning foreign DNA molecules have been developed in several different laboratories. The most highly developed systems utilize the E. coli phage lambda. Peter Philippsen described the λgt phages constructed at Stanford in Ron Davis's laboratory. These phage systems are particularly useful for the cloning of DNA of low biohazard potential. This, as well as all other, lambda phage system(s) takes advantage of the large amount of contiguous genes required for lysogeny but not for virulent growth. This region of the lambda genome is contained within the ECO RI restriction fragments λ•Band λ•Cand can be removed and replaced by foreign DNA. Removal of both λ•Band λ•C from a λ DNA molecule results in a DNA molecule too short after joining to give viable phage particles. For propagation of λ vectors it is necessary to insert one of the λ Eco RI fragments (λ•B or λ•C) or another piece of DNA which is subsequently removed and replaced by the cloned fragment. Another class of λgt phages are sufficiently long after joining that they do not require an inserted piece of DNA for their propagation. Detection of phages of this latter class carrying cloned DNA fragments is based on the much lower plating efficiency (10^{-5}) of phages of about 80% lambda length on certain mutant strains (pel^-) of *E. coli* as compared to phages of about 100% lambda length. In phages where the N gene, a gene involved in the regulation of transcription, has been removed it is necessary to also introduce the *nin* 5 deletion which confers on N^- mutants the ability to grow in the absence of N function. This deletion also provides additional space for the insertion of foreign DNA and prevents the formation of λdv plasmids.

Lambda vectors constructed by Williams and Blattner (University of Wisconsin, Madison) and by Donoghue and Sharp (M.I.T.) are amber in genes J and Z. These phages are also missing the lambda-specific recombination functions which are contained in the λ•C fragment. Two additional features of the Donoghue and Sharp phage vector system are mutations to virulence to render the genome insensitive to repressor already present in possible lambda lysogens which the vector might encounter in nature and a mutation near the J gene that causes the phage to adsorb poorly in liquid medium. Some of the Charon phages also contain an amber mutation in the S gene to prevent both late DNA replication and lysis.

Gerald Vovis described efforts in Zinder's laboratory (Rockefeller University) to develop a phage vector system using the filamentous phage fl. Their work indicates that pSC101 DNA can be inserted into fl replicative form DNA *in vivo*. At present these recombinant phage still require a helper phage for transduction of tetracycline resistance even though all phage genes are expressed. Further development of this system could provide a unique vector system since the DNA encapsulated in the virion is single stranded. This feature could be of considerable value for investigations requiring the isolated strands of a cloned DNA segment. It may also be possible to clone a wide range of sizes of DNA fragments with such systems since the length of filamentous virions seems to be determined by the length of DNA encapsulated.

MODEL RECOMBINANTS FOR THE DEVELOPMENT AND MANIPULATION OF EK2 PHAGE VECTOR SYSTEMS

Daniel J. Donoghue and Phillip A. Sharp

Center for Cancer Research
Massachusetts Institute of Technology
Cambridge, Massachusetts 02139

ABSTRACT. An attenuated bacteriophage λ has been prepared for use as an EK2 vector. This phage, designated *λgt vir Jam*27 *Zam*718-*λB'*, can accomodate up to 11x10^6 daltons of foreign DNA inserted through *Eco R1* ends. The virulence marker and *nin* 5 deletion are present to reduce the frequency of lysogen and/or plasmid formation. Two amber mutations have been introduced, requiring a suppressor in the bacterial host, and the phage recombination functions contained in the *Eco R1*-*λC* fragment have been deleted. In addition, this phage adsorbs at a significantly reduced rate to sensitive bacteria.

Model recombinants have been constructed by *in vitro* recombination with an *Eco R1* fragment coding for resistance to kanamycin. A simple replica plating technique has been employed to quantitate the frequency of *Km*^{R} transfer to infected *E. coli*. This test system can be employed to measure the frequency of gene transfer to virtually any strain of *E. coli*, including clinical isolates.

Another λ recombinant, containing the *mini-ColE1* factor, has been used as a model system to generate deletion mutants. Using EDTA selection and CsCl density fractionation, deletions of any desired size can be generated with randomly located endpoints. This simple procedure can potentially be applied to generate sets of overlapping deletions in any nonessential fragment of DNA inserted into a bacteriophage λ vector.

INTRODUCTION

The ability to insert specific fragments of DNA, whether prokaryotic or eukaryotic in origin, into bacterial vectors has ushered in a new era of molecular biology. The variety of problems amenable to analysis by recombinant DNA techniques is great, and is only slowly being realized. There is no one vector which is suitable for every recombinant DNA project; rather, each recombinant DNA analysis will have its own inherent goals and hurdles, and many different vectors, each responding to a slightly different situation, will be required.

At the present time, prokaryotic recombinant DNA vectors are typically of two types. The first comprises the plasmid vectors, which are generally derived either from R-factors (1,2) or from colicin-factors (3,4,5,6). The other major class of vectors includes the phage vectors, primarily phage λ and its derivatives (7,8,9). Different types of vectors within each class already exist -- some maximize the number of copies of the segment per cell, others maximize the levels of transcription from inserted fragments. A few systems optimize the screening for proteins coded for by the inserted fragment, while still others permit genetic manipulation of the inserted genes. The choice of vector is also influenced by considerations such as the size of the fragment, the type of joint used in linking the fragment to the vector, and not least of all, the assessed biological risk associated with a given experiment.

In order to encourage the development of improved vectors, it is essential to have clearly defined criteria for the acceptance and certification of these vehicles. Particularly in the development of new EK2 vectors, relatively straightforward tests are essential so that a potential vector can be rapidly screened and released to the scientific community for general use.

This communication describes an attenuated derivative of bacteriophage λ which is expected to be suitable for cloning at the EK2 level in *E. coli K*-12 and its derivatives. The working subgroup on phage vectors of the NIH Advisory Committee has established a series of tests of candidate EK2 vectors. To satisfy these tests model recombinants have been constructed with this vector, using a segment of DNA coding for an antibiotic resistance marker, and have been employed to quantitate (a) the frequency of gene transfer to infected *E. coli*, and (b) the frequency with which the vector can potentially be "rearmed" by replication in a heteroimmune lysogen. Some features of these tests will be important, by virtue of their simplicity, in the creation of new and more refined vectors for recombinant DNA use. Two other phage λ vectors have been proposed for use as EK2 vectors (10,11,12). Each possess some advantages over other vectors for certain types of analyses. We believe that a multiplicity of vectors is highly desirable if the full benefits of recombinant DNA technology are to be realized.

Although the complexity of λ genetics can be overwhelming, phage vectors in general provide a greater degree of experimental flexibility than plasmid vectors. For instance, phage vectors permit the rapid transfer of cloned DNA fragments from one bacterial strain to another. In addition, certain types of mutations in inserted DNA fragments can be easily obtained using phage vectors. In this communication, for instance, we describe the generation of overlapping deletion mutants in the *mini-ColE1* factor (3,4) by manipulation of a λ-*mini-ColE1* recombinant.

DESCRIPTION OF THE PHAGE VECTOR SYSTEM

The genotype of the proposed EK2 vector is λ*gt vir Jam*27 *Zam*718-λ*B'*. The construction of this vector has already been detailed elsewhere (13) and will only be summarized here.

λ*gt*-λ*C* (9) was used to construct λ*gt vir Jam*27 *Zam*718-λ*BC* by standard phage crosses. The DNA from this latter phage was then treated with *Eco R1*, followed by T4 DNA ligase, and transfected into C600 r_k^- m_k^-. The proposed vector was isolated from this *in vitro* recombination event; it lacks the *Eco R1*-λ*C* fragment, and contains only the *Eco R1*-λ*B* fragment inserted in the opposite orientation to that in wild type λ. The *Eco R1* restriction maps of these phages are shown in Figure 1.

The proposed vector contains two amber mutations in lambda tail genes *J* and *Z* and thus requires a bacterial amber suppressor for growth. The three cis-acting mutations V_1 V_2 V_3 comprising virulence, which lower the affinity of binding by λ repressor, are also present. The right end of the phage is identical to λ*gt*-λ*C*, containing the *nin* 5 deletion and two mutated *Eco R1* sites.

The phage adsorbs 5-10 fold more slowly to sensitive bacteria than a control phage, and is very dependent upon maltose for adsorption. Because of the slow adsorption of the phage, together with a reduced burst, it is easiest to propagate this phage by the plate stock method. However, titers of 10^{10}-10^{11} phage/ml crude lysate can be routinely obtained; thus, enough phage DNA for most biochemical analyses can be obtained from several infected plates.

A MODEL RECOMBINANT CODING FOR RESISTANCE TO KANAMYCIN

A model recombinant, coding for resistance to the antibiotic kanamycin, was constructed by an *in vitro* recombination event with the plasmid *pML* 21 (14). This phage, designated λ*gt vir Jam*27 *Zam*718-Km^R (see Fig. 1e), contains

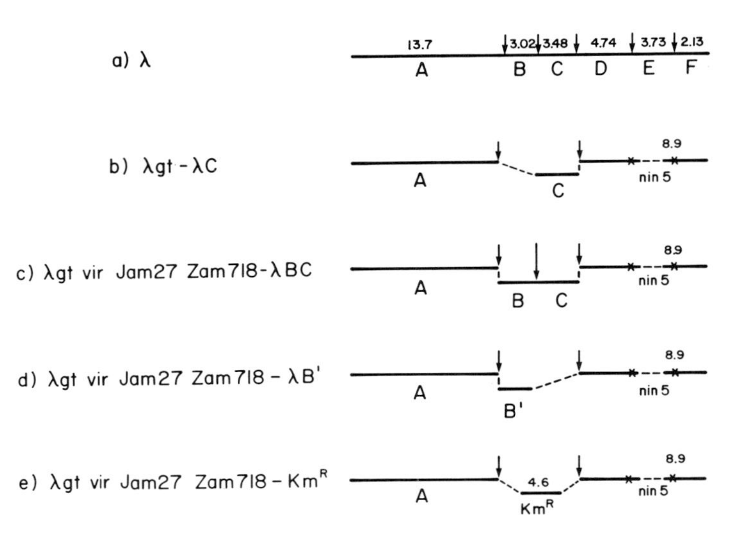

Figure 1. *Eco R1* maps of phage λ and the derivatives discussed in the text: a) The *Eco R1* map of phage λ as determined by Thomas and Davis (25). b) The *Eco R1* map of *λgt-λC* (9), which was the parent for the proposed EK2 vector. c) The *Eco R1* map of *λgt vir Jam*27 *Zam*718-*λBC*, which was an intermediate in the construction of the proposed EK2 vector. d) The *Eco R1* map of the proposed vector, *λgt vir Jam*27 *Zam*718-*λB'*; note the deletion of the λ-*R1-C* fragment, and the insertion in the opposite orientation of the λ-*R1-B* fragment. e) The *Eco R1* map of the model recombinant, carrying a 4.6×10^6 dalton fragment of DNA coding for resistance to kanamycin.

a 4.6×10^6 dalton fragment of DNA originally derived from R-factor R6-5. Selection for kanamycin resistance can be used to select over a 10^8-10^{10} fold range at a kanamycin concentration of 100 μg/ml. The gene responsible for inactivation of kanamycin apparently has its own promoter for expression (13). In addition, this segment of DNA contains an inverted repeated sequence, almost certainly due to the presence of insertion sequences which promote illegitimate recombination in bacteria (15,16,17,18).

Using this model recombinant, the frequency of gene transfer to various strains of *E. coli* has been measured. Since we routinely grow the proposed vector as plate stocks, we also used similar conditions to determine the frequency of gene transfer. Standard plate stocks were

TABLE 1

TRANSFER OF Km^R BY MODEL RECOMBINANT TO INFECTED *E. coli*

infected bacteria		transfer of Km^R
C600 $r_k^- m_k^-$	permissive	$5x10^{-11}$/phage out
C600 $r_k^- m_k^-$ (λgt-λC)	permisive lysogen	$7x10^{-10}$/phage out
WD5021	nonpermissive	$1x10^{-9}$/infected cell
WD5021 (λimm^{434}-*T Oam*29)	nonpermissive lysogen	$1x10^{-8}$/infected cell
493-1	clinical isolate	<$4x10^{-10}$/infected cell

Legend to Table 1. λgt *vir Jam*27 *Zam*718-Km^R was used to prepare plate stocks with the different strains indicated. In all cases, the infecting phage was *K*·modified. For the permissive strains, approximately 10^7 PFU of phage and $2x10^8$ cells were mixed and plated. For nonpermissive strains, approximately $1.25x10^{10}$ phage were plated with $2x10^8$ cells, for a final calculated moi of 5 (assuming a final cell density of $2.5x10^9$ cells/plate). In either case, plates were allowed to grow for 48 hr at 31°C, after which surviving colonies were replica plated to plates containing 100 µg/ml kanamycin sulfate. After an additional 48 hr at 31°C, kanamycin resistant colonies were scored. For permissive strains, the transfer of Km^R/phage out is calculated assuming approximately $2x10^{10}$ PFU/plate. This value is the minimum yield of PFU/plate as measured on duplicate plates after 24 hr at 31°C. In the case of non-permissive strains, the transfer of Km^R is calculated per infected cell. This assay is described in greater detail in Donoghue and Sharp (13).

prepared, using λgt *vir Jam*27 *Zam*718-Km^R as the infecting phage. After 48 hr at 31°C, surviving colonies were replica plated to plates containing 100 µg/ml kanamycin sulfate. After an additional 48 hr growth at 31°C, Km^R colonies were scored.

Some representative results obtained in this fashion are presented in Table 1. Transfer of Km^R is very low when either permissive su_{II} hosts or nonpermissive su^- hosts are used. Even when a homoimmune or heteroimmune λ prophage is present in the infected cells, the frequency of Km^R is still within the limits set by the NIH (19) for EK2 vectors. One clinically isolated strain of *E. coli* is included in Table 1; transfer of Km^R was very low, $<4 \times 10^{-10}$/infected cell. The ease of this replica plating technique, coupled with a strong selective force for kanamycin resistance, permits the rapid quantitation of gene transfer to virtually any strain of *E. coli*. For these reasons, drug resistance genes are the genetic markers of choice for conducting the tests currently required by the NIH for certification of new phage vector systems.

FREQUENCY OF "REARMING" BY A HOMOLOGOUS PROPHAGE

Any attenuated bacteriophage λ vector can presumably be "rearmed" by replication in a bacterium containing a homologous prophage. Accordingly, the NIH (19) have requested data concerning the frequency of this event for potential EK2 vectors. Because of the combination of two amber mutations in the left hand portion of our vector, plus the virulence mutations in the right hand portion, this frequency of "rearming" was easily determined.

Vector "rearming" was measured in two ways: 1) the model recombinant was plated on the nonpermissive heteroimmune lysogen as if preparing a plate stock, after which phage were harvested and titered for "rearmed" *vir* sus^+ phage, and 2) the model recombinant was titered directly on the nonpermissive heteroimmune lysogen and plaques produced by "rearmed" *vir* sus^+ phage were counted. Although both methods gave comparable results, only the latter will be described here. Nonpermissive heteroimmune lysogens carrying the λimm^{434}-*Tomizawa* prophage were used, as this lowers the rate of spontaneous induction, resulting in lower levels of free phage in the bacterial cultures (20). Similarly, in another experiment, the amber mutation *Oam*29 was incorporated in the heteroimmune prophage to further lower the level of free phage in the cultures. These precautions were necessary so that only "rearming" from an integrated prophage would be measured. It should be noted that λimm^{434} is highly homologous with phage λ and possesses a compatible late gene regulatory system.

As shown in Table 2, the value of rearmed phage/input phage was determined by titering the model recombinant on two different nonpermissive heteroimmune lysogens. The values obtained ranged between $1.6\text{-}3.2 \times 10^{-4}$ rearmed phage/ input phage. In these experiments rearming on the left side would produce the rearmed phage λZ^+ J^+ Km^R *vir nin* 5, and

TABLE 2

REARMING OF THE MODEL RECOMBINANT BY TITERING ON HETEROIMMUNE LYSOGENS

lysogen	rearmed phage/input phage
WD5021 su^- (λimm^{434}-T)	3.2×10^{-4}
WD5021 su^- (λimm^{434}-T $Oam29$)	1.6×10^{-4}

Legend to Table 2. The infecting phage was λgt *vir* *Jam*27 *Zam*718-Km^R and was *K*·modified. The "rearmed phage/input phage" number was determined by titering the phage on the indicated strains, and comparing the titer to the su^+ titer. All bacterial cultures used were in late log phase, and were confirmed at the time of the experiment to be su^-, λ^S, and imm^{434}. As a control, the efficiency of plating λgt *vir*-λC was checked on the strains and was identical with the titer on Q1 su_{II}. All plaques, regardless of morphology, were scored in the results reported above. Some plaques were of minute size, however, and probably did not reflect vector rearming during the initial replication of the infecting phage. All plates were grown at 31°C. The value reported for WD5021 su^- (λimm^{434}-T) is the average of two determinations, 5.5×10^{-4} and 2.2×10^{-4} rearmed phage/input phage. The value for WD5021 su^- (λimm^{434}-T *Oam*29) is the average of five determinations, ranging from 6.6×10^{-4} to 7.7×10^{-5} rearmed phage/input phage.

the reciprocal recombinant λZ^- J^- imm^{434} O^-. Similarly, rearming on the right side would yield the rearmed phage λZ^- J^- Km^R imm^{434} O^-, and the reciprocal recombinant λZ^+ J^+ *vir* *nin* 5. In the experiment described above, we scored the sum of λZ^+ J^+ Km^R *vir* *nin* 5 plus λZ^+ J^+ *vir* *nin* 5. The frequency of λZ^+ J^+ *vir* *nin* 5 produced will be equivalent to the frequency of λZ^- J^- Km^R imm^{434} O^-. Thus, the numbers represent the sum of rearming both on the left and right arms. Since this sum is less than 10^{-3} rearmed phage/input phage, then it follows that the frequency of rearming on each arm separately must be less than 10^{-3} rearmed phage/input phage, the limit required by the NIH for an EK2 vector.

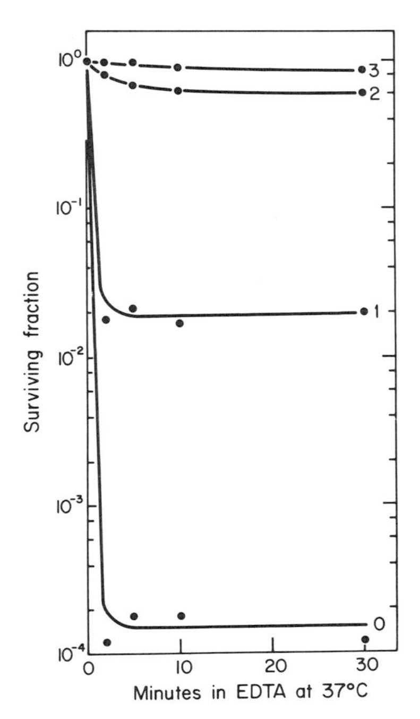

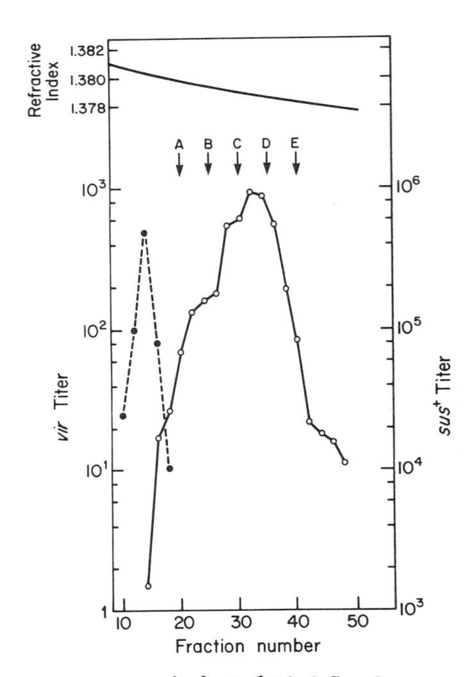

Figure 2 (left). Survival of *λgt- Km-ex mini-ColE1 λC* after treatment with 10 mM EDTA at 37°C for 30 min. The time course of EDTA inactivation is shown for the parent phage (line 0) and for phage stocks prepared after 1, 2, or 3 passages in EDTA. The parent phage is 102% of normal λ size; conditions for EDTA selection were essentially as described by Parkinson and Huskey (21). All phage stocks used for EDTA selection were prepared from plate stocks grown at 39°C in WD5021 *su*$^-$.

Figure 3 (right). Equilibrium CsCl centrifugation of EDTA selected phage. Phage taken after three EDTA selections (line 3,Fig. 2) were adjusted to a refractive index of 1.3790 with CsCl. Centrifugation was in a Ty65 rotor, 48 hr, 28,000 rpm, 15°C. 0.1 ml fractions were collected through the bottom. Deletion phages were assayed by titering *sus*$^+$ phage on WD5021 *su*$^-$. *λgt vir Jam27 Zam718 - Km-ex mini-ColE1 λC* was added as an internal marker to determine the density of the parent phage; this marker was titered for *virulence* on Q1 *suII (λcI857 Oam29)*.

GENERATION OF DELETION MUTANTS IN CLONED FRAGMENTS OF DNA

One advantage of using phage λ as a recombinant DNA vector is the high degree of flexibility offered in the selection of bacteria and conditions for proliferation. For example, growth conditions can be selected to rapidly generate sets of deletion mutants in inserted genes. This can permit the mapping of genetic functions in the inserted DNA, if any genetic functions are known; or deletion mutants can also be used to provide a set of specific hybridization probes with which to dissect the regulation of the inserted segment.

Since Mg^{++} is required for the stability of λ phage heads, treatment with chelating agents such as EDTA selects for spontaneously arising deletions, which are more resistant to the chelating agent (21). By varying the temperature at which the EDTA treatment is performed, one can select for larger or smaller deletions, as desired. Besides being extremely straightforward, this technique has the advantage that only small numbers of phage need be manipulated. Moreover, the endpoints of the resulting deletions are randomly located.

To demonstrate this, we chose an *in vitro* recombinant containing three inserted *Eco R1* fragments: 1) the λ-*R1-C* fragment, 2) the *mini-ColE1* factor, and 3) a fragment named *Km-ex*, produced from the 4.6×10^6 dalton fragment coding for Km^R by deletion of the inverted repeat sequence. The *Eco R1* map of this phage, and the molecular weights of the *Eco R1* fragments, are shown in Fig. 4.

In order to select for spontaneously arising deletions, the parent phage was treated with 10 mM EDTA at 37°C for 30 min. The survivors were used to make a new phage stock, which was subjected to a second EDTA treatment. This procedure was repeated a third time. The EDTA killing curves are shown for the parent (line 0, Fig. 2), and for the new phage stocks after 1, 2, and 3 passages in EDTA (lines 1, 2, and 3, respectively, Fig. 2). Each passage in EDTA produced a stock of increasingly resistant phage, so that after 2 or 3 passages, the phage were 95-100% EDTA-resistant.

When these phage were banded in an equilibrium CsCl gradient and fractionated, the profile shown in Fig. 3 was obtained. The majority of phage, as expected, banded at a less dense position of the gradient than the parent. In order to characterize the size of the resulting deletions, eight plaques were randomly picked from each of the fractions labelled A through E in Fig. 3. The DNA was prepared from each of these 40 phages and examined by cleavage with various restriction enzymes; in each case, the size of the deletion

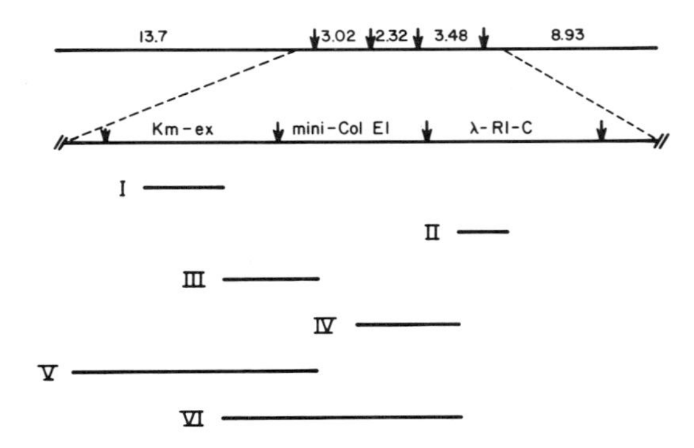

Figure 4. *Eco R1* map of the parent phage for the deletion mutants. The center portion, with three inserted *Eco R1* fragments, is expanded. Forty phages from gradient fractions A through E in Fig. 3 were found to fit into one of the Classes I through VI. Other classes are possible; however, these would involve deletions smaller on the average than those of fraction A, or larger on the average than those of fraction E. Although the entire λ-*R1-C* fragment could be deleted, there is a growth advantage in maintaining the λ*red* gene (22,23). Thus, the deletions isolated in Classes II, IV, and VI do not generally extend into the righthand portion of the λ-*R1-C* fragment. Sixteen different phages with endpoints in the *mini-ColE1 EcoR1* fragment were isolated; these belong to Classes II, III, and V.

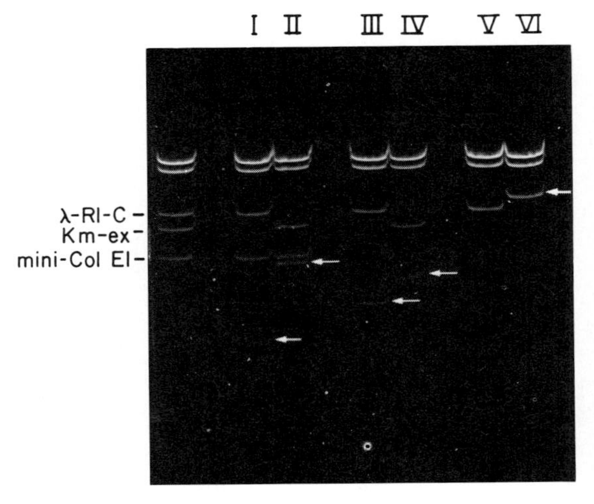

Figure 5. *Eco R1* digestion of representative deletion mutants. Each track shows the *Eco R1* digestion of the parent (far left) or an example from Classes I through VI, as described in Fig. 4. The parent phage in this case is λ*gt - Km-ex mini-ColE1* λ*C*, and contains the three inserted fragments as shown. 0.25 μg of DNA prepared from CsCl banded phage preparations was digested with *Eco R1* and electrophoresed in 1.4% agarose gels as previously described (13,24).

could be determined from the altered restriction fragment patterns obtained. As expected, the average size of the deletion increased from fraction A through fraction E. The average sizes of the deletion mutants examined, for fractions A through E, respectively, were as follows (in megadaltons): A 2.2±1.0, B 3.2±0.7, C 4.1±0.4, D 4.7±0.3, and E 5.1±0.7. Thus, deletions of any desired average size could quite easily be obtained by plating out the appropriate fraction of the gradient.

The forty phage DNA's examined could all be placed into one of the Classes I through VI, as shown in Fig. 4. These classes were determined by the location of the deletion endpoints in given *Eco R1* fragments of the parent phage. *Eco R1* digestions of representative phages from Classes I through VI are shown in Fig. 5.

These deletion mutants will be used to map the genetic functions encoded by the *mini-ColE1* factor. A similar deletion approach could be applied to any piece of DNA cloned in phage λ. This type of analysis, when applied to cloned fragments of eukaryotic DNA, represents one of the more exciting possibilities offered by EK2 phage vector systems.

DISCUSSION

Recombinant DNA techniques have made possible the rapid biochemical creation of specific transducing phages. Regardless of the origin of the inserted DNA fragments, once linked into a phage lambda vector, they become subject to the now-classic protocols for mutagenesis and genetic selection. For example, genetic selection has permitted the identification of specific segments of yeast DNA which are able to complement *IGPD* deficiency in *E. coli* (26). Even if no genetic function in an inserted fragment is identifiable, techniques such as the rapid creation of deletion mutants should make possible an analysis of transcription and translation from the inserted fragments.

The NIH guidelines for recombinant DNA research (19) proscribe certain types of genetic selections in phage vector systems at the EK2 level. Nonetheless, as improved EK2 vectors become available, some of the maneuverability of phage systems will surely be recovered.

This communication has described an attenuated bacteriophage lambda vector, which has been proposed for use at the EK2 level. In addition, straightforward tests have been described for satisfying the necessary criteria for certification as an EK2 vector; these tests should prove generally applicable in the development of other improved lambda vectors. One type of physical analysis -- the generation of deletion mutants in inserted DNA fragments -- has been described which is potentially applicable to any fragment of DNA in a phage lambda vector. Hopefully, the development of other general techniques such as this will permit phage vector systems to become powerful tools in the analysis of eukaryotic genetic function at the EK2 level.

ACKNOWLEDGEMENTS

The authors gratefully acknowledge the assistance of Margarita Siafaca for the thankless task of preparing this manuscript. In addition, D.J.D. wishes to acknowledge encouraging and stimulating discussions with Ellen Rothenberg and Susan Berget during the course of this work.

REFERENCES

1. Chang, A.C.Y. and Cohen, S.N. (1974) *Proc. Natl. Acad. Sci. U.S.A.* 71, 1030.
2. Morrow, J.F., Cohen, S.N., Chang, A.C.Y., Boyer, H.W., Goodman, H.M. and Helling, R.B. (1974) *Proc. Natl. Acad. Sci. U.S.A.* 71, 1743.
3. Hershfield, V., Boyer, H.W., Yanofsky, C., Lovett, M.A. and Helinski, D.R. (1974) *Proc. Natl. Acad. Sci. U.S.A.* 71, 3455.
4. Hershfield, V., Boyer, H.W., Chow, L. and Helinski, D.R. (1976) *J. Bacteriol.* 126, 447.
5. Polisky, B., Bishop, R.J. and Gelfand, D.H. (1976) *Proc. Natl. Acad. Sci. U.S.A.* 73, 3900.
6. Tanaka, T., Weisblum, B., Schnos, M. and Inman, R.B. (1975) *Biochem.* 14, 2064.
7. Murray, N.E. and Murray, K. (1974) *Nature (London)* 251, 476.
8. Rambach, A. and Tiollais, P. (1974) *Proc. Natl. Acad. Sci. U.S.A.* 71, 3927.
9. Thomas, M., Cameron, J.R. and Davis, R.W. (1974) *Proc. Natl. Acad. Sci. U.S.A.* 71, 4579.
10. Enquist, L.W., Tiemeier, D., Leder, P., Weisberg, R. and Sternberg, N. (1976) *Nature* 259, 596.
11. Tiemeier, D., Enquiest, L. and Leder, P. (1976) *Nature* 263, 526.
12. Williams, B.G., Moore, D.D., Schumm, J.W., Grunwald, D.J., Blechl, A.E. and Blattner, F.R. (1977) *Science*, in press.
13. Donoghue, D.J. and Sharp, P.A. (1977) *Gene*, in press.
14. Lovett, M.A. and Helinski, D.R. (1976) *J Bacteriol.* 127, 982.

15. Berg, D.E., Davies, J., Allet, B. and Rochaix, J.D. (1975) *Proc. Natl. Acad. Sci. U.S.A.* 72, 3628.
16. Kleckner, N., Chan, R.K., Tye, B. and Botstein, D. (1975) *J. Mol. Biol.* 97, 561.
17. Kopecko, D.J. and Cohen, S.N. (1975) *Proc. Natl. Acad. Sci. U.S.A.* 72, 1373.
18. Rubens, C., Heffron, F. and Falkow, S. (1976) *J. Bacteriol.* 128, 425.
19. National Institutes of Health (1976) *Federal Register* 41, 27901, and unpublished.
20. Pirrota, V. and Ptashne, M. (1969) *Nature* 222, 541.
21. Parkinson, J.S. and Huskey, R.J. (1971) *J. Mol. Biol.* 56, 369.
22. Enquist, L.W. and Skalka, A. (1973) *J. Mol. Biol.* 75, 185.
23. Cameron, J.R., Panasenko, S.M., Lehman, I.R. and Davis, R.W. (1975) *Proc. Natl. Acad. Sci. U.S.A.* 72, 3416.
24. Sharp, P.A., Sudgen, B. and Sambrook, J. (1973) *Biochem.* 12, 3055.
25. Thomas, M. and Davis, R.W. (1975) *J. Mol. Biol.* 91, 315.
26. Struhl, K., Cameron, J.R. and Davis, R.W. (1976) *Proc. Natl. Acad. Sci. U.S.A.* 73, 1471

BACTERIOPHAGE f1 AS A VECTOR FOR CONSTRUCTING RECOMBINANT DNA MOLECULES

Gerald F. Vovis, Mariko Ohsumi and Norton D. Zinder

The Rockefeller University, New York, New York 10021

ABSTRACT. DNA recombinants were obtained between the non-lytic, filamentous bacteriophage f1 and the non-conjugative plasmid pSC101 by screening for tetracycline transducing particles amongst the phage produced from f1 infected *Escherichia coli* harboring pSC101. An *E. coli* tetracycline resistant transductant was obtained that extruded filamentous particles of which as many as 17% contained the chimera molecule. Data are presented that are consistent with the DNA recombinant having been formed by the complete insertion of one of the parental molecules into the other. The use of such a recombinant vector for the construction of other chimera molecules is discussed.

INTRODUCTION

f1 is a male-specific, filamentous bacteriophage. Infected *Escherichia coli* are not lysed by the phage but rather continuously extrude phage through the membrane after a short eclipse period. The virion consists of several thousand molecules of the major coat protein covering a circular, single-stranded DNA molecule of about 1.8×10^6 daltons with one or two molecules of the minor coat protein at the end(s) of the phage particle (6). The existence of mini-phage particles, i.e. deletion mutants where as much as 80% of the genome has been deleted (2,3,4), demonstrates that it is the size of the DNA molecule which determines the size of the phage particle. Furthermore, the existence of multi-unit length phage particles, even though the DNA molecules within are unit length (6), suggests that filamentous phage containing DNA molecules much greater in size than that of the f1 genome should be able to be formed.

We have been investigating this question of how large a DNA molecule can be efficiently replicated and extruded as a filamentous phage particle. To this end, filamentous particles containing *in vivo* recombinants between f1 and the non-conjugative plasmid pSC101, which confers resistance to tetracycline (1), have been isolated. These recombinants are currently being characterized. This report concerns one such recombinant.

Isolation of in vivo f1-pSC101 recombinants

Tetracycline resistant (tet^r) transducing particles were detected amongst the phage produced by pSC101 containing *E. coli* infected with f1. All of the tet^r *E. coli* transductants that were examined were found to produce tet^r transducing particles at a very low frequency, i.e. $< 1 \times 10^{-8}$ tet^r transducing particles/plaque forming unit (10). An *E. coli* tet^r transductant was treated with N-methyl-N'-nitro-N-nitrosoguanidine and amongst the survivors were tet^r *E. coli* that produced tet^r transducing particles at a much higher frequency, i.e. $> 1 \times 10^{-2}$ (10). The particles produced from one such tet^r *E. coli* transductant were chosen for further investigation.

Characterization of an f1 lysate containing tet^r transducing particles

A culture of *E. coli* was infected with an f1 lysate that contained, by bioassay, about 1×10^{-2} transducing particles/plaque forming unit (pfu) and dilutions thereof were spread on the surface of agar underneath which tetracycline had previously been spread. A tet^r *E. coli* transductant was picked and used as inoculum for a culture. [^{3}H-methyl] thymidine and deoxyadenosine were subsequently added to label radiochemically the phage produced. The phage were concentrated by centrifugation after addition of polyethylene glycol (PEG) 6000 and NaCl (11). The concentrated phage stock was titered for pfu and tet^r transducing activity per ml. The results were 6×10^{11} and 1.6×10^{10}, respectively. Thus, the bioassay indicates that the tet^r transducing particles constitute approximately 3% of the phage particles.

An aliquot of the concentrated phage stock was subjected to velocity sedimentation on a preformed, neutral sucrose gradient. The gradient was fractionated and selected fractions were assayed for radioactivity, pfu, and tet^r transducing activity. The results are shown in Figure 1.

Two peaks of radioactivity were present. A previous analytical centrifugation of the ^{3}H labeled concentrated stock with ^{32}P labeled f1 phage showed that the slower sedimenting peak cosedimented with the ^{32}P labeled f1 marker, while the faster peak sedimented significantly faster than the 3-7% double-unit length particles normally found in an f1 phage stock (6). The shoulder ahead of the slower sedimenting peak in Figure 1 (fractions 40 through 42) corresponds to the expected position of double-unit length f1 particles. About forty-four percent of the radioactivity is

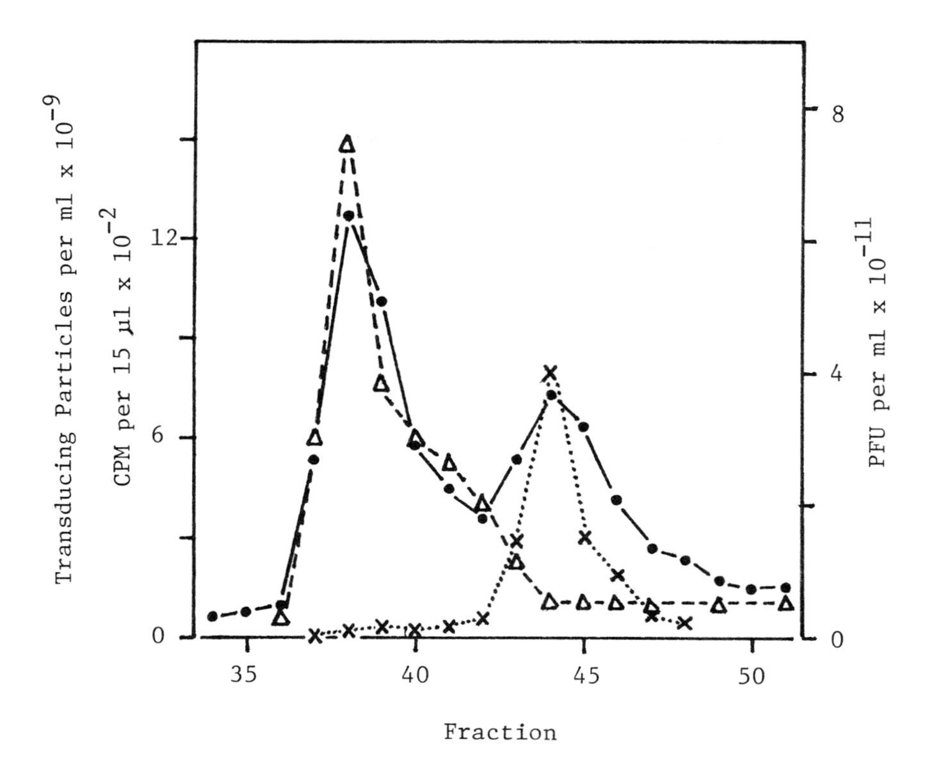

Fig. 1. Velocity sedimentation analysis of an ^3H labeled f1 phage stock containing tet^r transducing activity. An aliquot of the phage stock was layered on a neutral sucrose gradient (2) and centrifuged at 18°C for 7½ hr at 25K rpm in the Beckman SW27 rotor. Twenty-four drop fractions were collected from the top by pumping heavy sucrose into the bottom of the tube. Samples of the fractions were assayed for: ^3H cpm (●—●); pfu (Δ--Δ); and tet^r transducing activity (X····X). Only the relevant fractions are shown. Direction of sedimentation was from left to right.

present in fractions 43 through 49. If it is assumed that the particles in this region contain DNA molecules equal in size to the sum of f1 and pSC101 (see below), then such particles constitute approximately 17% of the total in the original concentrated phage stock. In the preparative centrifugation shown in Figure 1, the slower sedimenting peak contained the pfu, while the faster contained the tet^r transducing activity.

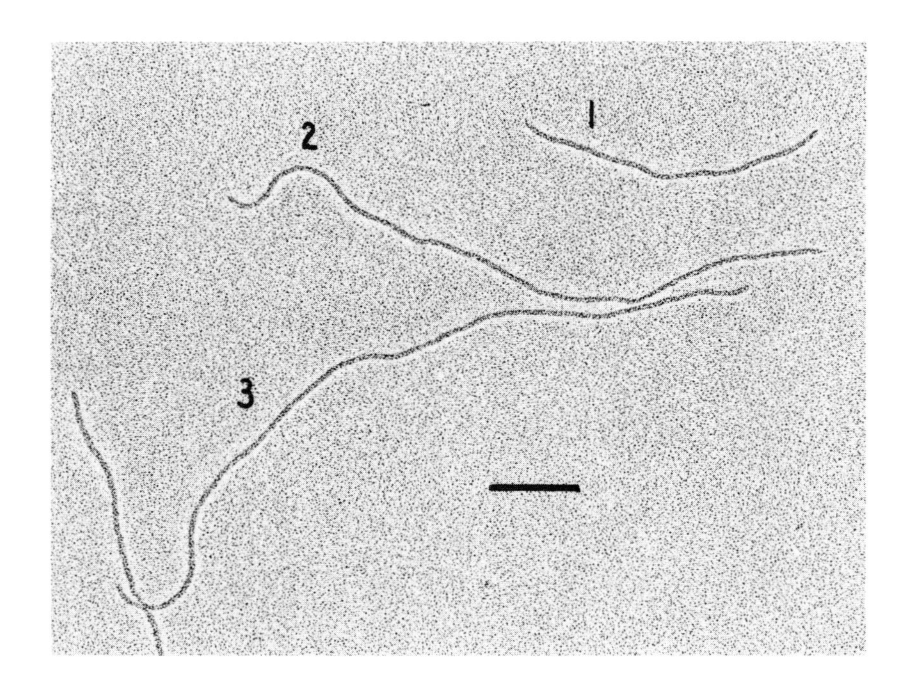

Fig. 2. Electron micrograph of filamentous particles found in the peak fraction of tet^r transducing activity (fraction 44, Fig. 1) from the sucrose gradient fractionation of a ^3H labeled f1 phage stock containing tet^r transducing activity. The phage were spread and stained as before (2). See text for a discussion of the numbered filamentous particles. The bar corresponds to 200 nm.

Figure 2 is an electron micrograph showing the three different size filamentous particles present in the peak fraction of tet^r transducing activity (fraction 44, Fig. 1). Particle number one corresponds in length to that of a unit length f1 particle, two to a double-unit length f1 particle, and three is approximately 2.7 times as large as a unit length f1 particle. This is the expected size of a filamentous particle containing a single-stranded DNA molecule equal in size to the sum of f1 and pSC101 (1,6).

The peak fraction of tet^r transducing activity (fraction 44, Fig. 1) was concentrated by PEG 6000 addition. The single-stranded viral DNA was purified by treatment with sodium dodecyl sulfate followed by extraction with phenol. The ^3H viral DNA was then subjected to velocity sedimentation in a neutral sucrose gradient with ^{32}P f1 viral DNA as an internal marker. The unit length, linear double-stranded DNA of pSC101, obtained by endo R·EcoRI treatment, was alkali-denatured, neutralized, and also subjected to velocity sedimentation with ^{32}P labeled f1 viral DNA. The results of this experiment are shown in Figure 3.

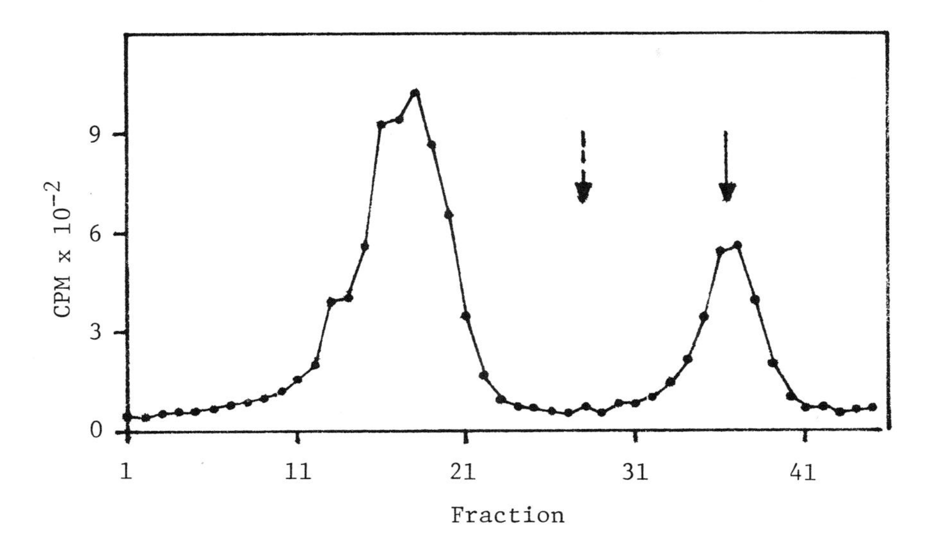

Fig. 3. Neutral sucrose velocity sedimentation analysis of ^{3}H labeled viral DNA. The purified viral DNA from fraction 44 (Fig. 1) was centrifuged for 5 hr at 36K rpm at 18°C in the Beckman SW40 rotor. Eleven drop fractions were collected from the bottom of the tube (2). The solid arrow marks the location of the ^{32}P labeled f1 viral DNA added as an internal marker. The dashed arrow marks the location of ^{3}H labeled linear, single strands of pSC101 in an accompanying gradient containing ^{32}P f1 viral DNA as an internal marker. Only the relevant fractions are shown. Sedimentation is from right to left.

The solid arrow indicates the location of the ^{32}P f1 marker and the dashed arrow the location of linear, single-stranded pSC101 DNA in the accompanying gradient. The molecular weight of the DNA in the faster sedimenting peak (fraction 17, Fig. 3) was calculated (8) to be about 4.2 x 10^{6} (the molecular weight of f1 was taken to be 1.8 x 10^{6}). The molecular weight obtained for single-stranded pSC101 DNA was 2.8 x 10^{6}, a number which is in good agreement with the published value (1).

The above results are consistent with the notion that the tet^r transducing activity present in this f1 lysate results from the filamentous phage particles that contain DNA molecules approximately equal in size to the sum of the molecular weights of f1 and pSC101 DNA and that as many as 17% of the particles in the lysate may contain such large DNA molecules.

The DNA molecules within E. coli that produce filamentous tet^r transducing particles

A tet^r *E. coli* transductant obtained as described above was used to inoculate a culture from which RFI (a covalently linked, circular double-stranded DNA molecule) was isolated. Two RFI species were present in such cells. The one that constituted approximately 60% by weight of the isolated DNA had an S value and a restriction enzyme pattern identical with f1 RFI (10). The other species had an S value greater than that of pSC101 RFI and a restriction enzyme pattern nearly identical to that expected with a mixture of f1 and pSC101 DNA (Fig. 4). Furthermore, this chimera DNA was apparently able to direct the synthesis in a "coupled system" of the six identified f1 gene products (7) and was as active per molecule as pSC101 DNA in transforming $CaCl_2$-treated *E. coli* to tet^r (10).

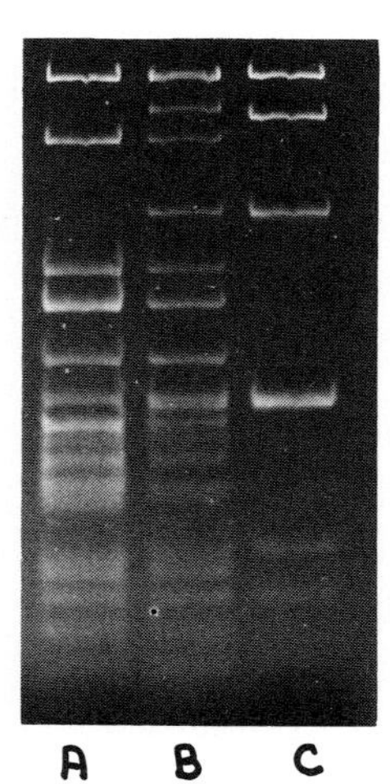

Fig. 4. Endo R·*Hae*III digestion of the f1-pSC101 recombinant DNA. Endo R·*Hae*III was used to digest: A) pSC101 RFI, B) f1-pSC101 recombinant RFI, C) f1 RFI. The digested material was layered on a polyacrylamide gradient gel (5). The restriction fragments were visualized by staining with ethidium bromide. Direction of electrophoresis is from top to bottom.

The results reported above describe the partial characterization of a recombinant molecule formed *in vivo* between the filamentous bacteriophage f1 and the non-conjugative plasmid pSC101. The size of the recombinant DNA molecule (Fig. 3) and the complexity of the restriction enzyme pattern (Fig. 4) are not inconsistent with the simple insertion of one of the constituent molecules into the other. Experiments currently in progress involving heteroduplex formation between the recombinant molecule and f1 and pSC101 DNA and further restriction enzyme digestion of isolated restriction fragments will locate the point of insertion of pSC101 into f1 on the physical and genetic map of f1 (5,9).

The recombinant DNA molecule contains an active tet^r gene (Fig. 1,3) and at least six active f1 genes (7). Abundant amounts of double-stranded and single-stranded DNA can be readily isolated from the same culture. The presence in the chimera of the single endo R·EcoRI restriction site of the pSC101 parent allows one to make potentially the chimera still larger and to select directly such larger recombinants (10). Experiments are currently in progress to test this potential and elucidate the role played by the presence of helper f1 phage.

ACKNOWLEDGEMENTS

The expert technical assistance of Michael Amer is gratefully acknowledged. This work was supported in part by grants from the National Science Foundation and the National Institutes of Health.

REFERENCES

1. Cohen, S.N. and Chang, A.C.Y. (1973) Proc. Nat. Acad. Sci., USA 70, 1293.
2. Enea, V. and Zinder, N.D. (1975) Virology 68, 105.
3. Griffith, J. and Kornberg, A. (1974) Virology 59, 139.
4. Hewitt, J.A. (1975) J. gen. Virol. 26, 87.
5. Horiuchi, K., Vovis, G.F., Enea, V. and Zinder, N.D. (1975) J. Mol. Biol. 95, 147.
6. Marvin, D.A. and Hohn, B. (1969) Bacteriol. Rev. 33, 172.
7. Model, P. and Vovis, G.F. (1976) unpublished.
8. Studier, F.W. (1965) J. Mol. Biol. 11, 373.
9. Vovis, G.F., Horiuchi, K. and Zinder, N.D. (1975) J. Virology 16, 674.
10. Vovis, G.F. and Ohsumi, M. (1976) unpublished.
11. Yamamoto, K.R., Alberts, B.M., Benzinger, R., Lawhorne, L. and Treiber, G. (1970) Virology 40, 734.

CLONING OF SATELLITE DNA SEQUENCES WITH THE λ VECTOR λgt-araB

Peter Philippsen[1], Chong S. Lee[2] and Ronald W. Davis[1]

[1]Department of Biochemistry
Stanford University School of Medicine
Stanford, California 94305

[2]Department of Zoology
University of Texas
Austin, Texas 78712

ABSTRACT. An *Eco*RI DNA fragment carrying the *E. coli* araB gene was inserted into the λ vector λgt. When DNA of this phage, λgt-araB, is used in cloning experiments, recombinant phage with reinserted araB gene can be distinguished from those containing new DNA by the plaque color on indicator plates. This technique allows the isolation of λ hybrid phage even if 99% of the recombinant phage consist of rejoined λgt-araB. This system was used to clone repeating units from satellite II DNA of *Drosophila nasutoides* which were obtained by cleavage with *Eco*RI endonuclease. The sizes of these *Eco*RI fragments are multiples of 110 to 120 nucleotide pairs. The result shows that at least some oligomers of the repeating units can become stably integrated into λ DNA and that the inactivation of restriction targets in satellite DNAs is not caused exclusively by methylation events.

INTRODUCTION

λ vectors for *in vitro* recombination of DNA consist of two DNA fragments representing those parts of the left and the right arm of λ DNA which carry the genes necessary for lytic growth. DNA fragments up to 40% λ length (20 kb) can become inserted into λ recombinant phage as calculated from the length of the vectors and the maximum packing capacity of λ heads. λ vectors are known for cloning of *Eco*RI fragments (1-13), *Hind*III fragments (4, 7, 8, 13) and *Sst*I fragments (13, 14).

λ vectors in general can be divided into two classes, replacement vectors and insertion vectors. Replacement vectors are too short (<74% λ) to become stably packaged into λ heads after rejoining of the ends (3). These vectors themselves are generated by cleavage of λ DNAs which contain a replaceable DNA fragment of sufficient length. Replacement vectors are very useful for shotgun experiments because each recombinant phage does contain extra DNA (5, 15, 16). Insertion vectors are sufficiently long to give viable phage particles after rejoining of the left and right vector end. Therefore, phage with inserted extra DNA constitute only a minor part of a recombinant phage pool. Selection for phage with DNA inserts is fairly easily accomplished by genetic or physical size selections (9, 14) or by screening for plaques of a specific phenotype (8, 11, 13). This classification of replacement and insertion vectors is also used by Murray et al. (8).

A special problem arises if small DNA fragments (<1 kb) are to be cloned because of the rapid self-cyclization of small DNA fragments (17). The yield of phage with small DNA inserts will be extremely low, even if a replacement vector of 74% λ length is used.

We wanted to clone EcoRI fragments from satellite II DNA of Drosophila nasutoides (18) in order to see whether repeated DNA can become stably integrated into λ DNA. In order to detect clones with satellite DNA, we used λgt-araB DNA, which consists of the vector λgt (3) and a replaceable center fragment containing the E. coli araB gene. When plated with E. coli araB⁻ cells, phage with other than rejoined λgt-araB DNA will give colorless plaques on indicator plates.

This paper describes the construction of the phage λgt-araB and the cloning of short fragments from Drosophila nasutoides satellite II DNA.

METHODS

DNA of λgt-λB (3) and λparaB114 (19) was prepared as described in (20). Satellite DNA from Drosophila nasutoides was isolated as described in (18). EcoRI and SalI restriction endonuclease were obtained from Marj Thomas and Steve Goff, respectively. C600rk⁻mk⁻ with defective L-ribulokinase gene (araB) was isolated as a spontaneous mutation; its reversion frequency is less than 10^{-8}. McConkey arabinose agar contains 40 g McConkey agar base (Difco) and 10 g arabinose per liter. In vitro recombination of DNA and transfection into SF8 cells followed the procedure described in (5). E. coli DNA ligase was a gift of Sharon Panasenko. Heteroduplex analysis in the electron microscope was performed according to (21).

RESULTS

Construction of λgt-araB.

Figure 1 outlines the procedure for the construction of λgt-araB. Equimolar amounts of EcoRI fragments from λgt-λB and λparaB114 DNA were mixed and incubated with E. coli DNA ligase at a total DNA concentration of 40 μg/ml for 12 h at 12°. 80 ng of the recombinant DNA gave 200 plaques after transfection, which corresponds to 5 plaques/ ng vector DNA. In control transfections, 500 plaques and 0.1 plaques were obtained per ng uncleaved and cleaved λgt-λB DNA, respectively.

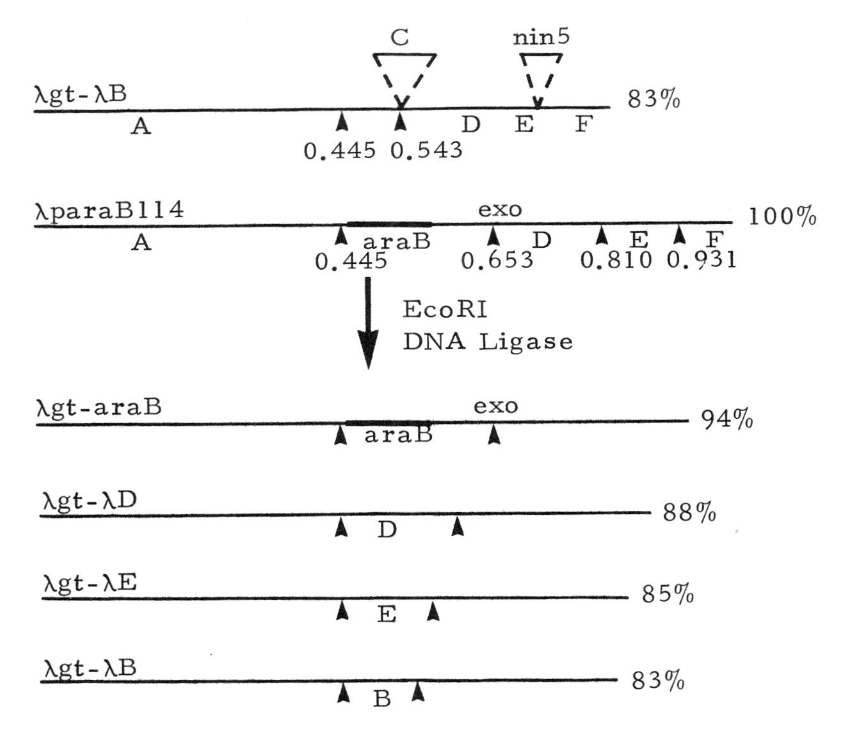

Figure 1. Reaction scheme for the construction of λgt-araB. The lines represent the structures of the phage genomes which were used and the structures of some of the possible recombinant DNAs. The positions of EcoRI sites are indicated by arrows and expressed as fractional λ length (20). The capital letters designate EcoRI fragments.

Twenty of the recombinant phage were analyzes by infection of E. coli araB⁻ and plating on McConkey plates containing arabinose. Three of the phage gave large and two gave small plaques with red centers after 25 h at 32°. Phage from one of the large plaques were picked and plaque purified. These λgt-araB phage were shown by heteroduplex analysis to have the expected structure. The phage in the small plaques with red centers contain an inverted araB fragment as shown by heteroduplex analysis with λparaB114 DNA. These so-called λgt-araB' phage have a decreased burst size because they are exo$^-$. The exo gene of λparaB114 is cleaved into two halves by EcoRI endonuclease, and it is restored only in λgt-araB. The observable slower expression of the araB phenotype in λgt-araB' as compared to λgt-araB is probably a direct consequence of the smaller burst size. From the other proposed recombinant DNA structures in Figure 1, only those with inserted EcoRI·E and EcoRI·B fragment could be identified in recombinant phage.

Cleavage of satellite II DNA with restriction nucleases.

60% of *Drosophila nasutoides* DNA consists of four AT-rich satellite DNAs. The sequence complexities obtained from DNA reassociation kinetics data are 5, 103, 2.3×10^6 and 46 nucleotide pairs for the satellites I, II, III and IV respectively (18). When a mixture of satellite I + II or satellite II DNA alone is cleaved with EcoRI or SalI endonuclease, a series of oligomers which are multiples of around 115 nucleotide pairs is formed. Purified satellite I DNA was not cleaved to a measurable degree under our conditions.

Figure 2a shows the cleavage pattern obtained from satellite I + II DNA with SalI endonuclease. Some interbands are seen which were not detected with EcoRI endonuclease. A mixed digestion with EcoRI and SalI endonuclease did not change the basic patterns. This means that the EcoRI and SalI cleavage sites belong to different parts of the satellite DNA or are located only a few nucleotides distant from each other. The bands representing the series of oligomers are fairly broad as compared to the bands in a HindIII restriction spectrum (Figure 2b). This indicates a considerable amount of length and/or sequence heterogeneity in the repeating units of satellite II DNA (Table 1).

The inactivation of some EcoRI or SalI cleavage sequences which causes the appearance of the oligomers can be due to random mutations, to methylation or to both. Assuming solely random mutations of a preexisting cleavage sequence as the source for this inactivation, the degree of sequence

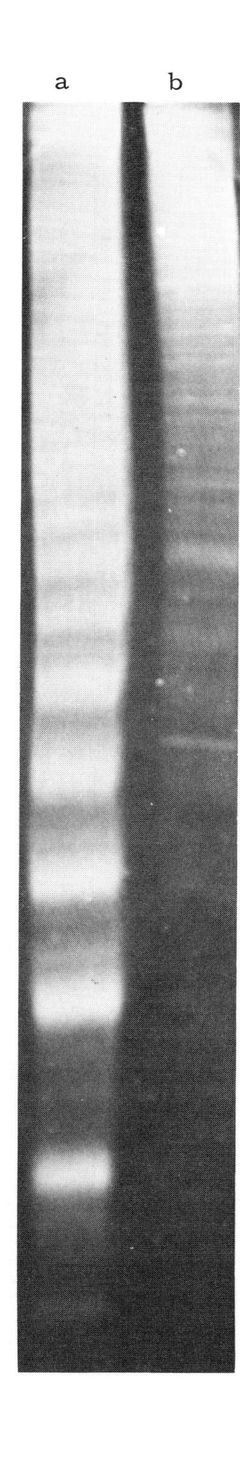

Figure 2. SalI (a) and HindIII (b) restriction spectrum of Drosophila nasutoides satellite I + II DNA in 5% polyacrylamide.

---

divergence can be estimated. We asked a computer to go through random mutations, i.e., base pair changes, including back mutations in a segment of 100 repeating units with 100 nucleotide pairs each and a 6 nucleotide pair-long cleavage sequence. A schematic representation of the expected cleavage patterns is shown in Figure 3. The pattern in Figure 2a is fairly close to the computer pattern obtained for 15% to 20% mutation in the studied segment. Cordeiro-Stone and Lee found 17 to 24% base pair mismatches for this satellite as concluded from the melting temperature of the native and the reassociated DNA (18).

Cloning of EcoRI fragments from satellite II DNA.

EcoRI fragments from λgt-araB and satellite I + II DNA were incubated with E. coli DNA ligase at a concentration of 10 μg vector DNA and 20 μg satellite DNA per ml. 180 μg of the recombinant DNA gave 350 plaques after transfection of SF8 cells which corresponds to 6 plaques/ng vector DNA. (Transfection of the araB$^-$ strain, which is recA$^+$,recBC$^+$, generated only 12 plaques.) In control transfections, 400 plaques and 0.1 plaques were obtained per ng uncleaved and cleaved λgt-araB DNA, respectively. The recombined phage were analyzed for the presence of

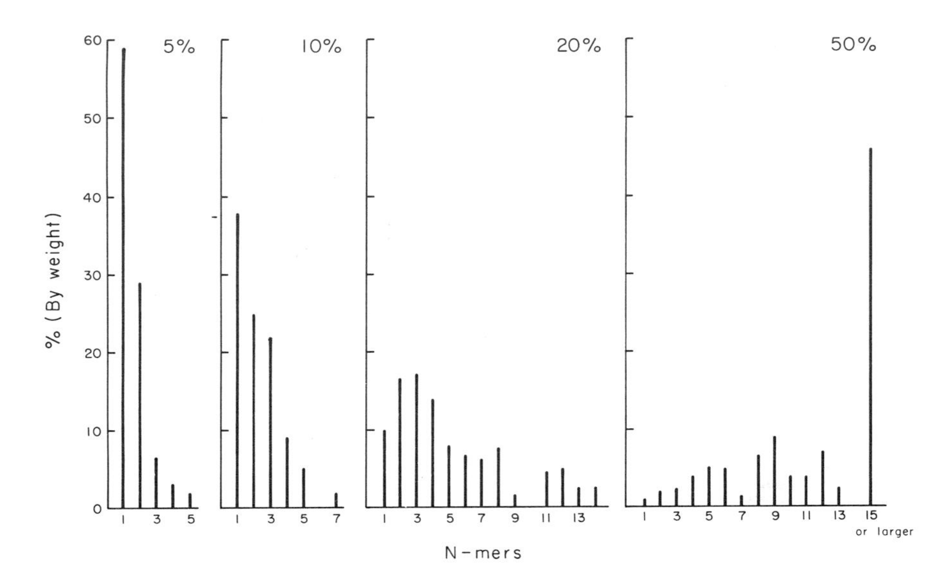

Figure 3. Effect of random mutations on the distribution of cleavage sites in satellite DNA composed of 100 nucleotide pair-long repeating units.

---

satellite DNA by spotting on McConkey arabinose plates containing a lawn of E. coli $araB^-$ cells. Most of the plaques had dark red colonies growing in their centers after 25 h at 33°. Two plaques had very small light red centers and six were clearly colorless. Phage from these eight plaques were plaque purified and further investigated. Three of them, λgt-Dn102, 105 and 107, contained EcoRI fragments from the series of oligomers of satellite II as summarized in Table 1. λgt-Dn107 carried, in addition, an inverted araB fragment. This phage was isolated from one of the two plaques with a light red center. Phage from the other plaque contained only the inverted araB fragment.

λgt-Dn106 had a very small fragment (around 70 nucleotide pairs long) inserted and carried a duplication of about 10% λ lengths in the right arm as shown by heteroduplex analysis. This made the phage genome sufficiently long to produce viable phage particles. The same duplication was found in another phage which had no DNA inserted. The remaining two phage contained a 6 kb and 8 kb long DNA fragment, respectively.

TABLE 1

Cloned EcoRI fragments[a] from Drosophila nasutoides satellite II DNA.

	Satellite II	λgt-Dn102[b]	λgt-Dn105	λgt-Dn107[c]
Monomer	110-130	—	120, 135	—
Dimer	220-260	—	250	—
Trimer	320-360	330	—	—
Tetramer	430-470	—	—	—
Pentamer	530-600	—	570	—
Hexamer	650-700	—	—	690

a. Sizes are expressed in nucleotide pairs.
b. Contains in addition a 7 kb fragment.
c. Contains in addition an inverted araB fragment.

DISCUSSION

The replacement vector described in this paper is not only useful for the cloning of small EcoRI fragments but also for the cloning of small quantities of DNA, e.g., DNA extracted from a band of an analytical agarose gel. This was demonstrated by the cloning of EcoRI fragments from yeast plasmid DNA after extraction from an agarose gel (John Cameron, personal communication). Similar replacement vectors containing other sugar metabolizing genes or suppressor genes have been constructed and used by other labs (8, 11, 13).

The cloning of satellite DNA was undertaken in order to answer part of the question of whether incomplete digestions of satellite DNAs with restriction nucleases are due to methylation or mutation events or both (23, 27, 31, 34).

Satellite DNAs with targets for restriction nucleases are either cleaved to completion producing fragments which represent repeating units or are cleaved only to a certain extent producing a series of fragments with multiple lengths of a repeating unit. An example for the first class is calf satellite I DNA, which is cleaved by EcoRI, SalI or SstI endonuclease into fragments of 1400 nucleotide pairs (25, 30 and P. Philippsen, unpublished). Examples for the second class were derived from studies with mouse satellite (23, 27), apodemus satellite (28), Drosophila melanogaster 1.688 satellite (31, 34) and monkey α satellite (32, 33, 35). There probably also exists a satellite in human DNA which belongs to this class (24).

Satellite II DNA from Drosophila nasutoides clearly belongs to the second class. But this satellite is unique with respect to the considerable amount of size heterogeneity found for the oligomers of the repeating unit (see Table 1). Restriction fragments from highly repeated DNA of many other organisms show no or very little size heterogeneity as judged from the band widths in polyacrylamide or agarose gels (22-37).

The analysis of the recombinant phage containing EcoRI fragments from satellite II DNA excludes methylation as the only source for the observed incomplete digestion. This is in agreement with the analysis of cloned fragments from the 1.688 satellite of Drosophila melanogaster, which also showed the presence of oligomers after cloning in E. coli (36). Evidence for at least a participation of methylases in the inactivation of restriction targets comes from experiments with calf satellite I DNA which is only cleaved with SmaI endonuclease after cloning in E. coli (37). A final answer concerning the understanding of the diversity of fragmentation patterns will be obtained only after extensive DNA sequence analysis with native and cloned satellite DNAs which is probably feasible for genomes with relative low complexity.

ACKNOWLEDGMENTS

We wish to thank Lynn Horn and Ruth Redse for preparing the manuscript, John Lis and Bob Schleif for sending their ara transducing strains prior to publication. The work was supported by grants from the National Institutes of Health, GM 21891, and from the National Science Foundation, BMS 75-05377. P. Philippsen thanks Studienförderung Cusanuswerk for generous support.

REFERENCES

1. Murray, N.E. and Murray, K. (1974) Nature 251, 476.
2. Rambach, A. and Tiollais, P. (1974) Proc. Nat. Acad. Sci. 71, 3927.

3. Thomas, M., Cameron, J. R. and Davis, R. W. (1974) Proc. Nat. Acad. Sci. 71, 4579.
4. Murray, K. and Murray, N. E. (1975) J. Mol. Biol. 98, 551.
5. Cameron, J. R., Panasenko, S. M., Lehman, I. R. and Davis, R. W. (1975) Proc. Nat. Acad. Sci. 72, 3416.
6. Enquist, L., Tiemeyer, D., Leder, P., Weisberg, R. and Sternberg, N. (1976) Nature 259, 596.
7. Williams, B. G., Moore, D. D., Schumm, J. W., Grunwald, D. J., Blechl, A. E. and Blattner, F. R. (1976) in Miles Symposium, R. Beers, ed., in press.
8. Murray, N. E., Brammar, W. J. and Murray, K. (1977) Mol. Gen Genet. 150, 53.
9. Cameron, J. R., Philippsen, P. and Davis, R. W. (1977) Nuc. Acid Res., in press.
10. Tiollais, P., Perricaudet, M., Petterson, U. and Philipson, L. (1976) Gene 1, 49.
11. Velten, J., Fukada, K. and Abelson, J. (1976) Gene 1, 93.
12. Donoghue, D. J. and Sharp, P. A. (1977) Gene, in press.
13. Blattner, F. R., Williams, B. G., Blechl, A. E., Denniston-Thompson, K., Faber, H. E., Furlong, L. A., Grunwald, D. J., Kiefer, D. O., Moore, D. D., Schumm, J. W., Sheldon, E. L. and Smithies, O. (1977) Science, in press.
14. Philippsen, P., Kramer, R. A. and Davis, R. W. (1977) in preparation.
15. Borck, K., Beggs, J. D., Brammar, W. J., Hopkins, A. S. and Murray, N. E. (1976) Mol. Gen.Genet. 146, 199.
16. Cameron, J. R. and Davis, R. W. (1977) Science, in press.
17. Jacobson, H. and Stockmayer, W. H. (1950) J. Chem. Phys. 18, 1600.
18. Cordeiro-Stone, M. and Lee, C. S. (1976) J. Mol. Biol. 104, 1.
19. Lis, J. T. and Schleif, R. (1975) J. Mol. Biol. 95, 395.
20. Thomas, M. and Davis, R. W. (1975) J. Mol. Biol. 91, 315.
21. Davis, R. W., Simon, M. and Davidson, N. (1971) in Methods in Enzymology, L. Grossman and K. Moldave,

eds. (New York, Academic Press) 21, 413.

22. Mowbray, S. L. and Landy, A. (1974) Proc. Nat. Acad. Sci. 71, 1920.
23. Hörz, W., Hess, I. and Zachau, H. G. (1974) Eur. J. Biochem. 45, 501.
24. Philippsen, P., Streeck, R. E. and Zachau, H. G. (1974) Eur. J. Biochem. 45, 479.
25. Botchan, M. R. (1974) Nature 251, 288.
26. Roizes, G. (1974) Nuc. Acid Res. 1, 1099.
27. Southern, E. M. (1975) J. Mol. Biol. 94, 51.
28. Cooke, H. J. (1975) J. Mol. Biol. 94, 87.
29. Mowbray, S. L., Gerbi, S. A. and Landy, A. (1975) Nature 253, 367.
30. Philippsen, P., Streeck, R. E., Zachau, H. G. and Müller, W. (1975) Eur. J. Biochem. 57, 55.
31. Manteuil, S., Hamer, D. H. and Thomas, C. A. (1975) Cell 5, 413.
32. Gruss, P. and Sauer, G. (1975) FEBS Letters 60, 85.
33. Fittler, F., unpublished.
34. Shen, C.-K. J., Wiesehahn, G. and Hearst, J. E. (1976) Nuc. Acid Res. 3, 931.
35. Musich, P., Brown, F., and Maio, J. (1977) J. Mol. Biol., in press.
36. Carlson, M. and Brutlag, D. (1977) Cell, in press.
37. Gautier, F., Mayer, H. and Goebel, W. (1976) Mol. Gen. Genet. 149, 23.

Workshop Summary on "Screening for Recombinant DNAs" by J.E. Dahlberg, University of Wisconsin

In this workshop several new methods and pieces of electrophoresis equipment were described which can facilitate screening for recombinant DNAs.

Dr. Ann kalka (Roche) described an _in situ_ (on the petri dish) lysis and antibody precipitation assay for detection of the protein products of cloned genes. The method is applicable to both phage and plasmid vectors. In control experiments to test the sensitivity of the method, Dr. Skalka has shown that she can detect the low level of β-galactosidase present in uninduced colonies of _E. coli_. The sensitivity of the method may be increased further by incorporation of an additional step, in which iodinated antibody is used.

Dr. Peter Philippsen (Stanford) described several of the elegant screening methods which are used in the laboratory of Dr. R. Davis. By overlaying phage plaques on a plate with a sheet of nitrocellulose, sufficient phage are absorbed to the membrane to allow for screening by filter hybridization. The phage are lysed and the DNA is denatured and baked onto the filter. Incubation of the filter with 10^5 - 10^6 c.p.m. of RNA probe allows one to detect which phage vectors carry the DNA of interest. The original plaque can then be located by placement of an autoradiogram over the plate. Using mutant phage, to keep plaque size small, it is possible to screen as many as 20,000 plaques per plate.

Dr. Wayne Barnes (MRC, Cambridge) described a rapid toothpick (2") method for stabbing colonies, to isolate plasmid DNAs and determine which vectors have altered sizes (resulting from insertions or deletions). The key to this method is the use of Tri-X film to photograph the ethidium bromide stained bands of DNA after gel electrophoresis.

Dr. Donald Kaplan (UCLA) spoke about a horizontal slab gel electrophoresis apparatus that he and Dr. G. Wilcox have developed. Ingenious features in the design allow one to run gels with as little as 0.18% agarose - low enough to make chromosome separation feasible. Besides being cooled to constant temperature, the apparatus has the feature that multiple concentrations of gels can be poured at the same time. Also, over 150 samples can be run in the same gel at one time. A nitrogen manifold permits one to use polyacrylamide gels in the same apparatus. Preparative gels up to two inches thick can be run.

The final presentation was by Dr. Edwin Southern (Edinburgh) who discussed a variety of techniques which he has developed for screening and production of restriction enzyme fragments. To illustrate the methods, he discussed the cloning of rabbit globin genes. The obvious first step in this procedure is to determine the sizes of DNA fragments carrying the genes of interest, after restriction of rabbit DNA with a variety of enzymes. This assay is done by blotting restriction fragments onto sheets of Millipore filters and hybridizing a probe RNA to the filters (the well known "Southern gel" method). Once the size of the fragment has been determined, large scale production of the fragment can be accomplished by electrophoresis of restriction fragments in an ingeniously designed circular gel apparatus. In this gel cell, electrophoresis is from an outer ring to a central electrode. The circular design allows for an increased concentration of DNA fragments as they migrate from the origin to the center. Samples are collected by periodically rinsing a dialysis membrane at the center. These methods have allowed Dr. Southern to screen and prepare clones of α and β globin genes on plasmids and they should be useful in other applications.

SCREENING FOR RECOMBINANT DNAS WITH IN SITU IMMUNOASSAYS

A. Skalka and L. Shapiro*

The Roche Institute of Molecular Biology
Nutley, New Jersey 07110

ABSTRACT. A series of *in situ* immunodiffusion assays have been developed for use in screening for the translation products of genes cloned in *in vitro* recombination experiments with either phage or plasmid vectors. The formation of antigen-antibody precipitates in a vector phage plaque can be used to detect the production of a specific protein from an amplified gene which is transcribed at normal efficiency. Immunodiffusion assays of individual bacterial colonies lysed *in situ* either by λ prophage induction or by biochemical means afford an even higher level of sensitivity than the plaque assay, probably adequate to detect the production of a few molecules of protein per cell.

INTRODUCTION

The ability to isolate specific cloned genes in a mixture of recombinants depends on the availability of methods to screen for their presence. We have recently reported the development of a series of immunoassay screens based on the detection of specific antigen-antibody complex formation within a vector phage plaque or surrounding vector containing bacterial colonies (4). These assays depend on gene expression, in that the elaboration of at least the antigenic segment of the gene's protein product is required. The assays do not require the gene product's function, however, and thus should be applicable to the isolation of part or all of many structural genes, including those for which there are at present no convenient assays and eukaryotic genes for which there are no counterparts in prokaryotes. Theoretically, the gene expression required in these assays need not depend on the presence of the cloned genes' promotor. At least in some cases, the DNA sequences which

*Permanent address: Dept. of Molecular Biology, Albert Einstein College of Medicine, Bronx, New York

code for antigenic sites could be linked either by chance or design *in vitro* to a promotor in the vector.

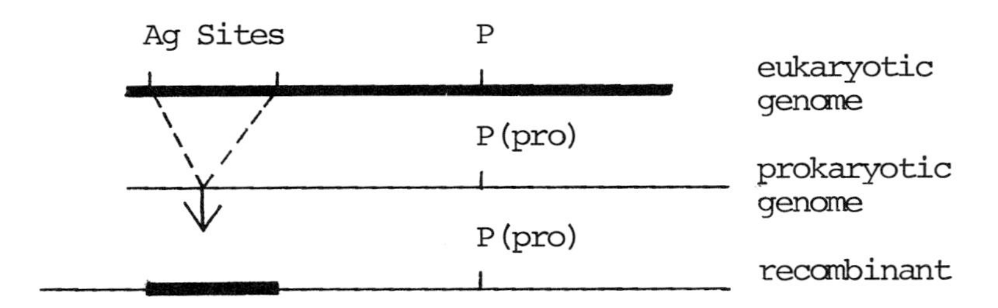

The lambda phage (λ*plac*5) employed in the development of these assays was constructed by Rambach and Tiollais (2) as a vehicle for cloning foreign DNA. The phage has only two EcoRI restriction endonuclease-sensitive sites and these flank a region that is nonessential for λ growth. This non-essential region can be replaced by foreign DNA *in vitro*. This vector already contains a segment of foreign DNA (the bacterial *lac* operon) which had been transduced into the non-essential region by conventional *in vivo* recombinant techniques.

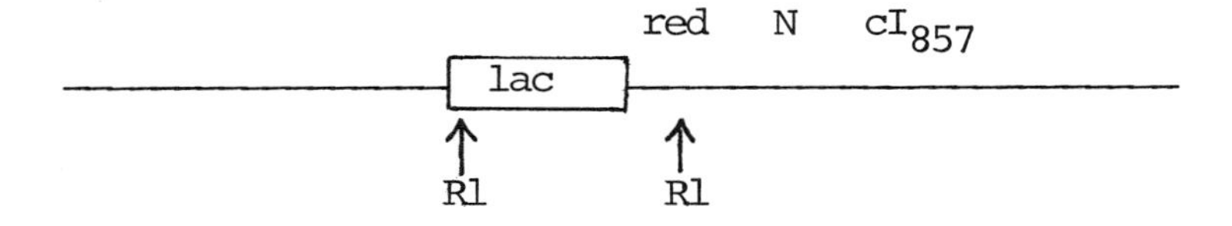

In one of the immunoassays, translation of the bacterial β-galactosidase gene in the *lac* DNA segment, served as a model for translation from a foreign cloned DNA fragment. In other assays, β-galactosidase produced from the *lac* operon within the bacterial chromosome was measured. In both cases for these model systems transcription originated at the bacterial *lac* promotor.

RESULTS AND DISCUSSIONS

I. The phage plaque immunodiffusion assay

This assay requires that the vector phage form clear plaques under the conditions to be used for screening. At the time of plating, antibody specific for the protein product of the desired "foreign" gene is mixed into the molten soft agar overlay containing recombinant phages and host bacteria. After overnight incubation, plaques from recombinant phages in which the foreign gene has been expressed can be identified by a turbid phenotype. This turbidity is due to the presence of immunoprecipitates which are formed as the infected host bacteria lyse and release the foreign protein or protein segment.

Figure 1(a) shows plaques from a mixture of a few λplac5 phages and many wild type lambda, plated with lac$^-$ (deletion) bacteria in the presence of anti-β-galactosidase antibody. Most of the plaques are clear (black) and a few are turbid (grey).

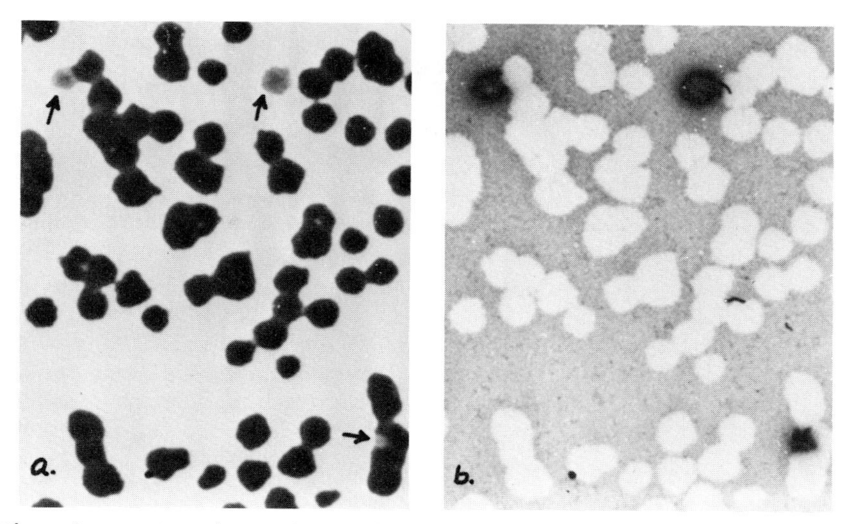

Fig. 1. Detection of β-galactosidase production in phage plaques by immunoprecipitate formation and subsequent confirmation by color development in the presence of the XG indicator substrate. (a) a mixture of λplac5cI$_{857}$ and λcI$_{857}$ phage were plated on tryptone agar plates with lac$^-$ (deletion) host bacteria (E. coli S90C) and anti-β-galactosidase antibody. Plaques were photographed against a black background after overnight incubation at 37°. Arrows indicate turbid plaques suspected to be due to release of β-galactosidase. (b) the same plates a few hours after plaques had been overlayed with soft agar containing 80 μg/ml of the β-galactosidase indicator substrate XG (5-bromo-4-chloro-3-indolyl-β, D-galactoside). Plaques were photographed against a white background and the blue plaques resulting from β-galactosidase enzyme activity appear black.

Direct assay for β-galactosidase enzyme activity verified that the turbid plaques were formed by the λplac5 phage (Fig. 1b).

Figure 2 compares these clear (a) and turbid (b) plaques as seen through a microscope. Antigen diffusion does not seem to interfere with the assay since the phenotypes are distinguishable even with plaques in direct apposition (c).

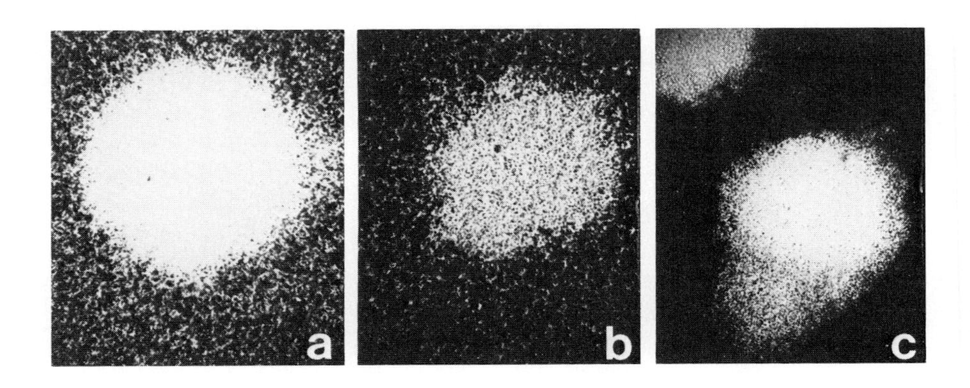

Fig. 2. Immunoprecipitates within phage plaques. Phage plaques formed in the presence of anti-β-galactosidase antibody, as described in Fig. 1, were photographed with a microflex photomicrographic attachment to a Nikon inverted scope using a 4 X objective (final magnification = 52 X) light transmitted through the microscope makes clear plaques appear white surrounded by a dense (and therefore dark) bacterial lawn. (a) λcI$_{857}$ plaque; (b) λplac5cI$_{857}$ plaque; (c) smaller λplac5cI$_{857}$ plaque adjacent to a λcI$_{857}$ plaque.

The presence of the antibody also has no apparent affect on the viability of the phage in the turbid plaques. Presumed recombinants can be picked and the phages stored in micro titer wells. The phenotypes can be verified at a later time by spot-checking samples from the micro wells on prepoured lawns containing specific antibody and host bacteria.

The λplac phage in our model system is i^- and thus derepressed for production of β-galactosidase. We estimate that each plaque could contain as much as 6 ng of β-galactosidase if production during the λ lytic cycle could reach the high levels made in a fully-induced normal E. coli, but it seems likely that infected cells contain less than that amount. It is possible to obtain a rough estimate of the sensitivity of this assay for any specific application by performing preliminary reconstruction experiments. Plaques are developed in the presence of specific antibody, and then various amounts of antigen are applied to them. The lowest concentration giving the turbid phenotype sets the limits of resolution for the assay (see Fig. 3). Antibody concentration can also be varied in such experiments.

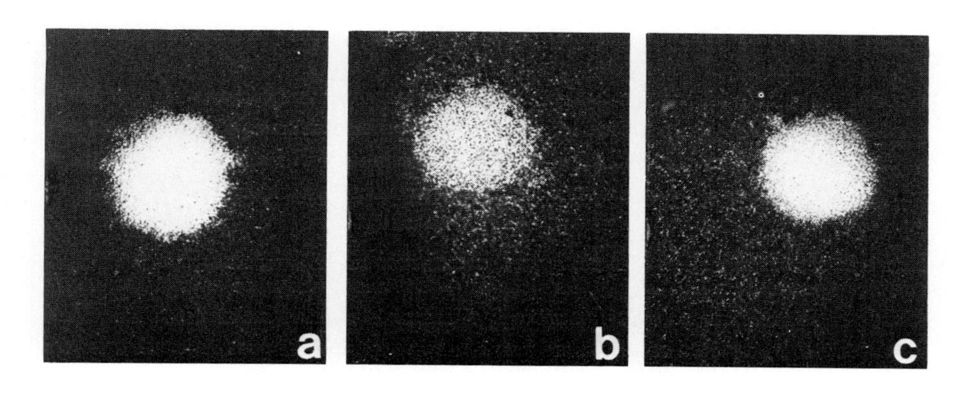

Fig. 3. C. crescentus flagellin immunoprecipitates within phage plaques; λplac5cI857 was plated in the presence of anti-C. crescentus flagellin antibody (undiluted antiserum 200 ul/ml). The resulting plaques were clear. After 16 hr at 37°C, purified C. crescentus flagellin was applied to 1/8 segments of the plate which contained a total of approx. 60 plaques. Incubation was continued for an additional 16 hr at 37°C. (a) No addition of flagellin; (b) addition of 0.5 μg of flagellin/phage plaque; (c) addition of 0.05 μg of flagellin/phage plaque. Photography was as in Fig. 2.

II. In situ colony immunodiffusion assays

These assays depend on the formation of specific antigen-antibody precipitin rings surrounding colonies which have been lysed in situ. Detection of the rings is easiest if the bottom agar of the petri dishes is poured thin (1 1/2 - 2 mm). Two methods have been developed; in the first lysis is by λ phage induction, in the second, biochemical lysis procedures are used.

(1) Lysis by induction of λ prophage.

In this method the host bacterium is a lysogen with a heat-inducible λ prophage. The foreign DNA could be carried in the prophage or on a plasmid in the host. Bacterial colonies are grown at 30° and induction is accomplished by shifting the plates to 42°. In the experiment shown in Fig. 4, lac$^+$ lysogens were grown in the presence of β-galactosidase inducer to simulate conditions which might be expected during maximum expression from an amplified gene. After 4 hours at 42°, to allow for lysis and diffusion of the antigen, a soft agar overlay containing anti-β-galactosidase antibody was poured over the colonies. After further incubation for 10 to 16 hours at 37° heavy immunoprecipitin rings were observed surrounding the colonies. Similar results were obtained with lac$^-$ (deletion) lysogens containing λplac prophages grown in the absence of β-galactosidase inducer. As expected lac$^-$ lysogens containing λ prophage gave no rings whether or not the β-galactosidase inducer was present (these data are not shown).

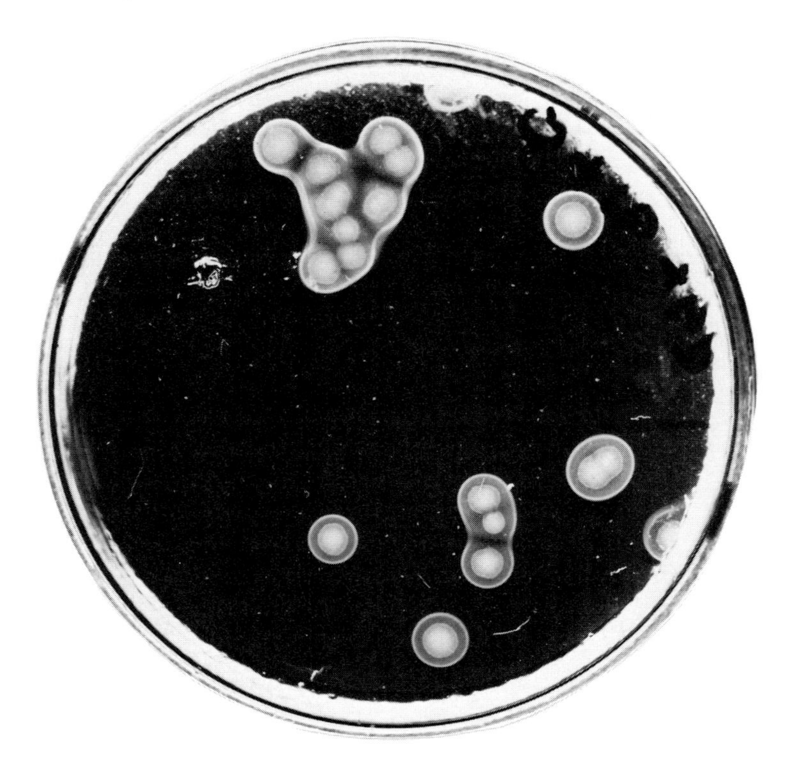

Fig. 4. Immunodetection of β-galactosidase produced by colonies of lac⁺ E. coli grown in the presence of IPTG (isopropyl thiogalactoside, 10^{-3}M). Colonies of E. coli W3110 (λcI_{857}) were grown at 30°C and were then lysed by prophage induction at 42°C for 4 hr. Immunoprecipitates were detected by the addition of soft agar containing 100 ul/ml undiluted antiserum. Photography was against a black background.

Figure 5 shows results from an experiment in which lac^+ lysogens containing λ prophage were grown without β-galactosidase inducer. In this case, following overnight incubation in the presence of antibody, the plates were stored at 4°. After two days, faint rings were visible around these colonies.

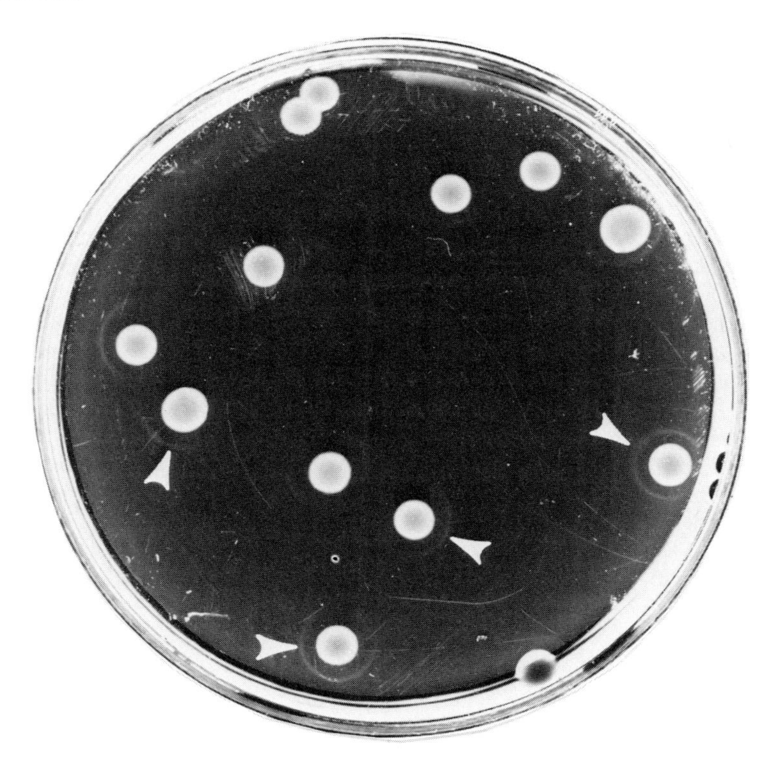

Fig. 5. Conditions as in Fig. 4 except IPTG was not present and incubation was continued at 4° for 2 days.

The basal uninduced level of β-galactosidase is known to be extremely low and can be accounted for by the production of less than one messenger RNA per cell per generation (3). Various estimates indicate that there are only about 10-20 molecules of β-galactosidase in an uninduced cell and that induction causes an approximately 1000 fold increase in the amount of this protein (1). These values indicate the range of sensitivity of this immunoassay.

(2) Lysis by biochemical procedures.

This method may be used when lysis by induction is inadvisable or inconvenient. Bacterial colonies are grown on medium in which 1.3% agarose is the solidifying agent. Antibody is added to the bottom agarose at the time of pouring. Bacteria can be grown at any temperature, and after growth a soft agarose overlay containing tris-EDTA and egg-white lysozyme is poured over the colonies. After the overlay solidifies, the plates are incubated at 37° for 1 hr. The colonies are lysed by pouring a small amount of 2% sarkosyl onto the surface of the overlay. Incubation is continued at 37° overnight to allow for development of the immunoprecipitin rings. Figure 6 shows results with a mixture of <u>lac</u>$^+$ and <u>lac</u>$^-$ bacteria grown in the presence of β-galactosidase inducer. The <u>lac</u>$^+$ colonies are surrounded by precipitin rings whereas the <u>lac</u>$^-$ colonies are not.

Rings could not be detected around colonies grown in the absence of β-galactosidase inducer which indicates that this assay is somewhat less sensitive than the one employing lysis by λ induction. It is possible that lysis is less efficient with this method. In addition, results of others (M. Scharff, personal communication) suggest that the detergent used for lysis could inhibit the antigen-antibody reaction by as much as 20%. If so, it may be possible to put this assay closer to the range obtained with lysis by λ induction by employing other detergents or lysis conditions.

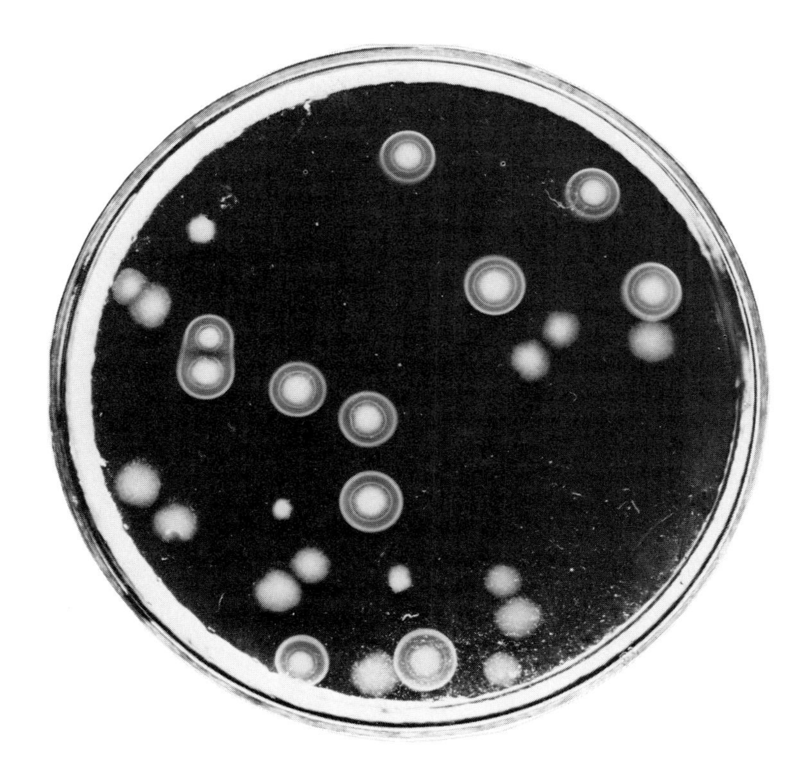

Fig. 6. Immunoprecipitates surrounding bacterial colonies lysed by biochemical procedures. Colonies of E. coli W3110 (lac+) plus E. coli S90C (lac- deletion) were grown at 37°C on plates which contained 100 ul/ml of anti-β-galactosidase antibody and IPTG (10^{-3}M). Colonies were then treated with lysozyme-EDTA and sarkosyl as previously described (4).

REFERENCES

1. Miller, J. H., In *Experiments in Molecular Genetics* (1972) Cold Spring Harbor Laboratory, New York, p. 398.

2. Rambach, A. and Tiollais, P. (1974) *Proc. Natl. Acad. Sci. USA*, 71, 3927.

3. Rotman, R, M. (1970) cited in Adamson, L., Gross, C. and Novick, A., *The Lactose Operon* (Zipser, D. and Beckwith, J. R. (Eds.) Cold Spring Harbor Laboratory, New York, p. 317.

4. Skalka, A. and Shapiro, L. (1976) *Gene* 1, 65.

MOLECULAR CLONING OF λh80dara RESTRICTION FRAGMENTS WITH NONCOMPLEMENTARY ENDS

Donald A. Kaplan, Lawrence Greenfield
and Gary Wilcox

Department of Bacteriology
and Molecular Biology Institute

University of California

Los Angeles, California 90024

ABSTRACT

The cloning of restriction fragments with noncomplementary ends eliminates the possibility of intramolecular reactions, thus increasing the efficiency of cloning. The efficiency of cloning is defined as the ratio of the number of clones with inserts to the total number. We have cloned restriction fragments of λh80dara in the cloning vehicle pBR317. A mixture of λh80dara and pBR317 DNA was simultaneously restricted with R.BamI and R.SalI and then ligated with T4 DNA ligase. The ligated material was used to transform E. coli RRI and a cloning efficiency of 83% was observed. In a similar experiment, λh80dara was simultaneously restricted with R.BamI and R.EcoRI* and ligated to pBR317 that had been restricted with R.BamI and R.EcoRI. The efficiency of cloning was 97%. The advantages of this type of cloning are (1) an increase in cloning efficiency, (2) the ability to orient restriction fragments by the appropriate choice of restriction enzymes, and (3) the ability to clone subsets of a previously cloned DNA fragment.

INTRODUCTION

Molecular cloning of bacterial and eukaryotic DNAs on various plasmids in E. coli is a powerful tool for the amplification and purification of DNA fragments (1,2). An essential step in molecular cloning is insertion of foreign DNA into the cloning vehicle. The experiments described in this paper were designed to improve the efficiency of inserting DNA into the cloning vehicle, thereby increasing the probability of cloning any fragment, even if it is available in only very small quantities. The cloning vehicle used in this study, pBR317, is a derivative of the plasmid *colE1* which contains genes conferring ampicillin and tetracycline resistance (Fig 1). The use of the appropriate pairs of restriction enzymes to digest both the plasmid and the DNA to be cloned eliminates the problem of self-closure (Fig 2) and increases the efficiency of cloning.

MATERIALS AND METHODS

Phage and Plasmid DNA: *λh80dara* was purified as described previously (3). *Escherichia coli* RRI *$(F^{-}leu^{-}pro^{-}thi^{-}lac^{-}str^{r}r_{K}^{-}m_{K}^{-})$* carrying pBR317 was given to us by R. Rodriguez. The plasmid DNA was isolated as described by Clewell and Helinski (4) after amplification in the presence of chloramphenicol (5).

Restriction enzymes: R.*Bam*I and R.*Pst*I were purified by conventional procedures (unpublished methods). R.*Eco*RI was purchased from Miles Laboratories. The other two enzymes were gifts from M. Grunstien (R.*Sal*I) and A. Ohtsuka (R.*Hind*III).

Cloning: Restriction, ligation, and transformation R.*Bam*I-R.*Sal*I: pBR317 (46 μg) and *λh80dara* (27 μg) were mixed (300 μl) and restricted with R.*Bam*I (100 units) and R.*Sal*I (25 units) in 100 mM Tris-HCl, pH 7.5, 50 mM NaCl and 5 mM $MgCl_2$ (R.*Eco*RI restriction buffer) for 24 hours at 37° C. The mixture was then heated at 65° C for 10 min to inactivate the restriction endonucleases. 75 μl of this mixture was added to an equal volume of ligase buffer (120 mM Tris-HCl, pH 7.5, 10 mM dithiothreitol, and 0.8 mM adenosine triphosphate). The mixture was cooled to 15.5° C before adding 0.4 units of T4 DNA ligase (Miles Laboratories). The reaction was maintained at 15.5° C for 24 hr; ATP was then added to a final concentration of 0.6 mM and 0.3 units of T4 DNA ligase was added and the temperature changed to 13.5° C. After 48 hr at 13.5°C the DNA was cooled to 0° C and used in transformation. For transformation, *E. coli* RRI was grown in TYE (tryptone 15g, yeast extract 10g, and NaCl 5g per liter) to 4.5×10^8 cells/ml. The cultures were harvested by centrifugation at 10,000 rpm in a Sorvall SS34 rotor at 4° C. The cell pellet was washed with 10 mM NaCl and then with 40 mM $CaCl_2$. The cells were then resuspended in 50 mM $CaCl_2$ (1.5 ml $CaCl_2/4.5 \times 10^9$ cells) and 0.25 ml of this cell suspension was mixed with 75 μl of the ligated DNA and stored at 0° C for 60 min. The mixture was heated for two min at 42° C and spread directly onto TYE plates containing 20 μg/ml of ampicillin (TYE Amp).

R.BamI-R.EcoRI* cloning: pBR317 (69 μg) was restricted with R.BamI (100 units) and R.EcoRI (100 units) in a total volume of 250 μl in R.EcoRI restriction buffer for 24 hr. λh80dara (68 μg) in 450 μl was first restricted with R.EcoRI* (100 units) in 25 mM Tris-HCl, pH 8.5, and 2 mM $MgCl_2$ (7) for 18 hr at 37° C and then with R.BamI (100 units) under R.EcoRI conditions for an additional 5 hr at 37° C. The restricted samples were separately heated to 65° C for 10 min. 75 μl of the restricted pBR317 was mixed with 225 μl of λh80dara and 300 μl of ligase buffer was added. The conditions for ligation and transformation were identical to those described above.

Screening procedure: The transformed mixtures were plated onto TYE Amp plates and

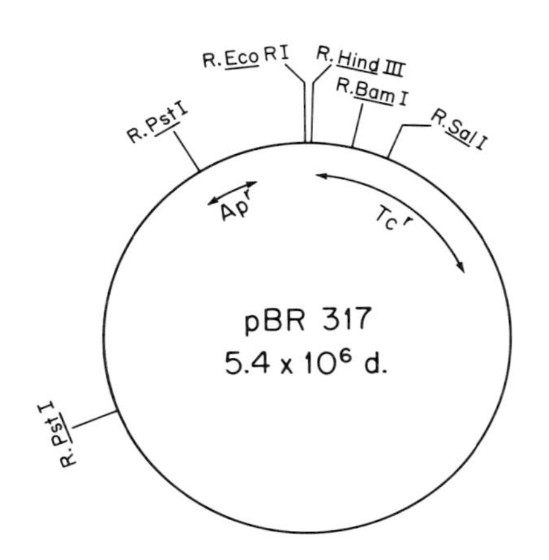

Figure 1

A schematic representation of the plasmid pBR317. Only those restriction sites that were used in this study are indicated. The abbreviations Ap^r and Tc^r stand for ampicillin resistance and tetracycline resistance respectively. The construction and properties of pBR317 have been described (6).

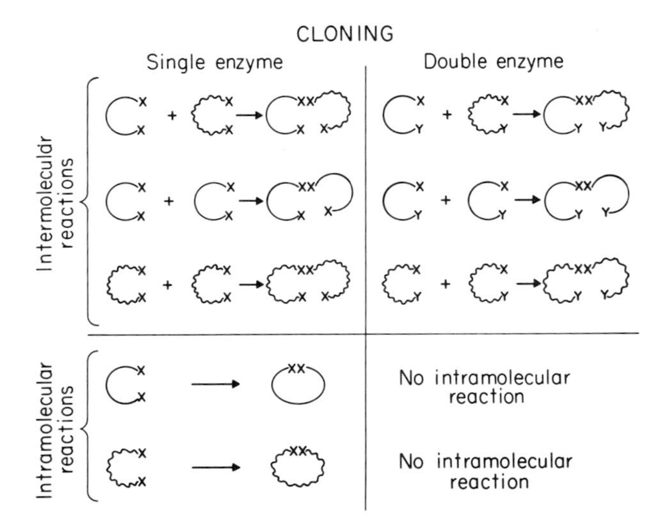

Figure 2

The left hand column displays the five initial reactions that are possible when cloning with plasmids and restriction fragments having the same ends. The smooth lined figures represent the cloning vehicle, the jagged ones the restricted fragments. The possible initial reactions for the double enzyme cloning of restriction fragments are shown in the right hand column.

incubated at 22° C for 16 hrs. Small colonies (∿ 2-3 mm) were easily seen on the plates at this time. 200-300 μl of tetracycline (Tet) (10 mg/ml) was sprayed onto the plates from a compressed air-driven aerosol unit (Nutritional Biochemicals Corporation). After an additional 16 hr incubation at 37° C large tetracycline-resistant clones were readily discernible from the smaller sensitive clones (Fig 3).

<u>DNA restriction analysis of chimeric plasmids</u>: 40 μl of chimeric plasmid DNA (5-10 μg) was restricted in 100 μl of R.*Eco*RI restriction buffer with R.*Bam*I (25 units), R.*Pst*I (25 units), R.*Hind*III (25 units), or R.*Eco*RI (25 units) for 16 hr at 37° C. The reactions were stopped by the

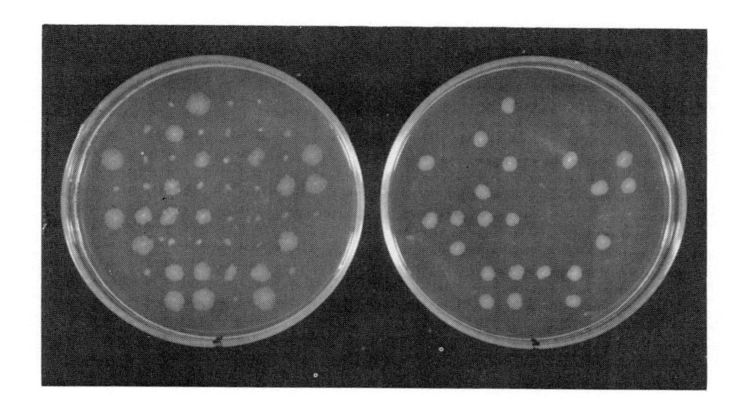

Figure 3

Left plate: Ampicillin resistant colonies were grown 16 hr at 22° C on TYE Amp plates. Small colonies (refer to photo) were easily seen after this time. The agar plate was then sprayed with tetracycline (200-300 μls, 10 mg/ml) and incubated an additional 16 hr at 37° C. The larger colonies in this photo represent those clones that are Ap^r and Tc^r, the smaller colonies are Ap^r and Tc^s.

Right plate: The identical set of ampicillin resistant colonies used in the first experiment (left plate) were picked to a TYE Tet plate. The position of each of the colonies was the same on both plates. The plates were incubated 16 hr at 37° C. Only those colonies that were Ap^r and Tc^r grew and in each case, correspond to the larger colonies on the left plate.

addition of 30 μl of glycerol-dye solution (50% glycerol, 10% sarkosyl, and 0.05% bromphenol blue).

Gel electrophoresis: Agarose gel electrophoresis was performed in a horizontal slab apparatus as described previously (8). A vertical apparatus was used for polyacrylamide gel electrophoresis (9).

RESULTS

Cloning with R.BamI and R.SalI: pBR317 and λh80dara DNA were mixed and restricted with R.BamI and R.SalI. The restricted DNA was then electrophoresed on a 1.2% analytical agarose slab gel. It was not possible to determine if both enzymatic reactions had gone to completion because the difference in size between pBR317 restricted only with R.BamI or R.SalI, or with both R.BamI and R.SalI is approximately 5%. Nevertheless, no intact pBR317 was detectable on the gels. After ligation, the DNA was electrophoresed on a 1.2% agarose slab gel along with the unligated material. After ligation none of the R.BamI-R.SalI fragments of λh80dara, or of restricted pBR317 could be seen. However, a smear of presumably ligated DNA was seen on the gel in the molecular weight range of 6-14 x 10^6 daltons. 75 μl of the ligated DNA was used to transform 250 μl of $CaCl_2$ treated heat-shocked *E. coli* RRI. Cells were spread on TYE Amp plates; the resultant colonies were picked and plated on TYE Tet plates. Of the Ap^r transformed cells 83% were found to be Tc^s. These transformants could have inserts with R.BamI-R.SalI ends, R.BamI-R.BamI ends, or R.SalI-R.SalI ends. A restriction analysis on the DNA of 8 clones showed that 7 had R.BamI-R.SalI ends and one had R.BamI-R.BamI ends only.

Randomly chosen colonies that were Ap^r and Tc^s were mass cultured and crude, cleared lysates were prepared and run on 0.7% horizontal agarose gels. Plasmid DNA was not detectable in 5% of these. Those crude cleared lysates containing plasmid DNA were extracted with phenol; the DNA was precipitated with ethanol and suspended in restriction buffer. DNA was treated separately with R.BamI; or with R.SalI; or with both enzymes simultaneously (Fig 4). 40% of the clones had more than one cloned fragment; of these 70% were a single R.SalI-R.SalI insertion along with a single R.BamI-R.SalI insertion. In addition, 30% of the clones had more than two insertions (Fig 4 lane 16).

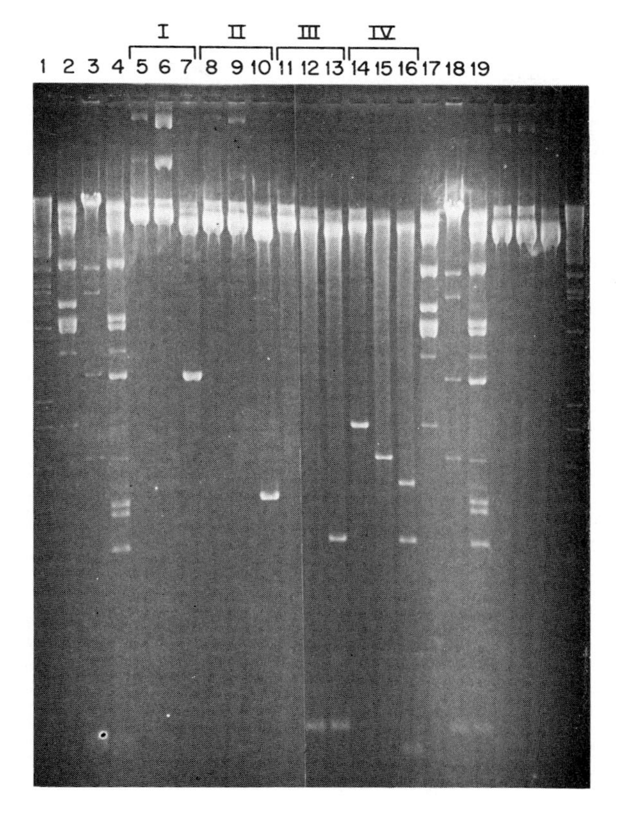

Figure 4

Four pBR317-*λh80dara* chimeras (labeled I, II, III, and IV) with R.*Bam*I-R.*Sal*I, R.*Bam*I-R.*Bam*I, and R.*Sal*I-R.*Sal*I insertions were restricted with R.*Bam*I alone (lanes 5, 8, 11, and 14), with R.*Sal*I alone (lanes 6, 9, 12, and 15), and R.*Bam*I and R.*Sal*I simultaneously (lanes 7, 10, 13, and 16). These were compared with *λh80dara* restricted with R.*Bam*I (lanes 2 and 17), R.*Sal*I (lanes 3 and 18) and with R.*Bam*I and R.*Sal*I simultaneously (lanes 4 and 19). Clones I and II had a single R.*Bam*I-R.*Sal*I insertion; Clone III had two insertions, one a R.BamI-R.SalI insertion and the other a R.SalI-R.SalI insert. Clone IV was difficult to analyze but probably has three R.BamI-R.Sal inserts.

Cloning with R.BamI and R.EcoRI*: Plasmid pBR317 DNA was restricted simultaneously with R.BamI, R.HindIII, and R.EcoRI. The restricted DNA was electrophoresed on a 1.2% agarose gel. DNA could not be seen migrating with unrestricted pBR317, but it was difficult to quantitate those that had been restricted with one, two, or three enzymes. Restricted pBR317 was mixed with λh80dara DNA that had been treated simultaneously with R.BamI and R.EcoRI* and ligated as described in Materials and Methods. The ligated material was electrophoresed on a 1.2% agarose gel with the unligated restricted DNAs. The unligated restricted DNA was no longer seen on the gels and the ligated material that remained migrated more slowly than restricted pBR317. 75 μl of ligated material was used to transform E. coli RRI. A total of 2,000 transformants were analyzed; 97% were ampicillin-resistant and tetracycline-sensitive. Crude, cleared lysates from 25 of the tetracycline-sensitive clones were analyzed on 0.7% agarose gels to determine the content and size of the plasmid DNA. This analysis was necessary to identify those clones that contained little or no plasmid DNA and those that might have two plasmids. One of 25 clones examined contained two different sized plasmids (Fig 5) and one of the clones did not have detectable amounts of plasmid DNA. A simple screening procedure was used to determine whether the remaining clones had a single insert into the R.BamI site of pBR317 (which would make the cell also Ap^r and Tc^s) and those which had the entire R.BamI-R.EcoRI region of pBR317 removed and DNA inserted. The chimeric DNA was restricted simultaneously with R.HindIII and R.PstI and the restriction fragments were electrophoresed on a 1% agarose gel with a R.HindIII and R.PstI digest of pBR317. If a simple insertion into the R.BamI site had occurred, and if the fragment between R.BamI and EcoRI of pBR317 had not been removed, only one of the two R.HindIII-R.PstI fragments of the

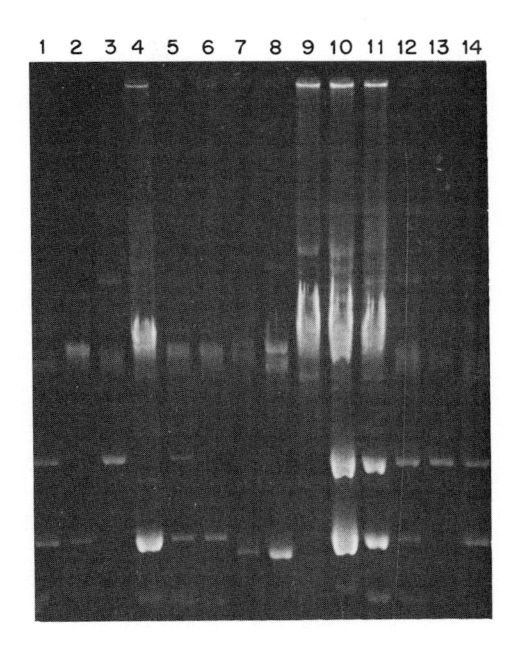

Figure 5

A single clone with two different sized plasmids was discovered among the 25 clones analyzed in the R.*Bam*I-R.*Eco*RI* cloning experiment. The clone was inoculated into TYE Amp media, incubated for 16 hr at 37° C and streaked on TYE Amp plates. Single colony isolates were picked and crude, cleared lysates were prepared from 12 clones for analysis on 0.7% agarose gels. Six of the clones contained two plasmids (lanes 1, 5, 10, 11, 12, and 14), three the smaller of the two plasmids (lanes 2, 4, and 6), two the larger one (lanes 3 and 13), and one contained neither (lane 9). Crude, cleared lysates of pBR317 were run in lanes 7 and 8.

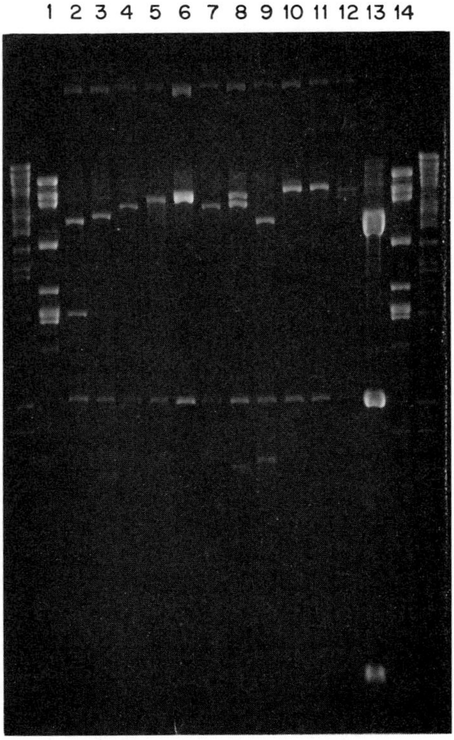

1 2 3 4 5 6 7 8 9 10 11 12 13 14

Figure 6

An analysis of the R.*Bam*I-R.*Eco*RI* inserts. Eleven of the pBR317-*λh80dara* clones from the R.*Bam*I-R.*Eco*RI* cloning experiment were analyzed to determine if the R.*Bam*I to R.*Eco*RI region of pBR317 had been removed and DNA had been inserted. pBR317 and plasmid DNA from each clone was analyzed by restriction with a mixture of R.*Hind*III and R.*Pst*I. Lanes 1 and 14 R.*Bam*I restriction of *λh80dara*, lanes 2-12 R.*Hind*III-R.*Pst*I restriction of pBR317-*λh80dara* DNA, and lane 13 R.*Hind*III-R.*Pst*I digest of pBR317.

chimeric restriction would be different from those of pBR317. However, if an insertion occurred between R.*Bam*I and R.*Eco*RI removing the R.*Hind*III site, both R.*Hind*III-R.*Pst*I fragments formed in pBR317 would be missing in the digestion products of chimeric DNA. Of the 11 clones analyzed, all revealed this latter pattern, demonstrating that the R.*Bam*I and R.*Eco*RI sites were the receptors in pBR317 (Fig 6).

Because a R.*Bam*I-R.*Eco*RI* digestion of the pBR317-*λh80dara* chimeras gave 20-30 difficult to resolve fragments, the plasmid DNAs were restricted with R.*Sal*I and R.*Pst*I. The resulting fragments, presumably containing the R.*Bam*I-R.*Eco*RI* insert, were isolated by horizontal slab gel electrophoresis (1.0% agarose). The isolated fragments were then restricted with R.*Bam*I and R.*Eco*RI* and run on 3.5-10% gradient polyacrylamide gels.

Preliminary results of these patterns compared with those of a R.*Bam*I-R.*Eco*RI* restriction of *λh80dara*, or R.*Bam*I-R.*Eco*RI* restrictions of pBR317-R.*Bam*I *λh80dara* chimeras or with partially purified R.*Bam*I restriction fragments of *λh80dara* restricted with R.*Eco*RI*, showed that the clones contained both R.*Eco*RI* to R.*Eco*RI* fragments and R.*Bam*I to *Eco*RI* fragments.

DISCUSSION

Jacobson and Stockmayer (10) have derived an equation describing the effective concentration (j) of one end of a molecule in proximity to the other end: $j = (3/2\pi lb)^{3/2}$ (ends/ml), where b is defined as the random coil segment and has a value of 7.7 x 10^{-2} μm (11) and l is the length of the molecule. The rate of intramolecular reactions is a function of the probability that one end of the molecule is in proximity to the other end, thus allowing hydrogen bonding (12,13). This is relevant to cloning experiments since the percentage of clones with inserts will be dependent on the relative rates of intramolecular reactions (self-closure) as compared to intermolecular reactions (reactions between molecules). Intermolecular reactions are directly related to i_t, the total concentration of ends. We will use the term "efficiency" to describe the ratio of clones with inserts to the total number of clones.

In a previous study, R.*Bam*I restriction fragments of *λh80dara* were cloned into the R.*Bam*I site of pBR317 with an efficiency of 30% when j_{pBR317}/i_t=.06 (Kaplan, D., L. Greenfield, T. Boone and G. Wilcox, manuscript in preparation). In this case, it is the intramolecular reaction of the fragments that must be counteracted, for they are smaller than pBR317 and have larger j values, to increase the probability of cloning an insert. To increase the probability of cloning the smaller fragments the concentration of pBR317 or the fragments can be increased. In theory, when $j<i_t$, the intermolecular reactions are favored. However, increasing the pBR317 concentration is not equivalent to increasing the concentration of the foreign DNA fragments. In the first instance, the total number of fragments cloned is increased but at the expense of efficiency. In the second instance, the amount of DNA cloned and the efficiency increases, but the number of multiple insertions also increases.

Although the use of single restriction enzymes for cloning is a valid and operable procedure, it is important to have a method which increases the efficiency of cloning. A simple way to increase the efficiency of cloning is to reduce, or eliminate, the possibility of self-closure of both the cloning vector and the DNA to be inserted. One of the advantages of the "connector method" (14,15) is that the ends cannot self-close. However, this method has the disadvantage that the cloned fragment cannot always be removed as homogenous DNA fragments from the chimera. The elimination of self-closure is also accomplished by restricting the DNAs with two different enzymes to generate two ends of noncomplementary base sequences. The cloning vehicle and the foreign DNA do not necessarily have to be restricted with the same enzymes. The only requirement is that the two single stranded sequences on the plasmid must be able to hydrogen bond with the two sequences on the restricted foreign DNA. For example, we restricted pBR317 with R.*Bam*I and R.*Eco*RI thus generating the single stranded sequences of GATC and AATT;

λh80dara was restricted with R.*Bam*I and R.*Eco*RI*, generating the same pair of single stranded ends.

To reduce possible intramolecular reactions in the plasmids to zero, the double restricted plasmid would have to be separated from those molecules not restricted or restricted only once. For most purposes this is probably unnecessary, however, it should be considered. In Fig 7 we have plotted the number of molecules restricted more than once, as a function of relative restriction enzyme activity. By including a third enzyme in the restriction of a cloning vehicle, having a single restriction site, between the two single sites to be cloned into, the number of plasmid molecules that have been restricted at least twice is greatly increased (Fig 7). For example, we restricted pBR317 with R.*Bam*I, R.*Hind*III, and R.*Eco*RI, where R.*Hind*III is between the R.*Bam*I and the R.*Eco*RI sites (see Fig 1).

Clearly, by decreasing the intramolecular reactions of the cloning vector and of the foreign DNA, the efficiency of cloning is greatly increased. In our experiments the R.*Bam*I-R.*Sal*I fragments of *λh80dara* were inserted into pBR317 with an efficiency of cloning of 83%. This is a much higher efficiency than the 30% efficiency obtained when the R.*Bam*I fragments of *λh80dara* were inserted into the R.*Bam*I site of pBR317. This increase in efficiency is even more pronounced when one considers that if a single enzyme cloning experiment were done under the conditions of the R.*Bam*I-R.*Sal*I experiment, the ^{j}pBR317/i_t ratio would be 0.18 as compared to ^{j}pBR317/i_t ratio of .06 for the single site cloning of the R.*Bam*I fragments. The increase in efficiency of cloning of the R.*Bam*-R.*Eco*RI* fragments into the R.*Bam*-R.*Eco*RI site of pBR317 was even more pronounced - 97%. The addition of a third restriction enzyme in the restriction of pBR317, to reduce intramolecular reactions, could account for this increase in efficiency.

There are, of course, some disadvantages to using two enzymes. Fragments with homogeneous ends will not be cloned alone and the probability of one of the restriction sites being within a given gene is increased. These problems may be alleviated by preparing partial digests or using different

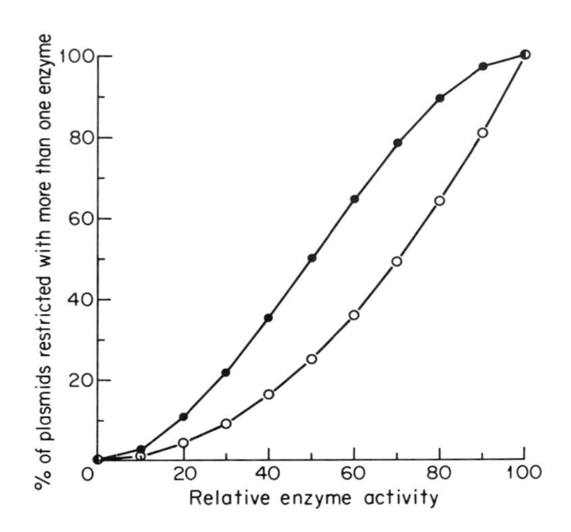

Figure 7

The probability that a plasmid will be restricted more than once in the presence of two or three restriction endonucleases. The relative enzyme activity represents the percentage of molecules which will be restricted by each enzyme. It is assumed that all the enzymes act independently at single sites on the plasmid and are present at identical activities. The percentage of plasmids which are restricted at more than one site in the presence of two (O) or three enzymes (●) is given on the abscissa. If two enzymes, R_1 and R_2, are present with activities α_1 and α_2, respectively, then the probability, ρ, that a plasmid will be restricted by both enzymes is given by $\rho = (\alpha_1)(\alpha_2)$. For three enzymes the corresponding equation is $\rho = (\alpha_1)(\alpha_2)(\alpha_3) + (\alpha_1)(\alpha_2)(1-\alpha_3) + (\alpha_1)(1-\alpha_2)(\alpha_3) + (1-\alpha_1)(\alpha_2)(\alpha_3)$.

combinations of enzymes. Beside being more efficient, the double enzyme cloning procedure has other advantages. First, it is possible to orient a fragment, for instead of having two identical ends on the plasmid and on the foreign DNA, there are two distinct ends that are not interchangeable. The orientation of a fragment produced by restriction with two enzymes has been reported using an SV40 vector but the efficiency and j/i_t ratios were not discussed (16). This orienting process, or directed cloning, is shown here with the R.*BamI* to R.*EcoRI** or the R.*BamI* to R.*SalI* cloning. A second advantage of the double enzyme method is that already cloned pieces (with identical ends) or fragments identified from restriction patterns with a single enzyme, such as with the Southern hybridization technique (17), can be restricted with a second enzyme and the resulting fragments cloned with high efficiency. We took advantage of this when we cloned the R.*BamI*-R.*EcoRI** fragments of *λh80dara* which were subsets of the R.*BamI* restriction fragments of *λh80dara*. In this way, small regions of the regulatory region of the arabinose operon have been isolated. A third advantage is that the cloned fragments can be recovered intact from the cloning vehicle.

ACKNOWLEDGEMENTS

We would like to thank Ralph Martinez and Dan Ray for their helpful comments on the manuscript. This study was supported by National Science Foundation Grant BMS 75-02805. GW is supported by an American Cancer Society Faculty Research Award and LG by Molecular Biology Training Grant GM 0158.

REFERENCES

1. Cohen, S.N., A.C.Y. Chang, H.W. Boyer and R.B. Helling. 1973. Proc. Nat. Acad. Sci. U.S.A. 70, 3240.

2. Hershfield, V., H.W. Boyer, C. Yanofsky, M.A. Lovett and D.R. Helinski. 1974. Proc. Nat. Acad. Sci. U.S.A. 71, 3455.

3. Wilcox, G, K. Clemetson, P. Cleary and E. Englesberg. 1974. J. Mol. Biol. 85, 589.

4. Clewell, D.B. and D.R. Helinski. 1969. Proc. Nat. Acad. Sci. 62, 1159.

5. Clewell, D.B. 1972. J. Bacteriol. 110, 667.

6. Bolivar, F., R.L. Rodriguez, M.C. Betlach and H.W. Boyer. Submitted to Gene.

7. Polisky, B., P. Greene, D.E. Garfin, B.J. McCarthy, H.M. Goodman and H.W. Boyer. 1975. Proc. Nat. Acad. Sci. U.S.A. 72, 3310.

8. Kaplan, D.A., R. Russo and G. Wilcox. 1977. Anal. Biochem. 78, 235.

9. Jeppesen, P.G.N. 1974. Anal. Biochem. 58, 195.

10. Jacobson, H. and W.H. Stockmayer. 1950. J. Chem. Phys. 18, 1600.

11. Hearst, J. and W.H. Stockmayer. 1962. J. Chem. Phys. 37, 1425.

12. Wang, J.C. and N. Davidson. 1966. J. Mol. Biol. 19, 469.

13. Dugaiczyk, A., H.W. Boyer and H.M. Goodman. 1975. J. Mol. Biol. 96, 171.

14. Lobban, P.E. and A.D. Kaiser. 1973. J. Mol. Biol. 78, 453.

15. Jackson, D.A., R.H. Symons and P. Berg. 1972. Proc. Nat. Acad. Sci. U.S.A. 69, 2904.

16. Nussbaum, A.L., D. Davoli, D. Ganem and G.C. Fareed. 1976. Proc. Nat. Acad. Sci. U.S.A. 73, 1068.

17. Southern, E.M. 1975. J. Mol. Biol. 98, 503.

HORIZONTAL SLAB GEL ELECTROPHORESIS OF DNA

Donald A. Kaplan and Gary Wilcox

Department of Bacteriology
and Molecular Biology Institute
University of California
Los Angeles, California 90024

ABSTRACT. A horizontal slab gel electrophoresis apparatus has certain advantages over its vertical counterpart: (1) rapid and simple preparation, (2) large sample capacity, (3) accommodation of low percentage agarose and polyacrylamide gels, and (4) versatility. An improved horizontal slab gel electrophoresis apparatus is discussed along with some of its applications to the separation of DNA.

INTRODUCTION

We have recently described an improved horizontal slab gel electrophoresis apparatus (1) which is very simple to use for either preparative or analytical purposes. In this paper we describe further improvements in the design of the apparatus and give examples of some applications. Experiments on the separation of large DNA molecules in low percentage agarose gels demonstrate that molecular weight estimates can be made on DNA as large as 84 x 10^6 daltons. Experiments using horizontal slab gels with a trapezoidal geometry show that the resolution of DNA molecules of similar size is greatly improved.

MATERIALS AND METHODS

The horizontal slab gel electrophoresis apparatus which we described recently (1) has been modified as follows: (1) the overall dimensions have been reduced 20% to minimize difficulties in handling, (2) the lid has been equipped with a manifold so that nitrogen can be flushed into the apparatus when acrylamide is being polymerized, (3) the combs have been reduced in thickness to 1 mm to improve resolution, (4) the trough is now made from a plastic which transmits ultraviolet light (Rohn and Haas UVT[R] plastic), and (5) the leveling

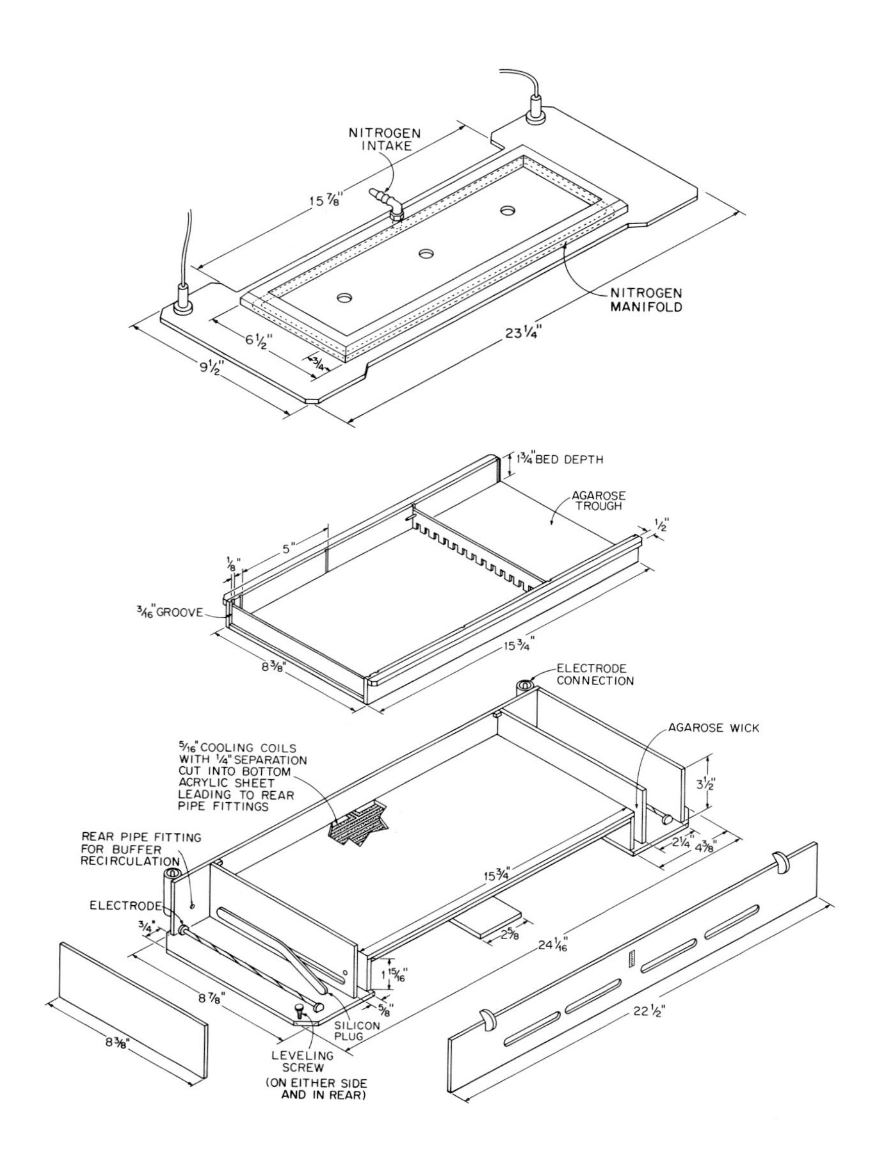

Figure 1. An exploded view of the horizontal slab gel electrophoresis apparatus.

bubble has been removed. A drawing of the modified horizontal slab gel electrophoresis apparatus is shown in Figure 1.

The apparatus is prepared for use as described (1) except that a circular level, which is placed in the trough, is used for leveling. The plastic troughs have a tendency to warp and for this reason are placed on a flat surface under a moderate amount of weight when not in use. A trough with a glass base has also been used. It has the advantage of not warping but the disadvantage of not transmitting UV light. The trapezoidal gels are made from a regular rectangular gel by cutting out part of the agarose slab with a scalpel and straight edge.

RESULTS AND DISCUSSION

We have described an improved horizontal slab gel electrophoresis apparatus (1). Perhaps the most important design feature of the apparatus was the use of vertical polyacrylamide or agarose wicks as the current carriers (see Figure 1). This eliminated the use of conventional paper wicks which have been a source of problems with the horizontal gel apparatus (2,3).

Two advantages of a horizontal slab gel apparatus are that a minimal amount of time is required for preparation and that it can handle a large number of samples. Preparation of the apparatus for use involves just three steps: (1) leveling the apparatus, (2) pouring the agarose or acrylamide solutions, and (3) placing the comb in the appropriate holder. This is much less time consuming than the comparable manipulations required in the preparation of a vertical slab gel and there are no leakage problems or difficulties in removing the comb. Both the versatility and sample capacity of the apparatus are demonstrated in the experiment shown in Figure 2. Three slabs of different agarose concentrations were poured sequentially. Electrophoresis of R.*Bam*I restriction fragments of various *λh80dara* phages containing different *ara* deletions was carried out. The larger molecular weight fragments are resolved best in the 0.5% agarose gel whereas the

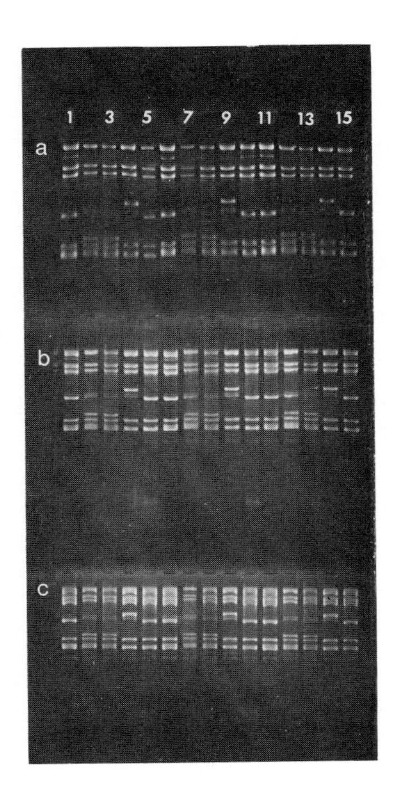

Figure 2. Analysis of DNA restriction fragments. R.BamI restriction fragments of $\lambda h80dara^+$ (lanes 1,6,11), λh80daraΔ766 (lanes 2,7,12), λh80daraΔ719 (lanes 3,8, 11), λh80daraΔ735 (lanes 4, 9,14), and λh80daraΔ718 (lanes 5,10,15) DNAs were run simultaneously on 0.5% (a), 0.75%(b), and 1.4%(c) agarose gels. The 1.4% agarose gel was closest to the anode. Electrophoresis was for 16 hours at 60 volts.

lower molecular weight fragments are resolved better in the 0.75% and 1.4% agarose gels. This protocol is very useful for mapping restriction endonuclease cleavage sites in DNA molecules because DNA restriction fragments which do not resolve at one agarose concentration may be resolved at another.

Horizontal Polyacrylamide Gels

Polyacrylamide gel electrophoresis provides the highest degree of resolution of DNA molecules in the size range of 20-1000 base pairs (4) and is used in DNA sequencing methods (5,6). We have adapted the horizontal gel apparatus so that it can be used for polyacrylamide gels. Acrylamide did not polymerize completely in our original horizontal gel apparatus probably because of quenching by molecular oxygen. Thus, the lid was equipped with a manifold through which nitrogen can be distributed into the apparatus (see Figure 1). Prior to use the apparatus was flushed with nitrogen for five minutes. Gel components, prepared as described by Dewachter and Fiers (7), were mixed, and poured through the hole in the lid. The flow of nitrogen was then adjusted so that it did not disturb the surface of the forming gel. After 45 minutes the nitrogen flow was shut off. After an additional 45 minutes the comb was removed, the sample wells filled with buffer and the samples layered underneath. Staining was carried out in the trough as described below. The resolution obtained using the horizontal gel apparatus was as good as that obtained with a conventional vertical apparatus.

Separation and visualization of large DNA molecules in low percentage agarose gels.

Large molecular weight DNAs can be analyzed on low percentage agarose gels. However, these gels can be difficult to handle during staining and visualization. The handling problem has been solved by making the tray bottom out of a plastic which transmits UV light (see Materials and Methods). The gel can then be stained and visualized in the tray without distortion of the gel. Staining and destaining solutions are added directly to the trough after the end plates have been sealed with melted agarose (the left end plate is shown in Figure 1). The trough with the gel in it is then placed on a UV light box for visualization.

The results of an experiment using a 0.2% agarose gel show that the mobility of a DNA molecule is proportional to the logarithm of its molecular weight over a range of 11×10^6 to 84×10^6 daltons (Figure 3). This result suggests that molecular weights of most bacterial plasmid and bacteriophage molecules can be estimated on such gels and that perhaps some chromosomal DNAs or chromosomes might also be isolated by electrophoresis on low percentage agarose gels.

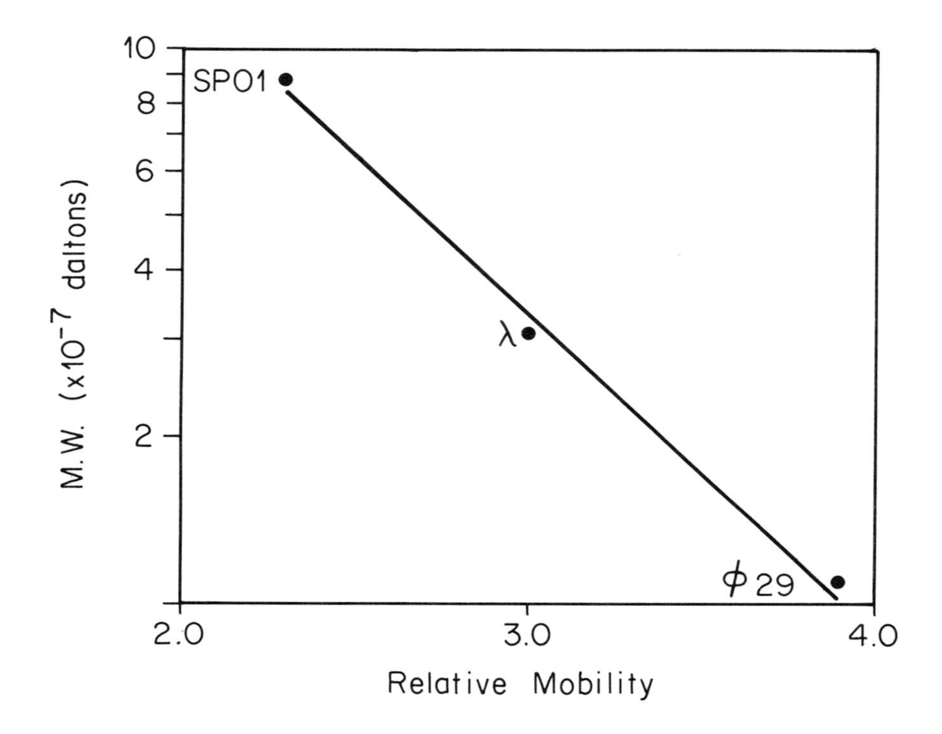

Figure 3. Electrophoresis in 0.2% agarose gels. Approximately 0.1 µg each of ϕ29 DNA ($MW=11x10^6$ daltons), λDNA ($MW=30x10^6$ daltons), and SP01 DNA ($MW=84x10^6$ daltons) were electrophoresed at 80 volts for 36 hours in a 0.2% agarose gel. The gel bed was kept at 0° C during electrophoresis. The data are plotted as log molecular weight vs. mobility.

Electrophoresis on gels with a trapezoidal geometry

Electrophoresis on horizontal slab gels with a trapezoidal geometry improves the resolution of DNA restriction fragments. The trapezoidal shape produces a voltage gradient because the resistance decreases as the cross-sectional area of the gel increases and the current remains constant. Thus, the leading edge of a band migrates slower than the trailing edge resulting in a decreased band width.

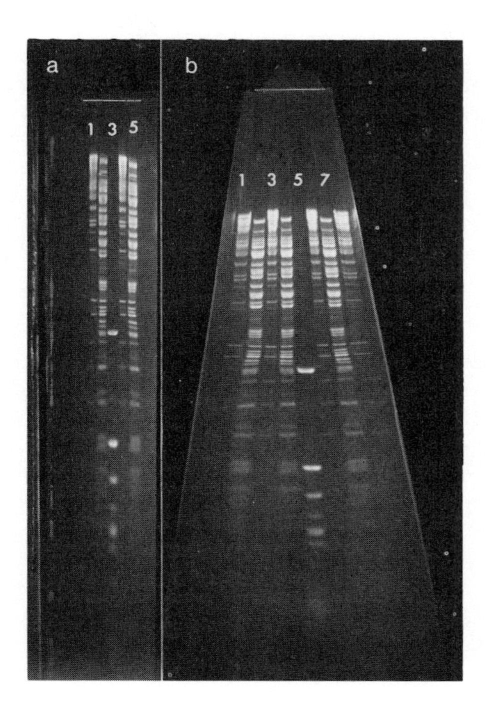

Figure 4. Electrophoresis using gels with rectangular and trapezoidal geometry. (a) 10 μl samples containing 1 μg of non-glucosylated T4 DNA restricted with R.EcoRI (lanes 1 and 4) or R.EcoRI* (lanes 2 and 5), 1 μg of M13 RF DNA restricted with R.HpaII (lane 3) were electrophoresed at 120 volts for 16 hours on a rectangular gel. (b) 10 μl samples containing 1 μg of non-glucosylated T4 DNA restricted with R.EcoRI (lanes 1,3,6,8) or R.EcoRI* (lanes 2,4, & 7) or 1 μg of M13 RF DNA restricted with R.HpaII (lanes 5) were electrophoresed at 120 volts for 16 hours on a trapezoidal gel. After staining with ethidium bromide, the gels were photographed with Plux X film using a UV transilluminator. Densitometer tracings of the negatives were made with a Joyce-Lobel densitometer.

Restriction endonuclease fragments of T4 and M13 DNA were electrophoresed on a horizontal gel with a rectangular and a trapezoidal geometry (Figure 4). The average width of a band on the trapezoidal gel was about 2/3 the width of the corresponding band on the rectangular gel. The decreased band width not only improves resolution but it can also increase the sensitivity of the method. Horizontal gels with a trapezoidal geometry should prove useful for the separation of RNAs, proteins, and perhaps DNA in the rapid DNA sequencing techniques.

ACKNOWLEDGEMENTS

The development of the horizontal gel apparatus was supported by UCLA Academic Senate Grant No. 1888. The research was supported by the National Science Foundation Grant BMS 75-02805 and an American Cancer Society Faculty Research Award to Gary Wilcox.

REFERENCES

1. Kaplan, D., R. Russo and G. Wilcox. 1977. Analytical Biochemistry 78:235.

2. Bartlett, R.C. 1963. Clin. Chem. 9:317-324.

3. Hoppe, H.H., W. Hennig and B. Brinkmann. 1972. Humangenetik 14:224-231.

4. Maniatis, T., A. Jeffrey, H. van deSande. 1975. Biochemistry 14:3787-3794.

5. Sanger, F. and A.R. Coulson. 1975. J. Mol. Biol. 94:441-448.

6. Maxam, T. and W. Gilbert. 1977. Proc. Nat. Acad. Sci. 74:560-564.

7. Dewachter, R. and W. Fiers. 1971. Methods in Enzymology 21:167.

WORKSHOP SUMMARY: Vehicles for Molecular Cloning in Mammalian Host Cells by G.C. Fareed, Molecular Biology Institute, University of California at Los Angeles, California

This workshop concerned the use of papovavirus DNA molecules for molecular cloning in permissive cells and in transformed, non-permissive cells. The first two presentations described SV40 vectors carrying prokaryotic DNA segments in the late gene region. The fate of a portion of the immunity region of bacteriophage λ DNA linked via poly dA-dT tails to a 60% segment of SV40 DNA was examined in monkey cells by Goff and Berg. The vector DNA harbored the SV40 origin for DNA replication and an intact early gene region. Although the hybrid virus could be efficiently cloned by its complementing a SV40 tsA helper *in vivo*, transcription of the prokaryotic DNA sequences was undetectable. This was not due to the dA-dT joints in the hybrid since reconstruction of SV40 genomes carrying dA-dT joints in the wild-type late gene region did not affect late gene expression. New SV40 vectors having shorter lengths and capable of expression of the *D* and *E* genes are being explored for future recombinant DNA studies.

Two helper virus-free cloning approaches with SV40 (or polyoma) vectors were analyzed by Upcroft, Upcroft, Skolnick, and Fareed. Purified SV40-*E. coli* hybrid DNA bearing the early gene region of SV40 was utilized to transform rat embryo cells and generate persistently-infected monkey kidney cells. Hybridization procedures demonstrated the preservation of vector and *E. coli* DNA sequences in free or unintegrated genomes from both mammalian cell populations.

Drs. John Jordan and Goran Magnussan described the properties of monkey cells persistently-infected with SV40 DNA and mouse cells persistently-infected with polyoma DNA, respectively. In both "carrier cultures" the block to virus multiplication appears to be a host cell defect(s) in virus maturation since the endogenous DNA retains infectivity. It was suggested that these host cells, when "cured" of free DNA by cell cloning in the presence of anti-virus antibody, may serve as useful host cells for the propagation of SV40 or polyoma vectors carrying foreign genetic information.

In the final presentation Shen and Hearst described the identification of sequences in SV40 DNA with approximate two-fold axis of symmetry (inverted repeats). These were fixed as hairpins in single-stranded SV40 DNA with a cross-linking agent and located by electron microscopy. Prominent hairpins appear at the origin and termination regions (0.67 and 0.17 map units, respectively) both for replication and transcription.

SV40 VECTORS FOR MOLECULAR CLONING IN MAMMALIAN CELLS

George C. Fareed

Department of Microbiology and Immunology
Molecular Biology Institute
University of California, Los Angeles
Los Angeles, California 90024

ABSTRACT. We have found that defective SV40 replicons can serve as efficient vectors for the propagation and molecular cloning of prokaryotic DNA segments in mammalian cells. The simplest SV40 vector DNA segments used were obtained from reiteration mutants and harbored the SV40 DNA replication origin. Segments of 520 and 2400 base pairs from the immunity region of bacteriophage lambda were linked *in vitro* to SV40 vectors and propagated in monkey kidney cells in the presence of added helper, wild-type SV40 DNA. In both recombinant genomes, the structure of the lambda DNA segment was largely unchanged after serial passage in monkey cells. An alternative method for molecular cloning with SV40 vectors takes advantage of the ability of the vector to express an early or late gene function and, thereby, complement a temperature-sensitive mutant helper genome. This approach has recently been employed (Hamer, Davoli, Thomas and Fareed, J. Mol. Biol., in press) to clone a SV40 genome bearing an *Escherichia coli* suppressor gene, $tRNA^{Tyr}su^{+}III$. The bacterial DNA segment was inserted in a unique orientation via *Eco*RI and *Hpa*II termini into a deletion in the late gene region of SV40. It was propagated with the aid of a helper tsA (early gene) mutant of SV40. We have employed this purified chimeric DNA, which contains the SV40 genetic information needed for cellular transformation and for viral DNA replication, for both the transformation of rat embryo cells and transfection of monkey kidney cells (Upcroft *et al*., this symposium).

INTRODUCTION

We have been concerned with the fate of prokaryotic DNA segments in mammalian cells and with the development of methods for molecular cloning of prokaryotic and eukaryotic DNA in such cells. The ability to propagate foreign DNA segments and specific host cell genes in appropriate eukaryotic host cells should facilitate the analysis of the mechanism and control of eukaryotic gene expression. Since a great deal of effort in the recombinant DNA field has focussed on the molecular cloning of eukaryotic gene regions in *Escherichia coli*, where these regions are not faithfully expressed as functional gene products except for certain yeast gene regions (1), we anticipated that one potential use of a mammalian vector system would be to examine in a eukaryotic environment the fate of a cloned eukaryotic gene that had been biologically amplified and physically characterized after the molecular cloning of *E. coli*. In addition, the mammalian vector system could potentially be used for the introduction of known functional genes of *E. coli* or its bacteriophages, which perhaps could correct a specific genetic defect in mammalian cells or be used to provide a product which the mammalian cells do not possess, such as a suppressor transfer RNA.

Our studies were initiated by an investigation of the fate in monkey kidney cells of recombinant DNA molecules containing specific segments of bacteriophage λ DNA and eukaryotic vector fragments from SV40 DNA. More recently we have cloned a SV40 genome bearing a suppressor tRNA gene from *E. coli* by mixed infection with a tsA helper and begun to study the properties of nonpermissive cells transformed by this suppressor-containing hybrid. In addition, the purified SV40-suppressor hybrid has been used to establish persistently infected, permissive monkey cells.

The approach for our obtaining mammalian transducing viruses arose from studies of reiteration mutants of papovaviruses (2). These mutants, derived from serial undiluted passage of SV40 or polyoma in permissive cells, characteristically have deleted a large part of the wild-type genome and reiterated in tandem a small part bearing the origin for viral DNA replication. Structural analysis has indicated that this origin is the only required *cis* function for replication. Thus, in analogy to the molecular cloning of foreign DNA segments in *E. coli* by enzymatic insertion into a suitable plasmid or phage λ replicon, one could use a segment of SV40 DNA containing its initiation site as the vector for a eukaryotic or prokaryotic DNA segment of appropriate size for encapsidation. About three years ago, we found that small fragments of SV40 DNA bearing the origin for DNA replication could be propagated and amplified in monkey cells in the presence of wild-type helper genomes (3). These fragments had been obtained from a class of evolutionary variants of SV40 which, as we refer to above, are called reiteration mutants. Such mutants contain tandem repeats of these unique monomer fragments which bear the origin for replication. We found that cleavage of mixed populations of such reiteration mutant genomes by certain restriction endonucleases, such as *Eco*RI or *Hin*dIII, generated the short monomer fragments which could be easily separated by gel electrophoresis. These monomer fragments could be recovered from gel slices, individually circularized in vitro using T4 ligase, and subsequently amplified in vivo after mixed transfection of monkey cells with added helper wild-type DNA. This amplification involved the oligomerization of monomers to yield dimers, trimers, tetramers, etc. Very recently, Shenk and Berg (4) have demonstrated a similar phenomenon using specific 10-13% fragments of wild-type SV40 DNA, which include the origin region. It occurred to us that the monomer fragments from these reiteration mutants might be able to serve as vectors for our propagating and cloning foreign DNA segments in monkey cells.

SV40-λ Chimeras

In our initial experiments (5), a 17% monomer segment from a pentameric reiteration mutant was isolated after HindIII cleavage. This mutant preserved in each monomer fragment a small portion of the SV40 genome, which included the origin for DNA replication and mapped between .62 and .72 on the SV40 genetic map. The mutant also possessed a host cell substitution of about 350 base pairs which was shown by reassociation kinetic studies to be from the low reiterated category of monkey cell DNA (6), a sequence which was reiterated approximately 10-20 times per diploid African green monkey cell. Since HindIII creates short cohesive termini at its site of cleavage, we isolated a 520 base pair fragment from the immunity region of phage λ, also by HindIII cleavage, and linked it in vitro to the 17% vector fragment from the pentameric DNA. We subsequently infected monkey cells with this ligation reaction product and helper wild-type SV40 DNA and found that the 520 base pair λ segment had been replicated and encapsidated in progeny virus (5). In Fig. 1 is a diagram which schematically shows the possible structures of the λ 520 base pair-SV40 vector fragment d hybrid. As illustrated here, endo R.HpaII, which cleaves within the SV40 vector fragment, generates a recombinant fragment carrying the λ substitution in the center of a 1400 base pair linear piece. We isolated this 1400 base pair fragment by gel electrophoresis after HpaII cleavage of the recombinant-containing viral DNA preparation. After prior ligation of the 1400 base pair fragment to form a monomeric circle of about 27% of SV40 mass, it was used to infect monkey kidney cells in the presence of helper SV40 DNA. The progeny from this infection of monkey cells in the presence of wild-type SV40 DNA yielded one major species which corresponded in mass to that expected for a trimer of the monomeric recombinant DNA fragment. When this DNA was cleaved with HpaII, it generated almost entirely a 1400 base pair monomer, and heteroduplex analysis of this recombinant DNA preparation showed that the region of homology between the λ immunity region fragment and the recombinant DNA

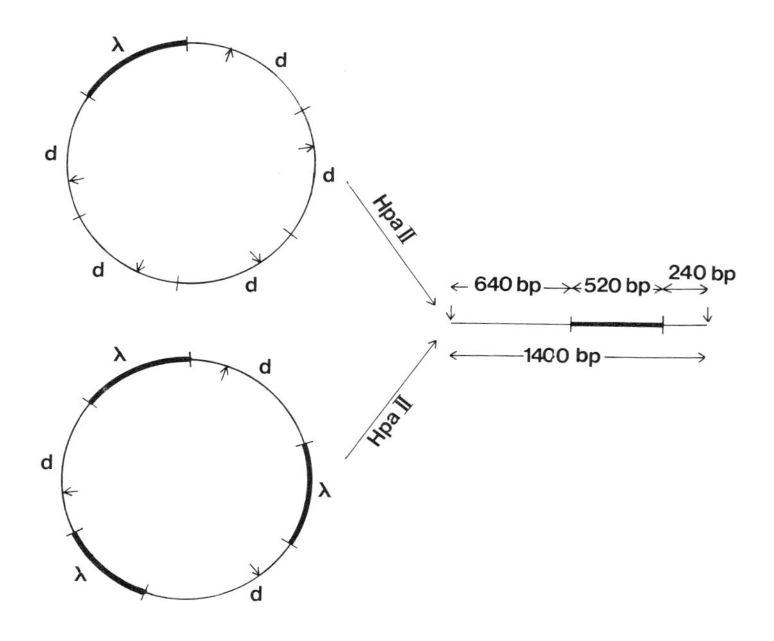

Fig. 1. Possible Structure of Hybrid Genomes.
Schematic diagram of possible structures of λ-d chimeric molecules in the serially passaged virus resulting from DNA infection with in vitro recombinants and wild-type SV40 DNA. The λ 520 base pair segment is represented by thick lines; straight lines denote endo R.HindIII sites; arrows indicate endo R.HpaII sites. Both types of molecules generate a 1400 base pair segment after endo R.HpaII cleavage. The upper circular molecule in the diagram is designated λ_1-d_5 and the lower one $(\lambda\text{-}d)_3$. (Reprinted, with permission, from Ganem et al., 1976, Cell 7, 349)

was preserved in this hybrid (Fig. 2). This was substantiated also by hybridization analysis where the 1400 base pair fragment and the trimeric $(\lambda\text{-}d)_3$ hybrid both annealed specifically with denatured λ DNA immobilized on nitrocellulose filters.

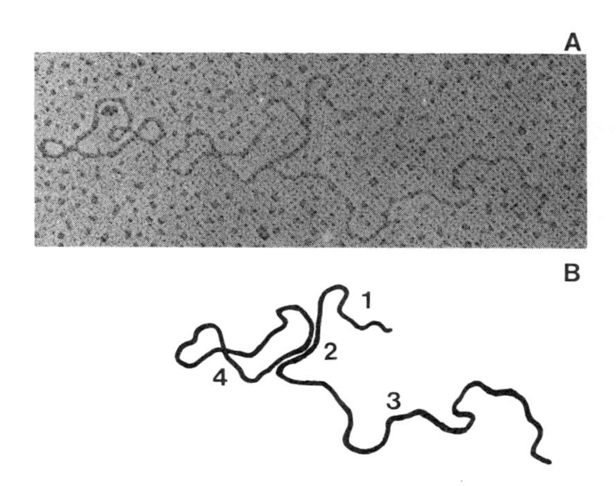

Fig. 2. Electron Microscopic Heteroduplex Analysis of Chimeric (λ-d)$_3$ Genomes.

The heteroduplex between open circular (λ-d)$_3$ and linear λ-EcoRI-B was prepared and measurements of 25 molecules were made. The region of homology between (λ-d)$_3$ and λ-EcoRI-B (designated 2 in the sketch) is 8.3% of the length of the λ-EcoRI-B segment and begins 21.7% from one end of λ-EcoRI-B (designated 1) and 70.0% from the other end (designated 3). The position of the duplex region in the heteroduplex is as predicted from the location of the λ 520 base pair segment within λ-EcoRI-B (Figure 3). The sum of segments 2 and 4 is approximately 83% the length of wild-type SV40 DNA (1.54 μm). (Reprinted, with permission, from Ganem et al., 1976, Cell 7, 349)

To establish the generality of this method and the feasibility for molecular cloning of hybrid genomes of this type we focussed our attention on a larger neighboring segment from the λ immunity region which included the leftward operator-promoter site, as shown in Fig. 3. This was excised as a 2400 base pair segment bounded by BamI and HindIII termini and purified by gel electrophoresis. The molecular cloning of this hybrid (7) in mammalian cells not only showed the efficacy of defective SV40 vectors, but also provided an opportunity for our examining a well-characterized prokaryotic genetic regulatory site, the leftward operator, in a eukaryotic environment. This region of λ seemed appropriate for the construction of novel recombinant DNA molecules, since it lacked the replication functions of λ and the leftward operator that it contained could be specifically identified in a hybrid molecule by λ repressor binding. The vector used to clone this λ segment was excised from a triplication mutant by EcoRI and BamI cleavages and was a fragment of the monomer segment from this reiteration mutant.

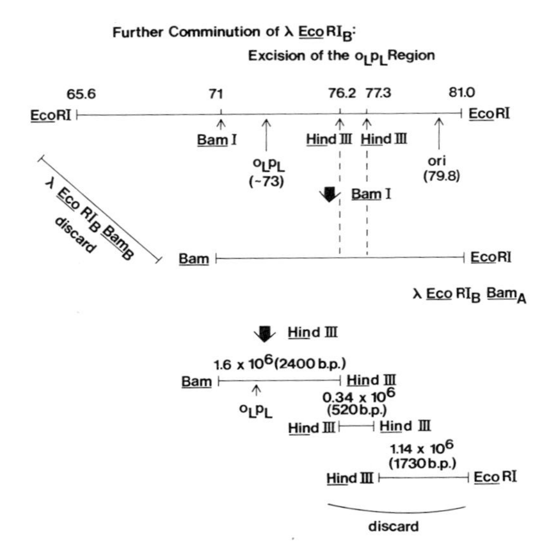

Fig. 3. Preparation of DNA fragments used for in vitro recombination. Schematic representation of the excision of the λ DNA fragment carrying the leftward operator, $O_L P_L$, from the immunity region located in λ-R.EcoRI fragment B.

EcoRI cleavage of this DNA generated the one-third wild-type SV40 monomer, whose physical structure had been previously determined (8). Subsequent BamI cleavage produced a 940 base pair segment bearing the origin and a smaller 730 base pair piece. We linked the 940 bp segment via the BamI end to the corresponding terminus of the λ 2400 base pair segment. After mixed infection of monkey cells with this ligation mixture and wild-type DNA, the progeny viral DNA was found to contain specific λ sequences by filter hybridization. This hybrid DNA was enriched with an infectious center plaquing procedure in which cells infected with helper and defective hybrid genomes are seeded onto a monolayer of uninfected cells. The single infectious center plaque out of about 40 tested for λ DNA sequences was subsequently used to generate a virus stock. The progeny intracellular DNA from this stock contained two different shortened genomes in addition to wild-type SV40 DNA as judged by agarose gel electrophoresis (Fig. 4). When these superhelical genome species were separately tested for homology to the λ immunity region fragment, the shortest genomes,

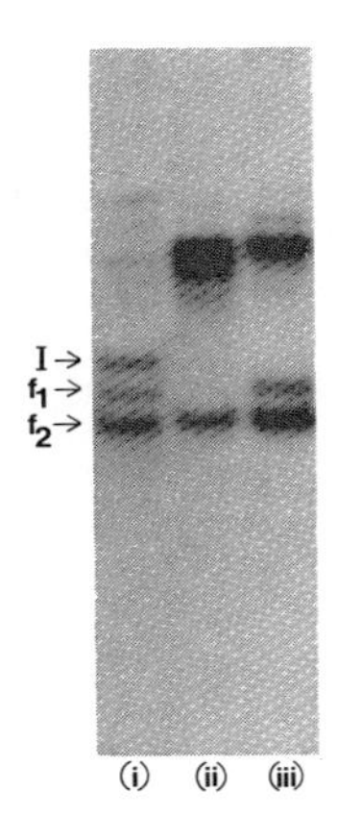

Fig. 4. Properties of the DNA derived from plaque *i.c.* 219. The infectious center method using DNA infection rather than virus infection was employed to clone the λ-SV40 hybrid. A DNA infection was carried out as follows: DNA (primarily viral) was extracted from cells infected with the original lysate created after infection of monkey cells with the recombinant DNA and WT SV40 DNA, and shortened viral DNA (1.2 μg) was purified by 1.4% agarose gel electrophoresis. This DNA, mixed with 0.6 μg of WT SV40 DNA, was used to infect a confluent monolayer of CV-1 cells in a 100 mm culture dish. After 18 hr, the cells were trypsinized and counted. Appropriate volumes of the diluted cell suspensions were seeded into 60 mm culture dishes along with about $2x10^6$ uninfected cells. After 24 hr at 37°, each dish was covered with a 1% agar overlay in Eagle's medium with 2% fetal calf serum, and additional overlays were made after 5, 9, and 12 days at 37° (the last one containing 0.01% neutral red). Plaques were readily evident at dilutions of 10^{-2} and 10^{-3} of the original infected cell suspension. These plaques were aspirated (0.2 ml) and 0.1 ml of each was used to infect a fresh culture of CV-1 cells in a 35 mm dish. After labeling with [^{3}H]thymidine, viral DNA was selectively extracted and tested for hybridization to λ DNA filters. One plaque aspirate, *i.c.* 219, proved to be positive and the remainder of the aspirate from this plaque was used to infect monkey cells in a 150 mm dish. The lysate resulting from this infection provided the viral stock for preparing intracellular viral DNA analyzed above. This shows an autoradiogram of *i.c.* 219 Form I (closed circular, superhelical) [^{32}P]DNA which had been subjected to electrophoresis in a 1.4% agarose slab gel before (i)

Fig. 4 (cont.) and after treatment with R.BamI (ii) or R.EcoRI (iii).

designated f2, annealed to the greatest extent. To be more certain that the hybrid genomes retained the leftward operator sequences, we next examined the affinity of λ repressor for these DNAs. Not surprisingly, as shown in Fig. 5, the f2 species bound λ repressor at levels where negligible binding was detected to wild-type SV40 or the f1 species. This

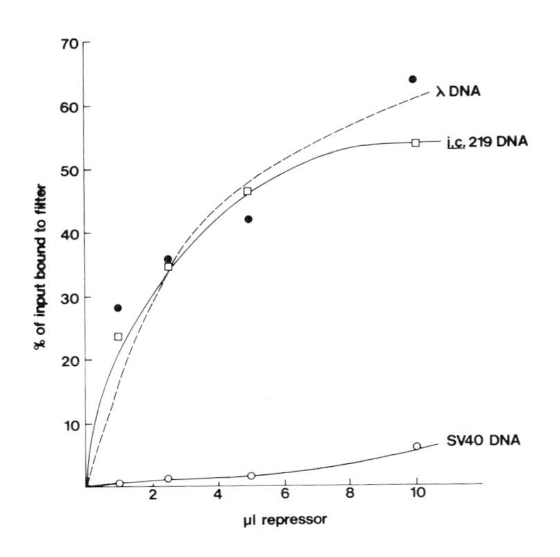

Fig. 5. Affinity of λ repressor for i.c. 219 DNA. Viral DNA was exposed to increasing amounts of λ repressor protein and passed through nitrocellulose filters. Material retained by filters is expressed as percent of input radioactivity. The binding assay mixture (0.6 ml) contained 0.02 μg of i.c. 219-f2 [^{32}P]DNA (10,000 cpm) and either λ [^{3}H]DNA (0.25 μg; 5000 cpm) or SV40 [^{3}H]DNA (0.3 μg; 5000 cpm) and the indicated volumes of λ repressor (3×10^{-8} M). Binding of f2 by λ repressor was abolished by addition of nonreadioactive λ DNA in excess.

observation was further refined by our asking whether λ repressor could selectively bind a specific restriction endonuclease cleavage fragment from the chimeric DNA. HpaII cleavage of the λ immunity region is known to produce two small fragments of 350 and 570 base pairs in addition to other fragments. Maniatis and Ptashne have shown that the leftward operator is located on the 350 base pair fragment (15). When the chimeric DNA was cleaved by HpaII and allowed to react with λ repressor, the 350 base pair fragment was bound by repressor. The gel electropherogram of the DNA fragments bound by λ repressor and subsequently eluted from a nitrocellulose filter by SDS treatment is shown in Fig. 6.

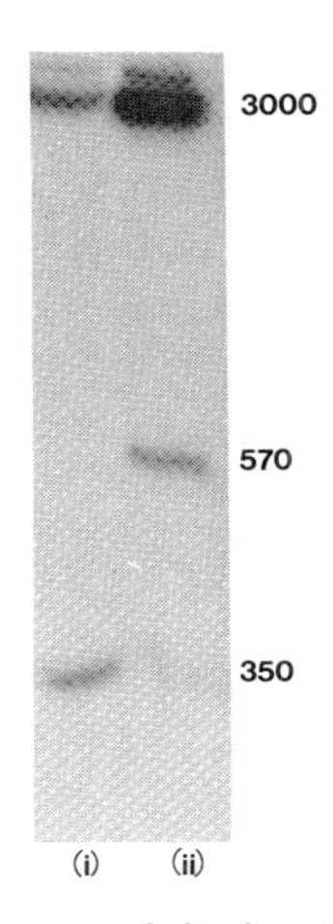

Fig. 6. λ Repressor binds the 350 bp fragment from the R.HpaII digest of i.c. 219-f2 [^{32}P]DNA. Unfractionated i.c. 219 DNA (2.3x10^6 cpm, 12.4 μg) was cleaved with R.HpaII and concentrated by alcohol precipitation in the presence of 200 μg of calf thymus DNA. It was then exposed to repressor protein (10^{-10} M) in a total volume of 2 ml. The resulting complex was retained on a filter and eluted with a buffer containing sodium dodecyl sulfate. A 10% aliquot of the eluate (7000 cpm) was subjected to electrophoresis and autoradiography (i). (ii): i.c. 219-f2 DNA was cleaved with R.HpaII. The 3000 bp segment in i was retained, probably due to nonspecific binding to the filter, whereas of the two small fragments only the 350 bp fragment was retained.

In channel 2 is the HpaII cleavage of f2 DNA which reveals the two small fragments and a large 3000 base pair fragment which contains the SV40 vector segment. In 1 is shown the DNA retained by λ repressor. Of the two small fragments, only the 350 base pair fragment was found as predicted. These structural findings were confirmed by electron microscopic heteroduplex analysis. The hybrid DNA was nicked, denatured, and allowed to anneal with denatured λ EcoRI immunity region fragment. The homology between the two DNAs was seen to be about 32% of the λ fragment, or 2300 base pairs, and was appropriately situated in the λ fragment to cover the region of the leftward operator. To map more precisely the λ segment in the hybrid, the λ EcoRI-B immunity region DNA fragment was cleaved by endo R.HindIII, as seen in Fig. 7.

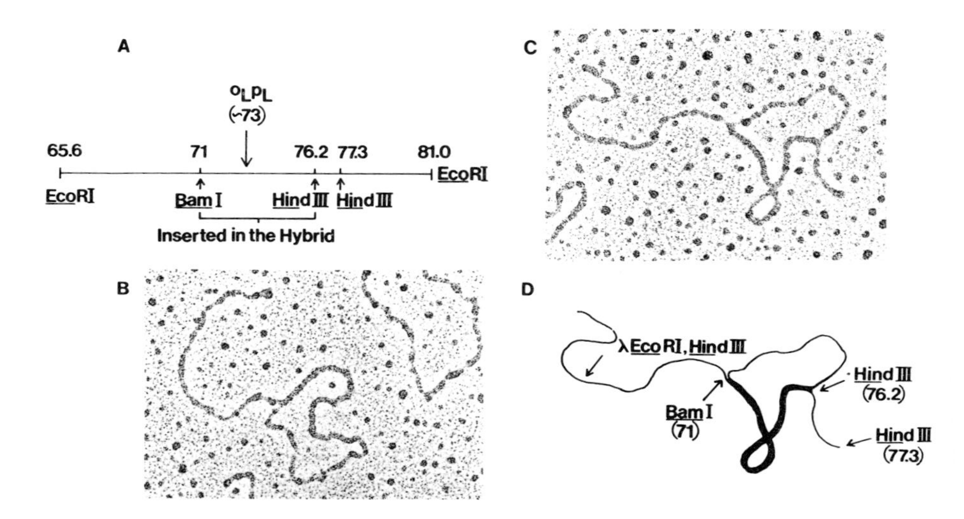

Fig. 7. (A) Schematic representation of the λ EcoRI-B fragment. The segment of λ EcoRI-B DNA extending from 71 λ map units (cleavage site for endo R.BamI) to 76.2 λ map units (one of the two sites for endo R.HindIII) is present in the SV40-λ hybrid as indicated. (B) Heteroduplex between the SV40-λ hybrid and λ EcoRI-B, HindIII fragment. The hybrid DNA was purified, nicked, and allowed to anneal to denatured λ EcoRI-B (further cleaved with endo R.HindIII). A partially duplex region is located in the circular portion of this representative heteroduplex molecule. The presence of one linear single-stranded segment extending from the circle proves that the hybrid DNA contains that portion of λ EcoRI-B up to the HindIII cleavage site at 76.2 map units. The single-stranded linear tail measures 0.359 ± 0.02 λ EcoRI-B lengths as expected for the part of the EcoRI-B, HindIII fragment between 65.6 and 71 λ map units that is not present in the hybrid genome. (C) Heteroduplex between the hybrid and λ EcoRI-B, HindIII (incomplete cleavage). A few heteroduplexes of this type were seen which probably were created from annealing of a hybrid molecule with the λ EcoRI-B fragment cleaved only once by HindIII at 77.3 map units. (D) Diagrammatic representation of the heteroduplex shown in (C), with the duplex region of homology drawn as a thick line. (Reprinted, with permission, from Davoli *et al.*, 1976, J. Virol. 19, 1100)

This produces three fragments, one of which, that spanning 65.6-76.2 λ map units, should form a heteroduplex with the hybrid. Such heteroduplexes were observed with one single-stranded branch extending from the circular region. A few heteroduplexes formed between the hybrid and the λ fragment, which had been cleaved only once by HindIII at 77.3 λ map units. These results indicated that no gross rearrangements of a region of λ DNA between 71 and 76 map units had occurred in the SV40-λ hybrid.

The final structural question concerned the SV40 vector sequences in the hybrid. Since the hybrid genomes were present in at least two-fold molar excess to helper genomes in the original virus stock, we anticipated that at least part of the vector sequences, indluding the replication origin, had been duplicated. This was substantiated (9) by an analysis of restriction endonuclease fragments known to arise from the vector segment and by the following heteroduplex analysis, as illustrated in Fig. 8. In the heteroduplex analysis the HpaII 3000 base pair linear fragment from the hybrid genome was denatured and annealed to single-stranded circular molecules of the triplication mutant from which the vector had been obtained. The heteroduplexes, such as that shown here, contained two regions of homology separated by a small deletion loop. These regions of homology indicate a tandem duplication of most of the vector sequences in the hybrid. This duplication occurred in vivo during the propagation of this hybrid. It provided two advantages: (1) the size of the hybrid genomes was increased to allow for efficient encapsidation since viral molecules less than 70% of wild-type size are not packaged, and (2) a selective advantage in DNA replication due to the duplicated origin for SV40 DNA replication.

These findings with small segments from phage λ DNA and short vector segments from SV40 reiteration mutants indicated that defective SV40 replicons could serve as vectors for molecular cloning and propagating foreign DNA in mammalian cells. However, since hybrid genomes constructed with non-complementing SV40 replicons, such as those used in these experiments, must be propagated and cloned with wild-type helper genomes, there is no simple means for selection for the hybrid genomes in a mixed population. To overcome this deficit, vector segments containing intact early or late gene regions could be employed. For example, excision of a suitable fragment from the late gene region of SV40 and insertion of a prokaryotic segment of similar size would create a defective hybrid capable of complementing early (tsA) mutants. In collaboration with Dean Hamer and Charlie Thomas at Harvard Medical School, we recently employed this approach to clone a bacterial transfer RNA suppressor gene in monkey cells (10). The basic features of the construction, propagation and

A

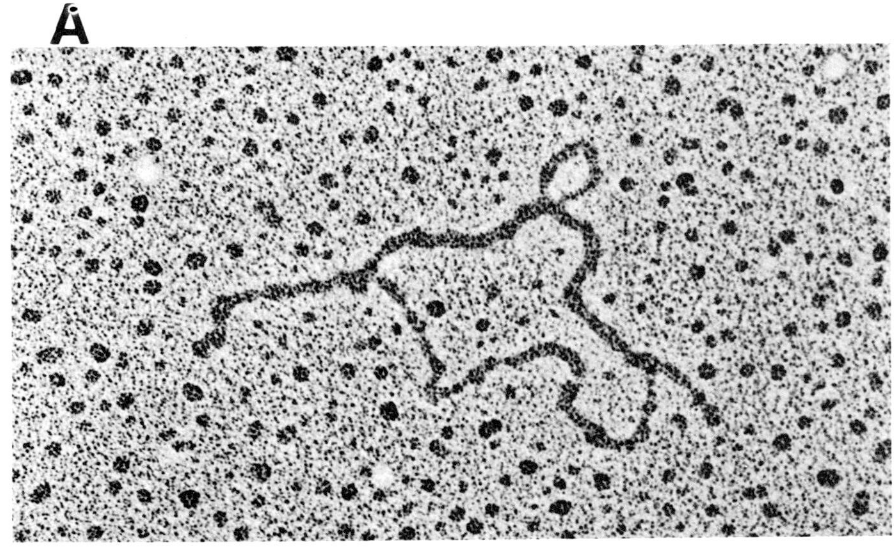

B

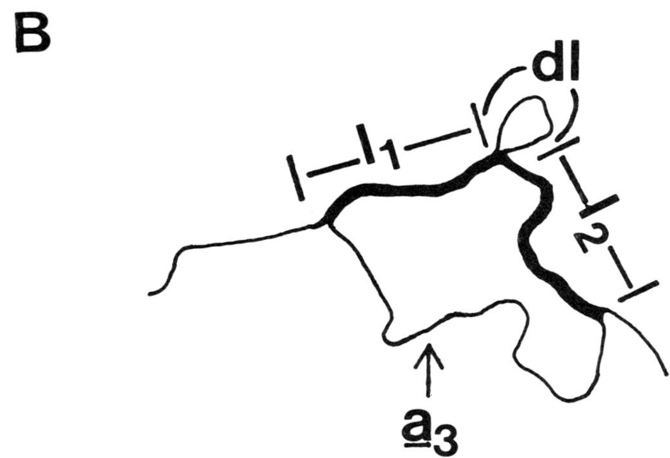

Fig. 8. Heteroduplex between reiteration mutant $\underline{a}_3$ and the HpaII-3000-base pair fragment. Reiteration mutant $\underline{a}_3$ DNA I was nicked and then denatured and renatured in the presence of the HpaII-3000-base pair fragment from the SV40-λ hybrid. This fragment had been purified after endo R.HpaII cleavage, as described in the legend to Fig. 6, and electron microscopy analysis was carried out. Measurements were made of 15 heteroduplexes, using single-stranded and duplex molecules of $\underline{a}_3$ as standards. The two duplex regions (shown as a heavy line in the diagram (B) and lettered I_1 and I_2), which

Fig. 8 (cont.) are 0.184 ± 0.006 and 0.182 ± 0.006 of the a_3 genome, respectively, represent a tandem duplication of the majority of the vector segment aBam-A preserved in the hybrid. The small, single-stranded deletion loop (dl in part B) is 0.160 ± 0.007 of the a_3 genome and is that part of fragment a not present in the hybrid. The single-stranded tails (thin lines in part B) are the remaining segments of λ DNA included in the HpaII-3000-base pair fragment. (Reprinted, with permission, from Davoli et al., 1976, J. Virol. 19, 1100)

biological characterization of one such complementing recombinant genome are reviewed by Hamer et al. (1977). The results of these findings and those of Goff and Berg (11) now have shown that deletions in the B, C and D gene regions of SV40 of 1300 base pairs from the HpaII site at .74 map units to the EcoRI site at 0 map units and of 2000 base pairs from the HpaII site to the BamI site at .15 map units can be replaced by foreign DNAs to produce genomes of over 90% of the original length. We had shown in studies of reiteration mutants that genomes of lengths of less than about 70% of SV40 size or greater than 100% of wild-type size were not encapsidated in progeny virions.

The bacterial nonsense suppressor gene we inserted into SV40 is su^+III from E. coli. This gene specifies a tRNA which translates the amber termination codon UAG as tyrosine. The source of the bacterial DNA was an E. coli plasmid constructed by Hamer called pCol-su^+III which carries a single copy of the tRNA tyrosine su^+III gene. Cleavage of this plasmid with the site-specific restriction endonucleases EcoRI and HpaII generated an 870 base pair fragment which included the suppressor tRNA structural gene sequence and its promoter and transcription termination regions. The viral vehicle was a 3700 base pair fragment of wild-type SV40 DNA and also produced by cleavage with EcoRI and HpaII. Since both EcoRI and HpaII produced DNA molecules with unique cohesive termini, the su^+III and SV40 fragments could anneal with one another in a single orientation and could then be covalently joined by treatment with polynucleotide ligase. This procedure yielded circular recombinant molecules which are 92% the length of wild-type SV40 DNA, and the su^+III fragment was inserted in such a way that the 5' end of the tRNA sequence was proximal to the 5' end of the 19S SV40 late region transcript (Fig. 9).

The SV40-su^+III recombinant DNA molecules were purified by two cycles of dye-density gradient centrifugation and, together with helper DNA from SV40 with a temperature sensitive mutation in gene A, were used to infect monkey cells at 41°C, which is nonpermissive for the helper. Under these

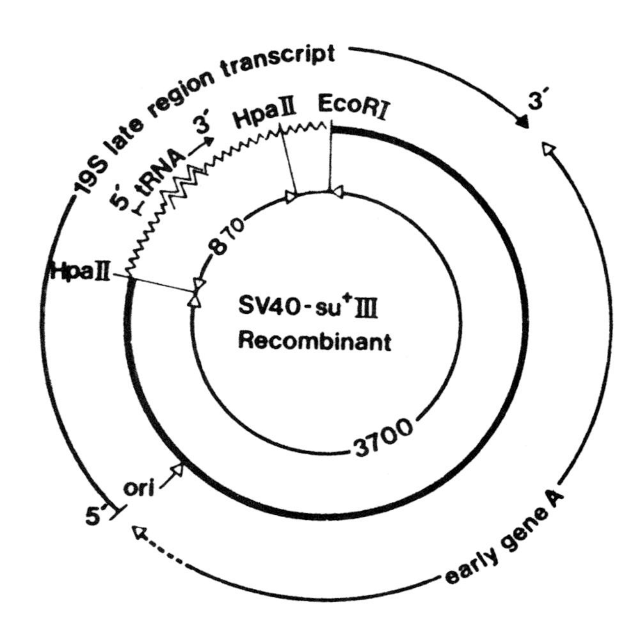

Fig. 9. Schematic diagram of the SV40-su$^+$III chimerical DNA.

conditions, virus was produced only be cells infected with both recombinant, which provides the early function, and the helper, which provides the late functions. Both the recombinant and helper genomes were encapsidated into virions, and this generated a mixed lysate which could subsequently be used to infect monkey cells for the preparation of virus-specific DNA and RNA. An analysis of the covalently closed circular viral DNA preparations from cells infected with the recombinant plus helper lysate by agarose gel electrophoresis showed two prominent viral species; about 70% of the viral DNA migrated at the position of helper, 5000 base pairs, and the remainder migrated at exactly the position expected for the SV40-su$^+$III recombinant, 4580 base pairs. Structural analysis of both the bacterial suppressor DNA sequences and the SV40 vector sequences revealed no significant alteration after this propagation in monkey cells (10). Furthermore, analysis of RNA synthesized during the late period after infection with the hybrid and tsA helper virus demonstrated that bacterial suppressor RNA species were produced, probably due to read-through during transcription of the SV40 late mRNA molecules. However, no functional su$^+$III tRNA^{Tyr} was detectable (10).

Transfection of Rat and Monkey Cells by the SV40-*E. coli* Hybrid DNA

Because of the conservation of the origin of replication and the A gene function, the hybrid SV40-*E. coli* DNA (SV40-$su^{+}III$) should be capable of transformation of non-permissive cells and of autonomous replication in permissive cells. We have transformed secondary rat embryo cells with the purified SV40-$su^{+}III$ DNA and constructed long term persistent infections of TC7 monkey kidney cells with this chimera. Various restriction enzymes, the Southern (12) elution procedure to transfer separated DNA segments from agarose gels to nitrocellulose filters and hybridization with probes specific for SV40 and $tRNA^{Tyr}su^{+}III$ have been used to analyze the fate of the recombinant DNA in both host cell systems. These studies are reviewed in the accompanying article by Upcroft *et al*. (this symposium).

CONCLUDING REMARKS

These experiments thus show that SV40 vectors are efficient for the propagation of specific prokaryotic DNA segments in mammalian cells. The clear advantage of a complementing vector, such as one which has a deletion in the late gene region, is that the hybrid recombinant can be easily cloned with a complementation technique. Furthermore, once the hybrid DNA has been purified free of the helper DNA, one can transform a variety of mammalian cells with it.

We have not yet undertaken the construction of a hybrid carrying a substitution in the early gene region of SV40, but, if one examines the SV40 map with respect to the known expression from the early gene region, it is likely that one could insert a foreign DNA segment between 0.5-0.55 and 0.15-0.2 map units. This would be an insertion very near the beginning of translation of the gene A protein or T antigen (17) and could place the foreign DNA segment under the control of the early gene region. Propagation of such a hybrid could be accomplished with various late conditional-lethal mutants of SV40 or by growth at the permissive temperature in the presence of a tsA helper. The SV40 tsA helper might allow for potentially valuable manipulation of the early region transcription. Reed *et al*. (18) have recently found that incubation at the non-permissive temperature results in a dramatic overproduction of the early mRNA species from SV40 tsA mutants. Thus, cells infected with a hybrid carrying a substitution in the early gene region and a tsA helper might be expected to overproduce the hybrid early mRNA at the non-permissive temperature.

ACKNOWLEDGEMENTS

This work was supported in part by a research grant from the National Cancer Institute, USPHS No. CA 20794-01. The expert technical assistance of Hagit Skolnik is gratefully acknowledged.

REFERENCES

1. Struhl, K., Cameron, J. and Davis, R.W. (1976) *Proc. Nat. Acad. Sci. USA* 73, 1471.
2. Ganem, D., Nussbaum, A.L., Davoli, D. and Fareed, G.C. (1976) *J. Mol. Biol.* 101, 57.
3. Davoli, D. and Fareed, G.C. (1974) *Nature* 251, 153.
4. Shenk, T. and Berg. P. (1976) *Proc. Nat. Acad. Sci. USA* 73, 1513.
5. Ganem, D., Nussbaum, A.L., Davoli, D. and Fareed, G.C. (1976) *Cell* 7, 349.
6. Davoli, D., Ganem, D., Nussbaum, A.L., Fareed, G.C., Howley, P., Khoury, G. and Martin, M.A. (1977) *Virology*, in press.
7. Nussbaum, A.L., Davoli, D., Ganem, D. and Fareed, G.C. (1976) *Proc. Nat. Acad. Sci. USA* 73, 1068.
8. Khoury, G., Fareed, G.C., Berry, K., Martin, M.A., Lee, T.N.H. and Nathans, D. (1974) *J. Mol. Biol.* 87, 289.
9. Davoli, D., Nussbaum, A.L. and Fareed, G.C. (1976) *J. Virol.* 19, 1100.
10. Hamer, D., Davoli, D., Thomas, C.A., Jr. and Fareed, G.C. (1977) *J. Mol. Biol.*, in press.
11. Goff, S. and Berg, P. (1977) *Cell*, in press.
12. Southern, E.M. (1975) *J. Mol. Biol.* 98, 503.
13. Ketner, G. and Kelly, T.J., Jr. (1976) *Proc. Nat. Acad. Sci. USA* 73, 1102.
14. Botchan, M., Topp, W. and Sambrook, J. (1976) *Cell* 9, 269.
15. Maniatis, T. and Ptashne, M. (1973) *Nature* 246, 133.
16. Maniatis, T., Kee, S.G., Efstratiadis, A. and Kafatos, F.C. (1976) *Cell* 8, 163.
17. Shenk, T.W., Carbon, J. and Berg, P. (1976) *J. Virol.* 18, 664.
18. Reed, S.I., Stark, G.R. and Alwine, J.C. (1976) *Proc. Nat. Acad. Sci. USA* 73, 3083.

CONSTRUCTION OF SV40 VECTORS AND EXPRESSION OF INSERTED SEQUENCES

Stephen P. Goff and Paul Berg

Department of Biochemistry, Stanford University
Stanford, Ca. 94305

ABSTRACT. Cleavage of SV40 DNA with restriction endonucleases HpaII and BamI yields a fragment of the genome (0.14 to 0.735 on the SV40 map) suitable for cloning DNA segments in cultured CV-1 monkey cells. This vector (termed SVGT-1) has been joined to a fragment of phage lambda by the poly (dA): poly (dT) method. The resulting hybrids were propagated in CV-1 cells in the presence of a temperature-sensitive SV40 helper, tsA58 at 41°C. The structures of the cloned hybrid molecules were determined by analysis of restriction endonuclease digests and by heteroduplex analysis.

The RNA produced by CV-1 cells infected with hybrid genomes contained virtually no lambda-specific RNA, although the hybrids replicated as well as the helper. The reason for the lack of stable lambda RNA is unknown; perhaps the linker segments of poly (dA): poly (dT) block transcription or render the transcripts highly unstable. To test this possibility we have joined the two fragments of SV40 DNA produced by HpaII and BamI cleavages, using the poly (dA): poly (dT) method. The resulting molecule is a wild-type genome except for short sequences of poly (dA): poly (dT) at the HpaII and BamI cleavage sites. Complementation tests were used to determine whether these sequences block the expression of the intact late genes between them (the D and E genes).

INTRODUCTION

Recently techniques have been developed which permit the biochemical construction and propagation of novel transducing phages (2, 16, 18) and plasmids (4, 15, 25) carrying specific genetic markers. The generality of these procedures permits the introduction, into bacteriophage genomes and plasmids, of DNA segments from organisms that do not ordinarily interact genetically with such phages and plasmids.

We here describe the successful construction and propagation of a transducing animal virus. Approximately 2 kilobases (kb) of DNA were removed from the late region of the SV40 genome by sequential cleavages with HpaII and BamHI endonucleases (at 0.735 and 0.14, respectively, on the SV40 DNA map) and a segment of about 1.5 kb of λ phage DNA was inserted in its place. The resulting hybrid DNA was cloned and propagated in CV-1 monkey kidney cells by mixed infections (14) with tsA58, an early mutant of SV40 (22, 23). A more detailed description of these experiments has been published elsewhere (7). Similar mutants, but with wild-type SV40 as helper have also been described (6, 17).

We have examined the RNA isolated from monkey kidney cells infected by the hybrid virus λ-SVGT-1. Although the hybrid DNA replicated nearly as well as the helper virus, and under conditions when large amounts of SV40-specific RNA was produced, virtually no λ-specific RNA was detected. One possible explanation for this observation is that the short poly (dA): poly (dT) linker segments act to block transcription proceeding into the inserted DNA segment. To test this possibility, we have constructed SV40 mutants which contains two poly (dA): poly (dT) inserts at the HpaII and BamI endonuclease recognition sites at 0.735 and 0.14 SV40 map units. These mutants were then tested genetically to determine whether transcription could proceed into the late region between the linker segments. The late D gene of these mutants was found to be expressed normally, indicating that poly (dA): poly (dT) sequences alone do not block late transcription in SV40. We conclude that some feature of the λ DNA segment of the λ-SVGT-1 hybrids probably precludes transcription of the insert or renders the transcripts highly unstable in the infected cell.

RESULTS

Preparation of SV40 DNA Vector (SVGT-1)

A vector suitable for cloning and propagating foreign DNA segments was prepared by excising virtually the entire late region of the viral DNA by two successive cleavages with HpaII and BamHI restriction endonucleases; these enzymes cleave SV40 DNA at map position 0.735 (20) and 0.14 respectively. The large segment of the viral DNA (0.6 SV40 genome length), hereafter referred to as SVGT-1, contains the origin of SV40 DNA replication (at 0.67 (5, 19)); it was separated from the 0.4 SV40 genome length segment by two sequential electrophoreses in agarose.

SVGT-1 was modified for joining to the λ phage DNA segment by digestion with an excess of λ-exonuclease (12) to remove approximately 50 nucleotides from the 5' end of each strand (10, 13) followed by incubation with deoxynucleotidyl terminal transferase (11) and ^{3}H-dATP to add about 200 deoxyadenylate residues per exposed 3' end (10, 13). The modified fragment recovered after these two reactions was uncontaminated with intact circular or full-length linear SV40 as judged by its low infectivity.

Preparation of λ Phage Insert

Lambda phage DNA contains five EcoRI endonuclease cleavage sites(24) and six HindIII endonuclease cleavage sites (1); digestion with both enzymes generated 12 fragments ranging in size from 0.3 to 23 kb. The longer and intermediate size fragments were separated readily by electrophoresis in agarose. By comparison with SV40 DNA marker fragments of known length, the sizes of the intermediate λ fragments 6, 7, 9, and 10 were found to be 1.91, 1.81, 1.28, and 0.91 kb (±0.05 kb), respectively.

Fragment 8 is 1.48 kb in length and contains ORI, the origin of λ DNA replication, and two structural genes, CII and cro, as well as four transcriptional promoters: Pre and Prm promoters regulate CI gene transcription, a promoter near ORI controls transcription of the short OOP RNA transcript, and Pr, the repressor-regulated promoter, controls late gene transcription (8). After elution of fragment 8 from the gel it was treated with λ phage exonuclease and then with deoxynucleotidyl terminal transferase and ^{32}P-dTTP as mentioned above.

Construction and Propagation of λ-SVGT-1 Hybrid Viruses

SVGT-1 containing ^{3}H-poly (dA) termini and the λ DNA insert with ^{32}P-poly (dT) termini were mixed and annealed. The annealed DNA was used without further purification or treatment to infect monolayers of CV-1P in the presence or absence of tsA58. At 41°C, the tsA58 DNA alone produced no plaques, the annealed DNA alone gave no plaques ($<$200 PFU/μg DNA), but infection with the two DNAs together produced approximately 2.5 x 10^3 PFU/μg annealed DNA. DNA isolated (9) from cells infected with 9 of 34 plaques caused a striking increase in the reannealing rate of labeled λ DNA fragment 8. As is shown below, each of these nine plaques contained, in addition to the helper virus, particles with a recombinant virus genome in which the λ phage DNA segment was covalently joined by short poly (dA): poly (dT) segments to SVGT-1.

Structure of the Putative λ-SVGT-1 Hybrid DNAs

Each of the putative λ-SVGT-1 hybrid DNA preparations contains tsA58 DNA; consequently, when such DNA samples are digested with HindIII endonuclease, electrophoresis in agarose should reveal all of the fragments expected from SV40 DNA. In addition, a new fragment (about 0.6 SV40 fractional length) should be generated containing the DNA and poly (dA): poly (dT) segments with short SV40 DNA "tails" (Fig. 1).

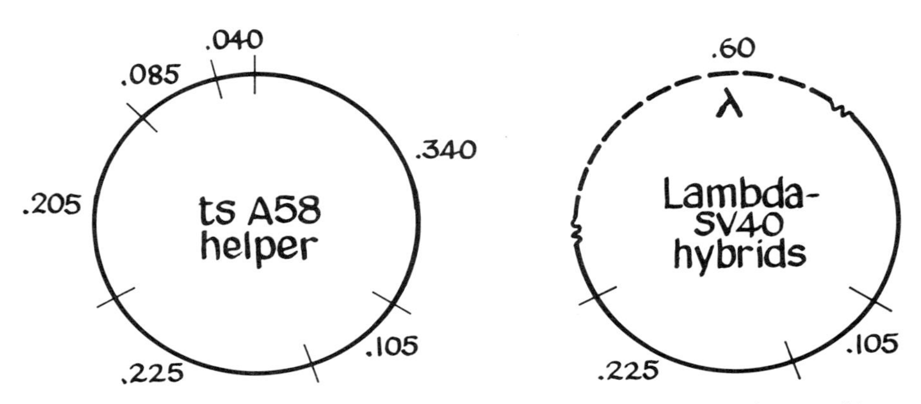

Fig. 1. Fragments produced by HindIII endonuclease digestion of helper and hybrid virus genomes. The radial slashes indicate the sites of cleavage and the numbers the length of the fragments expressed in SV40 fractional lengths. tsA58 DNA yields all the fragments of wild-type SV40; λ-SVGT-1 hybrid DNA yields two of these fragments plus a novel fragment of 0.60 SV40 fractional length.

Figure 2 shows that this expectation was fulfilled; HindIII endonuclease digests of each of the nine putative λ-SVGT-1 hybrid DNA preparations gave an electrophoretic pattern qualitatively similar to the ones shown in Fig. 2A.

Using a procedure developed by Southern (21), which permits transfer of denatured DNA from such gels to nitrocellulose sheets without disturbing their relative positions, one can detect the λ DNA-containing band by *in situ* hybridization with a radioactively labeled probe. ^{32}P-labeled λ cRNA, prepared with whole λ DNA and *E. coli* RNA polymerase, was hybridized to such nitrocellulose "imprints", and, after suitable washing, autoradiograms of the nitrocellulose sheets were prepared (Fig. 2B). Only the large HindIII endonuclease-generated fragments contained λ DNA as judged by their hybridization to the labeled λ cRNA.

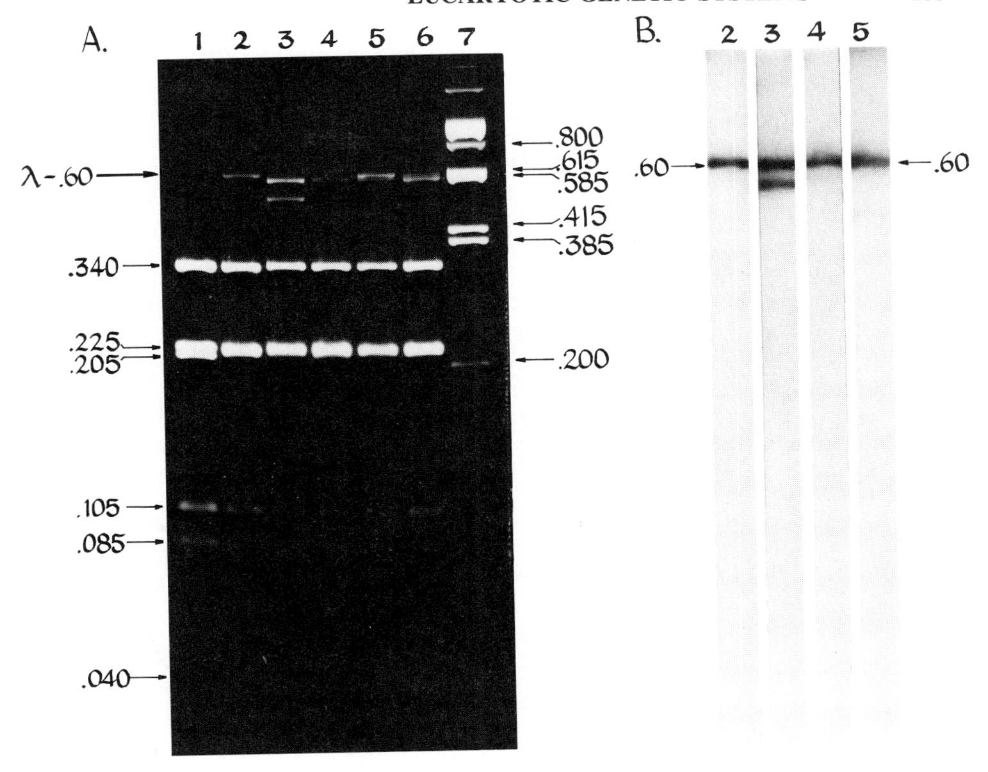

Fig. 2. Agarose gel electrophoresis of DNA fragments produced by HindIII endonuclease digestion of hybrid DNAs.

A: DNA bands were visualized by staining the gel with ethidium bromide. Track 1: Wild-type SV40 DNA. Tracks 2 through 6: λ-SVGT-1 DNA prepared from plaque isolates 7, 9, 10, and 18, and 32. The DNA used in these digestions was obtained from infections with virus from the plaques. The new fragments produced in the digestions of the hybrid DNAs are 0.60 fractional SV40 units in length. Each of the nine isolates containing λ DNA sequences yielded similar digestion products. Track 7: length standards prepared by partial cleavage of SV40 DNA with HpaI endonuclease.

B: Autoradiogram after hybridization of a replica of the gel in Fig. 2A with ^{32}P-labeled λ cRNA. The DNA in each gel track was transferred to nitrocellulose strips, hybridized _in situ_ to the labeled cRNA, and then autoradiographed. DNA from plaque isolates 7, 9, 10, and 18 are shown.

Heteroduplex Analysis of the Putative λ-SVGT-1 Hybrid DNA

A novel and characteristic heteroduplex structure should be formed when full-length single strands from EcoRI endonuclease cleaved, wild-type SV40 DNA are annealed to the strands of the large fragment formed by cleavage of λ-SVGT-1 hybrid DNA with HindIII endonuclease; in that fragment, the λ DNA segment is joined by a poly (dA): poly (dT) bridge to SV40 DNA (Fig. 3). The "tails" of SV40 DNA anneal to a full-length linear SV40 DNA strand generating a circular structure that is part single stranded and part duplex DNA. Because the λ DNA sequence is flanked by poly (dA) at one join and by poly (dT) at the other, that segment occurs as a loop joined by a short "neck" of dA:dT duplex to the SV40 DNA heteroduplex region (Fig. 4). The size of the inserted segment as determined by length measurement of the single-stranded heteroduplex loop is 1.46 kb (±0.06), identical to the size of λ fragment 8. The existence of these heteroduplex structures is thus quantitatively consistent with the predicted structure of the large HindIII endonuclease-cleavage product and, therefore, of the λ-SVGT-1.

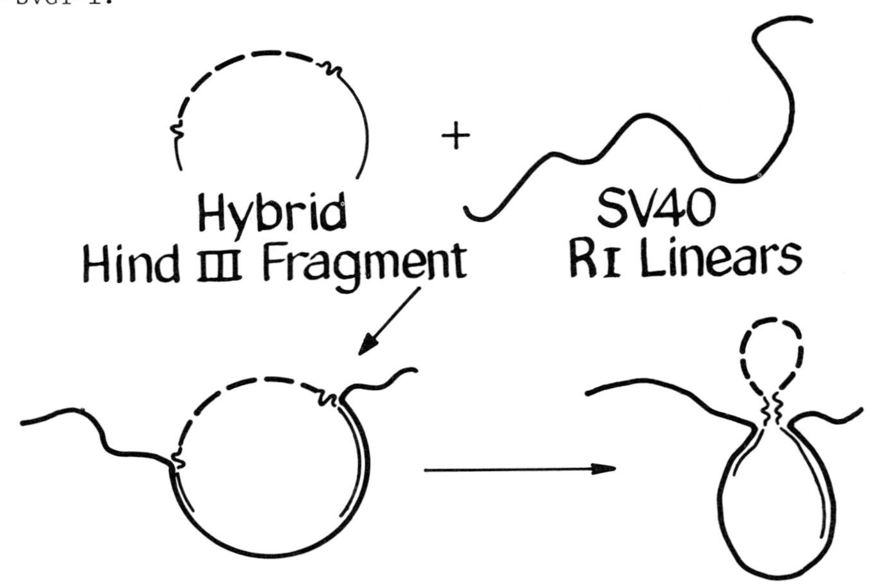

Fig. 3. Schematic representation of the heteroduplexes expected from EcoRI endonuclease-cleaved SV40 linear DNA and the λ DNA-containing segment produced by HindIII endonuclease cleavage of λ-SVGT-1 hybrid DNA. Heteroduplexes are formed between the linear SV40 DNA and the SV40 DNA tails of the HindIII-generated fragment. When the poly (dA) and poly (dT) sequences on the same strand anneal to form a short "snapback" region, the λ DNA segment appears as a single-stranded loop; with higher formamide concentrations, the λ DNA, poly (dA), and poly (dT) segments are extended.

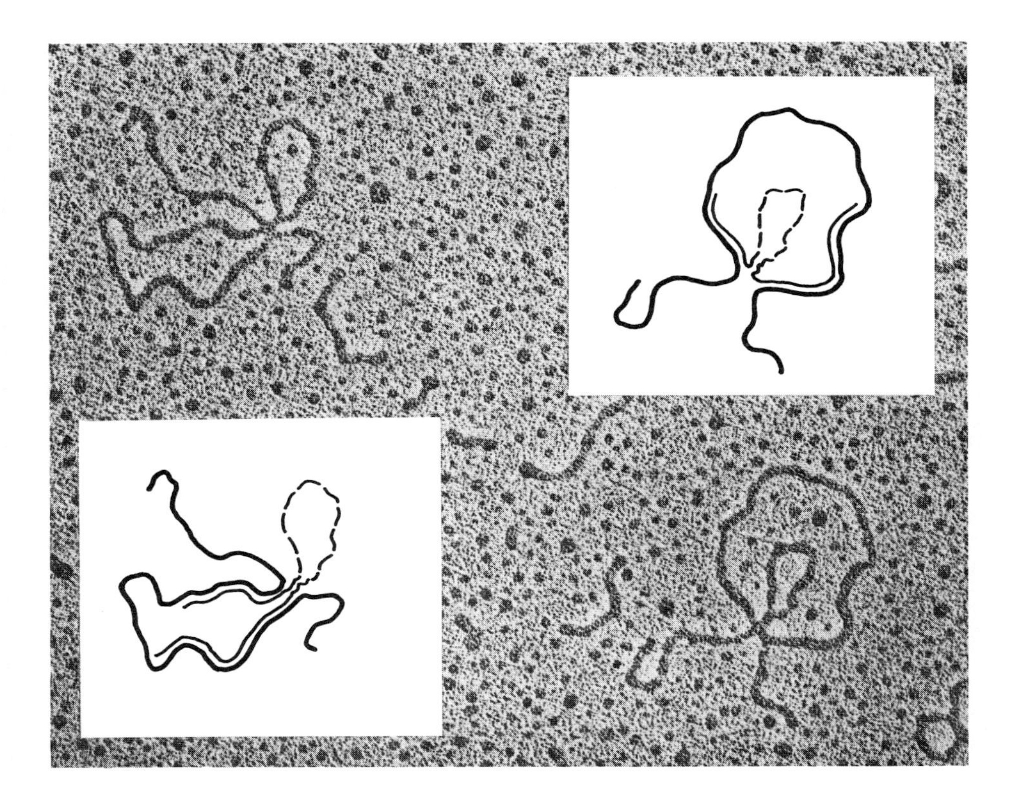

Fig. 4. Electron micrograph of heteroduplexes diagrammed in Fig. 3. In the tracings, heavy lines represent the linear SV40 DNA, the thin lines represent the SV40 tails of the fragment derived from hybrid DNA, and the dashed lines represent λ DNA segments.

By analysis of restriction digests with HincII and by further heteroduplex analysis (data not shown), it was found that hybrids with the λ insert in each of two possible orientations had been isolated.

Expression of the λ DNA Segment During Propagation of the λ-SVGT-1 Hybrid DNA

Our experiments establish that the λ-SVGT-1 hybrid DNA replicates during infection. Of interest is whether the λ DNA sequence is transcribed. To answer this question, we examined the RNA produced following infection of monkey cells with a mixture (approximately 1:1) of cloned λ-SVGT-1 DNA (either clone 9 or 18) and _tsA_58 DNA.

During the infection, the λ-SVGT-1 hybrid DNA replicated nearly as well as the tsA58 helper DNA; the ratio of hybrid to helper DNA, as estimated by agarose gel electrophoresis, was between 0.3 and 1.0. But, at the same time, when substantial amounts of SV40-specific RNA were being synthesized, there was no formation of λ-specific RNA (as judged by filter hybridization) and virtually none by measurement of reassociation kinetics (Table 1). These results indicate that although the λ-SVGT-1 hybrid DNA replicates nearly as well as SV40 DNA itself, transcripts of the λ DNA sequence either are not synthesized or do not persist in the infected cells; substantially the same results were obtained with hybrids having either of the two possible orientations of the inserted λ DNA sequence.

Table 1. Synthesis of SV40- and λ-specific RNA following infection of CV-1 cells with λ-SVGT-1 and SV40 virus.

RNA from cells infected with	% of ^{32}P-RNA hybridized to		% of total RNA homologous to	
	SV40 DNA	λ DNA	SV40 DNA	λ fragment 8 DNA
Nothing	<0.05	<0.02	<0.02	<0.0001
SV40 alone	1.97	<0.05	0.47	<0.0001
tsA58 plus hybrid 9	2.21	<0.02	0.49	0.0003
tsA58 plus hybrid 18	1.20	<0.03	0.10	<0.0001
λ cRNA		35.2		0.86

Analysis of RNA isolated from infected CV-1 cells. Cells were infected as indicated, labeled with ^{32}P-inorganic phosphate, and total RNA isolated. The first pair of columns shows the percentage of the input RNA that hybridized to filters containing SV40 or λ DNA. No λ-specific RNA was labeled in the hybrid-infected cells above the background of 0.05%. The second pair of columns shows the fraction of the total RNA homologous to SV40 or λ fragment 8 DNA as measured by the acceleration of the annealing of the labeled DNA probe by the added RNA. The values for SV40 RNA have been calculated taking into account that the predominant SV40 RNA (the 16S late mRNA) can increase the annealing rate of only 24% of one strand of the labeled DNA.

Construction of Novel SV40 Mutants: Do poly (dA): poly (dT) Linkers Block Transcription?

One explanation for the lack of stable transcripts made from the inserted λ DNA sequences is that the presence of the poly (dA): poly (dT) linkers which flank the insert somehow block the production of stable RNA. We have constructed mutants of SV40 which were designed to test this hypothesis. SV40 form I DNA was cleaved with restriction endonucleases HpaII and BamHI as described above; the two resulting fragments (the "early" fragment, 0.6 SV40 fractional units in length, and the "late" fragment, 0.4 SV40 units in length) were separated by electrophoresis in agarose. Aliquots of each of these preparations of DNA were treated with λ-exonuclease and deoxynucleotidyl terminal transferase in the presence of ^{3}H-dATP or dTTP, as described above. The "early" fragment containing poly (dA) termini and the "late" fragment containing poly (dT) termini were mixed, annealed, and used to transfect CV-1P monkey kidney cells in the presence of tsA58 helper virus DNA at 41°C. Plaques arising from these transfections were picked and after plaque-purification, the virus was used to prepare viral DNA stocks in CV-1 monkey kidney cells. The stocks each consisted of a mixture of tsA58 helper DNA and a mutant DNA similar to WT SV40 but containing two poly (dA): poly (dT) linker segments, one at the HpaII endonuclease recognition site, and one at the BamI endonuclease recognition site. These mutants were thus identical to the λ-SVGT-1 hybrids, but with the SV40 late region inserted instead of the λ DNA segment. The lengths of the poly (dA): poly (dT) linker segments of all mutants tested were in the range of 20-90 nucleotides as judged from the decreased mobility of the HindII + III restriction endonuclease fragments C and G (see Fig. 5). All the mutants examined had been rejoined with the "late" fragment inserted in the normal orientation (data not shown).

The mutant DNAs could be separated from the helper DNA by cleavage of the mixture with restriction endonuclease HpaII followed by equilibrium centrifugation in CsCl and ethidium bromide. The mutant DNAs were then tested by the standard complementation test (14) to determine if the late D gene, contained within the poly (dA): poly (dT) linker segments, could be expressed. In all cases, the mutant DNAs gave normal complementation with the mutant tsD202 and were thus genotypically D+ (data not shown). We conclude that the presence of the poly (dA): poly (dT) linker segments alone does not block expression of genes located between linkers.

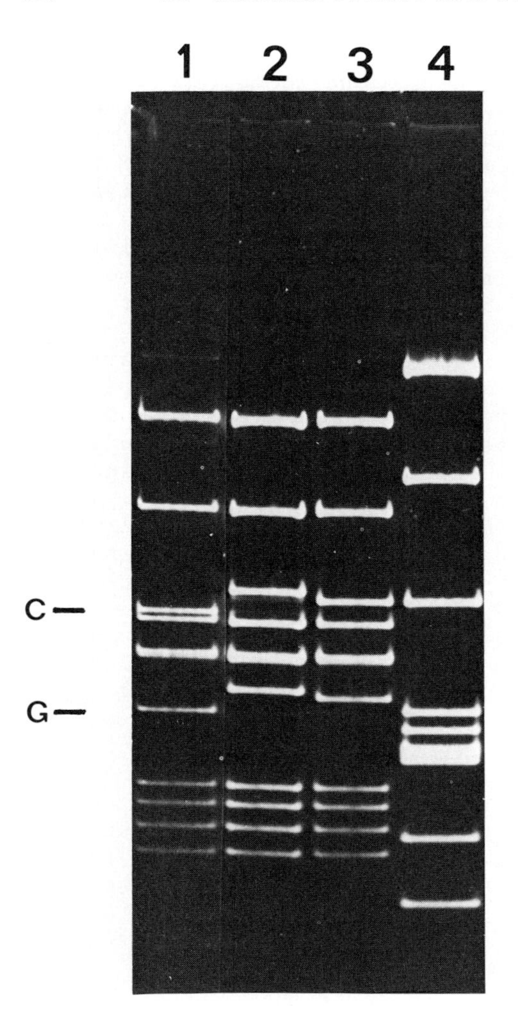

Fig. 5. Agarose gel electrophoresis of Hind II + III restriction endonuclease digests of SV40 mutant DNAs. Track 1: wild-type SV40 DNA. Tracks 2 and 3: mutant SV40 DNAs containing poly (dA): poly (dT) linker segments at the HpaII and BamHI endonuclease recognition sites. Track 4: marker DNA fragments produced by endonuclease HaeIII digestion of wild-type SV40 DNA. In the two mutants shown fragments C and G, containing the HpaII and BamHI recognition sites respectively, have a reduced mobility relative to the parental fragments.

DISCUSSION

There are several limitations to the use of SVGT-1 for cloning foreign DNA segments. One stems from the inability to encapsidate SV40 DNA molecules larger than about 5 kb; since SVGT-1 is itself about 3 kb, the insert must be 2 kb or smaller to be propagated as a virion. To clone larger DNA segments requires development of additional SVGT vectors. Presently, we are exploring smaller SVGT vectors that contain only the origin of DNA replication or that region plus a single gene that could complement an appropriate ts or defective helper virus. An alternative way to reduce, or possibly eliminate, the size limitation of the cloned DNA segments is to construct vectors that can be propagated as plasmid-like elements in animal cells.

A puzzling outcome of our attempts to assess λ DNA expression during multiplication of λ-SVGT-1 genomes was the failure to detect more than a trace of λ-specific RNA sequences, irrespective of the orientation of the inserted sequence. One explanation is that the poly (dA): poly (dT) linkers surrounding a DNA segment in some way prevent the synthesis of a stable transcript from that segment. However, mutants of SV40 constructed to test this hypothesis demonstrate that the poly (dA): poly (dT) linkers alone do not block expression of the segment between them; the SV40 late D gene is expressed normally even when surrounded by linkers. In addition, it should be noted that fully viable mutants of SV40 have been constructed (3) with one of the two poly (dA): poly (dT) inserts present in these mutants. We postulate that some feature of the λ DNA segment in λ-SVGT-1 either prevents transcription or renders the transcript produced highly instable. Experiments to test these hypotheses are underway.

ACKNOWLEDGMENTS

These experiments were supported by research grants from the U.S. Public Health Service and the American Cancer Society. S.G. is a Smith, Kline and French predoctoral scholar as well as a Public Health Service Trainee.

REFERENCES

1. Allet, B., and Bukhari, A.I. (1975) J. Mol. Biol. 92, 529.
2. Cameron, J.R., Panasenko, S.M., Lehman, I.R., and Davis, R.W. (1975) Proc. Nat. Acad. Sci. U.S.A. 72, 3416.
3. Carbon, J., Shenk, T.E., and Berg, P. (1975) J. Mol. Biol. 98, 1.
4. Clarke, L., and Carbon, J. (1975) Proc. Nat. Acad. Sci. U.S.A. 72, 4361.
5. Danna, K.J., and Nathans, D. (1972) Proc. Nat. Acad. Sci. U.S.A. 69, 3097.
6. Ganem, D., Nussbaum, A.L.,Davoli, D., and Fareed, G.C. (1976) Cell 7, 349.
7. Goff, S.P., and Berg, P. (1976) Cell 9, 695.
8. Hershey, A.D. (Ed.) (1971) The Bacteriophage Lambda. Cold Spring Harbor Laboratory, Cold Spring Harbor, N.Y.
9. Hirt, B. (1967) J. Mol. Biol. 26, 365.
10. Jackson, D., Symons, R., and Berg, P. (1972) Proc. Nat. Acad. Sci. U.S.A. 69, 2904.
11. Kato, K.I., Goncalves, J.M., Houts, G.E., and Bollum, F.S. (1967) J. Biol. Chem. 242, 2780.
12. Little, J.W., Lehman, I.R., and Kaiser, A.D. (1967) J. Biol. Chem. 242, 672.
13. Lobban, P.E., and Kaiser, A.D. (1973) J. Mol. Biol. 78, 453.
14. Mertz, J.E., and Berg, P. (1974) Virology 62, 112.
15. Morrow, J.F., Cohen, S.N., Chang, A.C.Y., Boyer, H.W., Goodman, H,M. and Helling, R.B. (1974) Proc. Nat. Acad. Sci. U.S.A. 71, 1743.
16. Murray, N.E., and Murray, K. (1974) Nature 251, 476.
17. Nussbaum, A.L., Davoli, D., Ganem, D., and Fareed, G.C. (1976) Proc. Nat. Acad. Sci. U.S.A. 73, 1068.
18. Rambach, A., and Tiollais, P. (1974) Proc. Nat. Acad. Sci. U.S.A. 71, 3927.
19. Salzmann, M.P., Fareed, G.C., Seebring, E.D., and Thoren, M.M. (1974) Cold Spring Harbor Symp. Quant. Biol. 38,257.
20. Sharp, P.A., Sugden, B., and Sambrook, J. (1975) Biochemistry 12, 3055.
21. Southern, E.M. (1975) J. Mol. Biol. 98, 503.
22. Tegtmeyer, P. (1972) J. Virol. 10, 591.
23. Tegtmeyer, P. (1974) Cold Spring Harbor Symp. Quant. Biol. 39, 3.
24. Thomas, M., and Davis, R.W., (1974) J. Mol. Biol. 91, 315.
25. Wensink, P.C., Finnegan, D., Donelson, J.C., and Hogness, D.S. (1974) Cell 3, 315.

EPISOMAL STATES OF AN SV40-*ESCHERICHIA COLI* RECOMBINANT GENOME IN DIFFERENT MAMMALIAN CELL LINES

P. Upcroft, J.A. Upcroft, H. Skolnik and G.C. Fareed

Department of Microbiology and Immunology
and Molecular Biology Institute
University of California, Los Angeles
Los Angeles, California 90024

ABSTRACT. The structure and fate of SV40 genomes carrying in the late gene region an *E. coli* suppressor gene have been examined both in transformed rat embryo cells and in persistently infected permissive monkey cells. High molecular weight DNA obtained from cloned lines of rat embryo cells which had been transformed by the purified hybrid viral DNA was cleaved with various restriction endonucleases and fractionated by agarose gel electrophoresis. DNA fragments were denatured *in situ* and transferred to nitrocellulose filters. Both SV40 vector and *E. coli* suppressor DNA sequences were localized on specific DNA fragments by hybridization with radiolabeled SV40 DNA or suppressor plasmid DNA probes. An unexpected observation with certain cloned rat cell lines was the presence of substantial amounts of free or unintegrated hybrid viral DNA molecules. Analysis of the total cellular DNA from persistently-infected monkey cell cultures revealed large amounts of free viral DNA which contained both SV40 and suppressor DNA sequences as judged by hybridization analysis.

INTRODUCTION

Several studies have recently shown that portions of the simian virus 40 (SV40) genome can be replaced by specific segments of bacterial or bacteriophage DNA, and these recombinant molecules can then be propagated by lytic infection in African green monkey kidney cells with helper virus to complement the deleted functions (Fareed, this volume; Goff and Berg, this volume). However, lytic propagation generally results in the loss of viability of the cells carrying the introduced viral genomes and, therefore, precludes any study of the long term effects of new genetic information in such cells. We have been investigating the introduction of foreign DNA into mammalian cells in a stable manner without the need for helper virus. The recombinant molecule that we have utilized, SV40-*su$^+$III*, was constructed by D. Hamer and his colleagues (D. Hamer, D. Davoli, C.A. Thomas, and G.C. Fareed, J. Mol. Biol., in press) to investigate the possible action of a bacterial suppressor tRNA gene in monkey cells. SV40-*su$^+$III* has a 1300 base pair segment encoding for portions of the late genes B/C and D of SV40 deleted. This region has been replaced by an 870 base pair plasmid segment carrying the Escherichia coli tRNA^{Tyr}su^{+}III gene (Fig. 1). The details of the propagation of this hybrid in monkey kidney cells are summarized in the accompanying paper by Fareed (this volume).

Because of the conservation of the origin of replication and the A gene function of SV40, the hybrid SV40-E. coli DNA (SV40-su^{+}III) should be capable of transformation of non-permissive cells and of autonomous replication in permissive cells. We have transformed secondary rat embryo cells with the purified SV40-su^{+}III DNA and constructed long term persistent infection of TC7 monkey kidney cells with this chimerical DNA.

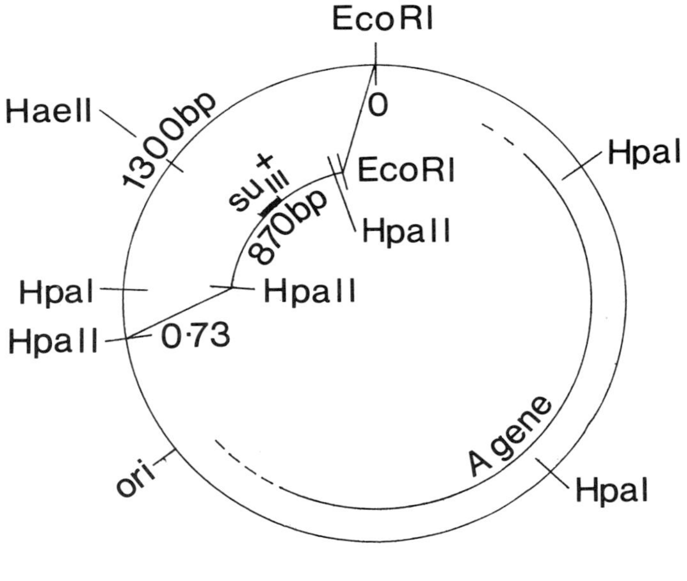

Fig. 1. Diagrammatic Representation of the Genome of the SV40-Escherichia coli Recombinant, SV40-su^{+}III. The SV40 segment of SV40-su^{+}III comprises all of the SV40 genome except a 1300 base pair fragment between the HpaII and EcoRI restriction sites at 0.73 and 0 respectively on the SV40 map. This fragment has been replaced by an 870 base pair segment carrying the E. coli tRNA^{Tyr}su^{+}III gene from the plasmid pColsu^{+}III excised by EcoRI and partial HpaII cleavage. [For construction, see Fareed, G.C., this volume]. Thus the SV40-su^{+}III genome is 92% of wild-type SV40 in length. There are two HpaII sites

Fig. 1 (cont.). in the recombinant DNA, one at either end of the plasmid insert. The single *HaeII* site in SV40 has been deleted. Also shown are the *HpaI* sites in the SV40 genome

CHIMERICAL (SV40-*E. COLI*) DNA IN TRANSFORMED CELLS

Transformation of secondary rat embryo cells by the purified hybrid DNA was accomplished with the calcium transfection technique of Abrahams *et al*. (1). Initially, the SV40 and suppressor DNA sequences in uniformly T antigen positive transformed rat cells were examined by reassociation kinetic analysis (2). Radiolabeled SV40 DNA was cleaved by restriction endonucleases *EcoRI* and *HpaI* to generate four fragments (3). Three of these fragments (A, C and D) were expected to be represented in the transformed cellular DNA, and one fragment (B), mapping between 0.76 and 0 on the SV40 physical map, was not expected to be represented in the transformed cellular DNA. When these radiolabeled SV40 DNA probes were separately denatured and allowed to reassociate in solution in the presence of a large concentration of denatured total cellular DNA from the uncloned transformed rat cells, the SV40 *EcoRI-HpaI* fragments -A, -C and -D were found to be represented at about three copies per diploid DNA content, whereas fragment -B sequences were essentially absent from the total cellular DNA. When the radiolabeled 870 base pair suppressor *su$^+$III* DNA segment was denatured and allowed to reanneal in the presence of denatured total cellular DNA from the transformed rat cell culture, the acceleration of its reannealing corresponded to approximately 2-3 copies of the suppressor DNA region per diploid DNA content. In a control analysis, cellular DNA from a wild-type SV40-transformed rat cell line produced no apparent acceleration of reassociation of the *su$^+$III* DNA fragment. (D. Solomon, G. Khoury, unpublished results).

In order to examine the arrangement of hybrid viral DNA sequences in the transformed rat embryo cells, we have utilized various restriction endonucleases to cleave high molecular weight DNA obtained from cloned lines of the transformed rat embryo cells. The cleavage products were fractionated by agarose gel electrophoresis and then transferred to nitrocellulose filters by the Southern procedure (4). Both SV40 and *E. coli su$^+$III* DNA sequences were localized by hybridization on specific DNA fragments with ^{32}P cRNA as described by Ketner and Kelly (5) or with nick-translated DNA (6). An unexpected observation from these analyses of the total cellular DNA from cloned lines of rat embryo transformed cells has been the presence of free, or unintegrated, as well as integrated hybrid viral DNA sequences. Because of this observation, the estimation of the copy number for the integrated hybrid viral DNA in the transformed cells from the reassociation kinetic analyses is uncertain at present. The cleavage patterns of total cellular DNA from one cloned rat embryo cell line (C1B) are illustrated in Figs. 2 and 3. Cleavage of the

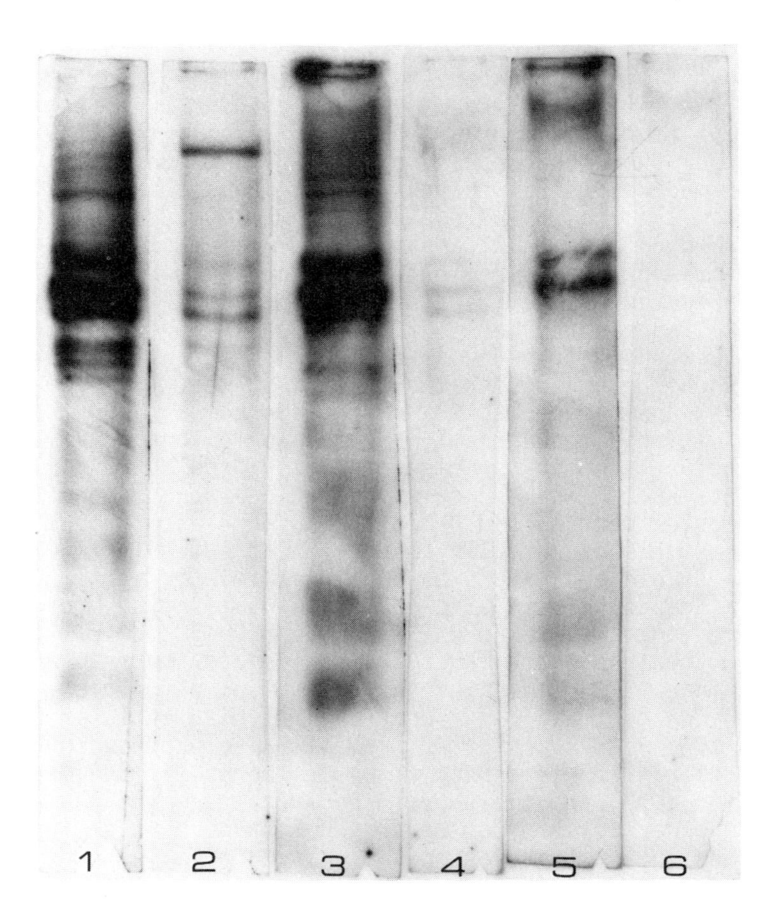

Fig. 2. Hybridization analysis of cloned secondary rat embryo cells transformed by purified SV40-su^{+}III DNA (C1B). The SV40-su^{+}III hybrid DNA was separated from the helper tsA DNA based upon its shorter size and increased electrophoretic mobility through 1% agarose gels (Hamer et al., in press). Subconfluent secondary rat embryo cells in a 75 cm^2 Falcon flask were transformed with 0.9 μg of the SV40-su^{+}III DNA using the calcium co-precipitation technique (1). Several weeks after the DNA transfection foci of transformed cells appeared and these were trypsinized and transferred to

Fig. 2 (cont.) new culture flasks. Individual clones of transformed rat cells were obtained by serial cell dilution and plating in multi-well dishes. One clonal line was expanded for isolation of total cellular DNA (8) and hybridization analysis. Twenty microgram samples of total cellular DNA per gel lane were subjected to electrophoresis in 1% agarose in a horizontal slab gel apparatus (Upcroft *et al.*, manuscript in preparation). Each lane was excised and, following denaturation *in situ*, the DNA fragments were eluted onto nitrocellulose filters as described by Southern (4). Each filter was hybridized with either ^{32}P-labeled, "nick-translated" (6) SV40 DNA or pCol*su$^+$III* DNA at a specific activity of 5×10^7 cpm/μg and autoradiography was performed. In the autoradiograph shown above the gel lanes (from left to right) were loaded with the following: (nos. 1 and 2) *Eco*RI-cleaved DNA, (nos. 3 and 4) *Bam*I-cleaved DNA, (nos. 5 and 6) *Bal*I-cleaved DNA. Nos. 1, 3 and 5 were hybridized with the SV40 probe and 2, 4 and 6 with the pCol*su$^+$III*. Exposure time was 60 hr.

total cellular DNA with *Bal*I (an enzyme which does not cleave the hybrid DNA) and hybridization with the ^{32}P-labeled SV40 DNA probe revealed two bands migrating in the vicinity of linear SV40 DNA. Cleavage of the total cellular DNA from this cloned cell line with *Eco*RI or *Bam*I, which each cleave the original SV40-*su$^+$III* hybrid DNA once, leave both of these bands in the same proportion as the *Bal*I digest. Furthermore, cleavage with *Sal*I, another enzyme which does not cleave the hybrid, or with *Bal*I+*Sal*I reveals both DNA species unaltered (data not illustrated). Cleavage with a vast excess of *Sal*I, sufficient to cause conversion of supercoiled SV40, by nicking to greater than 50% open circles, under similar migration conditions still reveals the same two bands. These bands also hybridize the nick-translated ^{32}P-labeled *su$^+$III* tRNA gene probe, and the simplest interpretation is that they correspond to one (open circular and linear forms) or possibly two free hybrid viral DNA species which are not integrated in this particular cloned cell line.

Cleavage with *Eco*RI or *Bam*I and hybridization with the SV40 probe produces a considerable number of bands, including the two revealed in DNA cleaved by *Bal*I and *Sal*I. Those new fragments which migrate more rapidly than the two main bands common to all digests are not identical in the two restriction endonuclease reaction products. These latter bands hybridize the tRNA gene probe, and we interpret these to be the products of integrated SV40-*su$^+$III* genomes. They are not generated by simple cleavage of the two major bands common to all the cleavage patterns illustrated and the yield is too great to

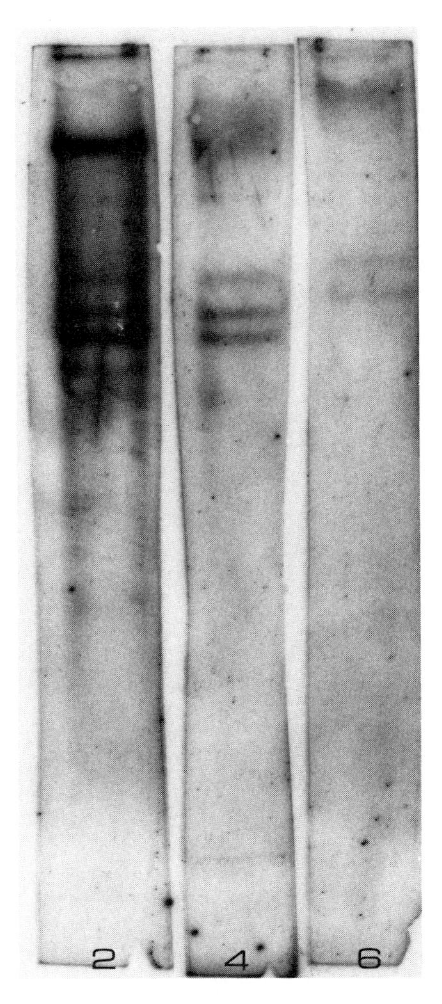

Fig. 3. Hybridization analysis of transformed rat embryo cells. This autoradiograph is a one week exposure of gel slots 2, 4 and 6 of Fig. 2 (i.e., EcoRI-, BamI- or BalI-cleaved DNA, hybridized with the pColsu^{+}III probe).

be products of higher oligomers which correspond to certain of the species migrating more slowly than unit length linear SV40 DNA in the patterns generated by EcoRI and BamI cleavages.

Fig. 4 compares the EcoRI cleavage patterns of total cellular DNA from two more isolates (A2 and A7) from the transformed rat embryo cells with the isolate shown in Fig. 2. Some fragments identified with the SV40 probe are common to the two transformed lines, A2 and A7, but many are different. Both patterns are different from the first isolate C1B and there is considerably less monomeric free viral DNA than in C1B.

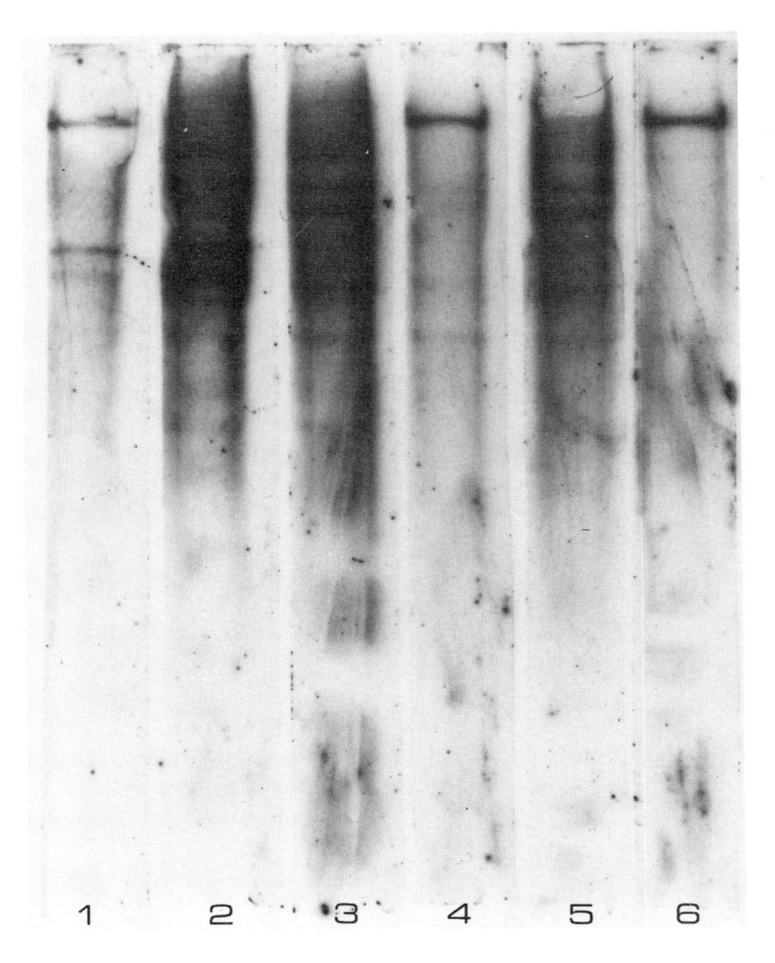

Fig. 4. Hybridization analysis of three different cloned secondary rat embryo cell lines transformed by purified SV40-su^{+}III DNA. This autoradiograph compares the EcoRI cleavage patterns of total cellular DNA from two cell lines, A2 and A7, with C1B shown in Fig. 2. These different lines were cloned from transformed populations of a common rat embryo cell culture. Details of analysis were as described in Fig. 2. Gel lanes 1 and 2 are of C1B, 3 and 4 of A2 and 5 and 6 of A7 DNAs. Lanes 1, 4 and 6 were hybridized with the su^{+}III-

Fig. 4 (cont.) specific probe and lanes 2, 3 and 5 were hybridized with the SV40 probe.

The single HaeII site in SV40 has been deleted in SV40-su$^+$III and there is no HaeII site in the plasmid insert (unpublished data). Cleavage of total transformed cell DNA with HaeII and analysis by the Southern procedure should yield a pattern of open circular viral DNAs and integrated species (supercoils hybridize poorly under these conditions). The analysis of cell lines C1B and C3E compared with EcoRI restricted C1B DNA showed three species in common, a major band migrating at the open circular position, and two higher molecular weight components, consistent with dimers and trimers of SV40-su$^+$III. Moreover, two bands were seen in the HaeII cleavage of C1B migrating in the vicinity of unit length open circular SV40-su$^+$III. Cleavage of these two species by EcoRI or BamI would generate the two common species migrating at the expected linear position seen in Fig. 2.

The SV40-su$^+$III hybrid contained two HpaII cleavage sites in the suppressor gene region (Fig. 1). Cleavage of the original hybrid by HpaII generated an 870 base pair fragment carrying only bacterial DNA sequences. When the total cellular DNA from the cloned transformed rat cells (C1B and C3E) was cleaved by HpaII, an 870 base pair fragment, which comigrated with the expected fragment from pColsu$^+$III, was identified with the radiolabeled su$^+$III DNA probe (unpublished results).

CHIMERICAL DNA IN PERSISTENTLY INFECTED MONKEY CELLS

We have also employed this purified chimerical DNA which carries the SV40 genetic information needed for viral DNA replication in permissive cells to generate persistently infected monkey kidney cells. The TC7 subline of CV-1 monkey kidney cells was infected with a preparation of the SV40-su$^+$III DNA using the DEAE-Dextran procedure (7), and 48 hours after DNA transfection the cells were subcultured and serially diluted for cell cloning. Individual clonal isolates were expanded and tested for viral DNA by incorporation of ^{3}H-thymidine into the low molecular weight DNA fraction of cell extracts (8). Of a large number of cloned or partially cloned lines from the original DNA-infected cell culture, two cell populations were identified which carried substantial amounts of free DNA. The cloned lines identified by this approach were expanded and further investigated. Analysis of the total cellular DNA from these cloned lines of TC7 cells persistently infected with the SV40-su$^+$III by ethidium bromide staining of agarose gels revealed fluorescent bands in the region of superhelical SV40 DNA. One of the two persistently infected cell lines contained a complex pattern of superhelical viral DNAs. This was explained in part by rearrangements that had occurred in the input SV40-su^{+}III. By the Southern procedure, each of these major superhelical DNAs from the total cellular DNA hybridized to nick-translated SV40 DNA as illustrated in Fig. 4. When the suppressor plasmid DNA probe was used for hybridization with the total cellular DNA from this persistently-infected TC7 cell line, all of the 3-4 major free viral DNA species were found to hybridize with the tRNA probe. Three of the four species were EcoRI resistant, the fourth partially resistant (Fig. 5). All were BamI sensitive (data not shown). Subcloning of cells from this persistently-infected cell culture has revealed the persistence of free viral DNA in the majority of subclones and expression of SV40 intranuclear T antigen. A number of these subclones contained a much simpler pattern of superhelical viral DNAs with 1-2 main free DNA species in contrast to the original clonal isolate.

CONCLUDING REMARKS

These studies thus show that both the SV40 vector and bacterial DNA sequences remain associated with transformed rat embryo cells and persistently infected monkey kidney cells. In the two transformed rat embryo lines tested the bacterial sequence is totally conserved. An unexpected observation with the rat cells has been the presence of both free or unintegrated and integrated hybrid viral DNA sequences. Both of these host cell vector systems have two advantages over the lytic infectious system for SV40: (1) no infectious virus is produced (assessed by co-cultivation with permissive monkey cells) and (2) the size of the non-viral, prokaryotic or eukaryotic DNA inserted in the late gene region of the SV40 vector may be quite large since no requirement for encapsidation in progeny virus is imposed. Furthermore, expression of the newly-introduced gene(s) can be followed in a long term, stable cell system as compared with a lytic infection, which generally terminates the viability of the host cell prematurely.

ACKNOWLEDGEMENTS

This work was supported in part by a research grant from the National Cancer Institute, USPHS No. CA 20794-01. All experiments with recombinant virus and cells carrying hybrid viral genomes were performed under conditions of P3 physical containment, as described in the National Institutes of Health "Guidelines for Research Involving Recombinant DNA Molecules" (1976). We thank D.H. Hamer for the *E. coli* strain carrying

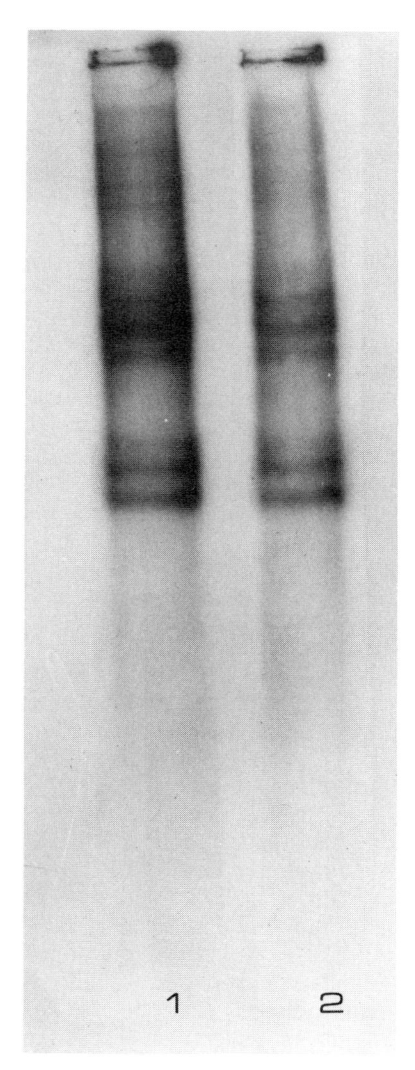

Fig. 5. Hybridization analysis of hybrid viral DNA from a persistently infected monkey cell line. The purified SV40-*su$^+$III* DNA was used to transfect TC7 monkey kidney cells and a chronically infected cell population was subsequently cloned (Fareed and Upcroft, in preparation). The total cellular DNA from this culture was purified and analyzed in the Southern (4) technique before (#2) and after (#1) cleavage by endo R.*EcoRI*. The hybridization probe was ^{32}P-labeled cRNA from SV40 DNA (9).

plasmid RSF2124 su$^+$III (which is termed pColsu$^+$III in this paper) and for the plasmid RSF2124 DNA.

REFERENCES

1. Abrahams, P.J., Mulder, C., Van De Voorde, A., Warnaar, S.O. and van der Eb, A.J. (1975) *J. Virol.* 16, 818.
2. Gelb, L.D., Kohne, D.E. and Martin, M.A. (1971) *J. Mol. Biol,* 57, 129.
3. Danna, K.J., Sack, G.H. and Nathans, D. (1973) *J. Mol. Biol.* 78, 363.
4. Southern, E.M. (1975) *J. Mol. Biol.* 98, 503.
5. Ketner, G. and Kelly, T.J. (1976) *Proc. Nat. Acad. Sci. USA* 73, 1102.
6. Maniatis, T., Kee, S.G., Efstratiadis, A. and Kafatos, F.C. (1976) *Cell* 8, 163.
7. McCutcheon, J. and Pagano, J.M. (1968) *J. Nat. Cancer Inst.* 41, 351.
8. Hirt, B. (1967) *J. Mol. Biol.* 26, 365.

MAPPING OF SEQUENCES WITH TWOFOLD SYMMETRY ON THE SV40 GENOME*

Che-Kun James Shen and John E. Hearst

Department of Chemistry,
University of California,
Berkeley, California 94720

ABSTRACT. Sequences with approximate twofold axes of symmetry have been detected in SV40 DNA and mapped by electron microscopy by the ability of SV40 single strands to form hairpins which are stabilized by the photochemical cross-linking reagents trioxsalen (4,5',8-trimethylpsoralen). SV40 I was digested with restriction enzymes EcoRI or HpaII and the resulting linear SV40 DNA molecules were denatured and photochemically reacted with trioxsalen at 16.0 ± 0.5°C at different ionic strengths. In 20 mM NaCl, one specific hairpin 100-150 base pairs long was detected at 0.17 ± 0.02 map units on the EcoRI map of SV40 DNA which is an "in vitro" promoter site for *E. coli* RNA polymerase and is near the termination site of DNA replication. The 3' ends of the three SV40 mRNA's have been mapped to this region as well. In 30 mM NaCl, five more hairpins appeared on these denatured and cross-linked SV40 DNA molecules. Four of these were found to be at 0.26 ± 0.02, 0.68 ± 0.03, 0.84 ± 0.02, and 0.94 ± 0.01 units on the EcoRI map, respectively. The fifth one is located on or near the EcoRI cleavage site of SV40 DNA. Of these five additional hairpins the one at 0.68 map units is near the replication origin and the hairpin at 0.94 map units is near the 5' end of the 16S late messenger RNA. The possible functions of these sequences are discussed in terms of the nature of the promoter sites, the replication origin, the processing of RNA precursors, and regulation at the translational level.

*A more detailed description of these experiments and the accompanying discussion can be found in Ref. 40.

INTRODUCTION

Direct sequencing data have shown that many regulating segments of prokaryotic DNA contain sequences with approximate twofold symmetry: the operators (1,2), the promoters (3-8), and the origin of replication (9). This property may allow the formation of hairpins in these regions after strand separation and, possibly, the mapping of these secondary structures on the genome if the hairpins can be visualized directly in the electron microscope.

However, it has been difficult to detect hairpins on single stranded or denatured viral DNA molecules using the protein monolayer-formamide spreading technique (10). Two possibilities could explain this failure. First, the hairpins, if there are any, are too short (< 50 base pairs) to be discerned. Secondly, an examination of the regulatory sequences mentioned above indicates that few of them consists of perfect twofold symmetry. It may be that the hairpin forms with the lowest free energy are not stable under the formamide spreading conditions because of base mismatching in the foldback duplex. The latter problem can be solved by photochemically cross-linking the single stranded or denatured virus DNA with trioxsalen. Trioxsalen molecules intercalate between base pairs of DNA and, upon irradiation with long wavelength UV, form covalent interstrand cross-links (11-15). Using this technique we have shown previously that as many as eight hairpins can be stabilized by trioxsalen and visualized on fd DNA in the electron microscope. Furthermore, these hairpins appeared at specific sites on the fd genome and most of them are located near or in the *in vitro* fd promoter regions (16).

We report here that at least six regions of the SV40 genome are capable of forming discernable hairpins and most of them coincide with positions of biological importance in the genetic map of SV40.

MATERIALS AND METHODS

Restriction of SV40 I. Supercoiled SV40 DNA (SV40 I) was digested by EcoRI in 100 mM Tris•HCl (pH 7.6), 50 mM NaCl, 5 mM $MgCl_2$, 0.2 dithiothreitol and 0.1 mM EDTA at 37°C for 3 hours. For the HpaII reaction, SV40 I was incubated with the enzyme at 37°C in 60 mM Tris•HCl (pH 8.0), 7 mM $MgCl_2$ for 1 hour. After incubation, the reaction mixtures were first extracted with 100 mM Tris•HCl (pH 7.0) saturated phenol and then with ether. The aqueous phase was dialyzed against 10 mM Tris•HCl, 1 mM EDTA, pH 8.0.

Gel electrophoresis. Electrophoresis of DNA samples in 1% agarose gels was performed as previously described (18,19). EcoRI cut λ DNA fragments were co-electrophoresed as molecular weight markers (20).

Denaturation of restricted SV40 I DNA and photochemical cross-linking of denatured SV40 DNA. The linear SV40 DNA resulting from the restriction enzyme reactions was denatured in alkali, neutralized, and dialyzed at 4°C against 1 mM Tris•HCl, 0.1 mM EDTA, pH 7.0 as described before (16).

A typical DNA solution to be irradiated had a final volume of 100 μl and contained 5 μg/ml of denatured SV40 DNA and the appropriate NaCl concentration. This solution was mixed with 1 μl of 1 mg/ml trioxsalen (in 100% ethanol) and sealed in a glass pipet. The pipet was irradiated at 16.0 ± 0.5°C with long wavelength UV light in the apparatus described by Issacs et al. (17).

After irradiation for 10 minutes, the sample solution was mixed with another microliter of the trioxsalen-stock solution and the irradiation was continued for another 10 minutes.

Electron microscopy. The photochemically cross-linked DNA solution was spread directly for electron microscopy using the 40% formamide spreading technique (10). Sample grids were examined in a Philips 201 Electron Microscope. Lengths of DNA segments were measured as described before (16).

RESULTS

Formation of hairpins on denatured SV40 II DNA. In order to see whether hairpins form on single-stranded SV40 DNA molecules, nicked SV40 DNA

(SV40 II) was denatured and photochemically cross-linked with trioxsalen at 16°C and 30 mM NaCl. It was observed that essentially all of the molecules have hairpins on them. Fig. 1 shows an example in which as many as 6 hairpins appear on a single-stranded circular SV40 DNA molecule.

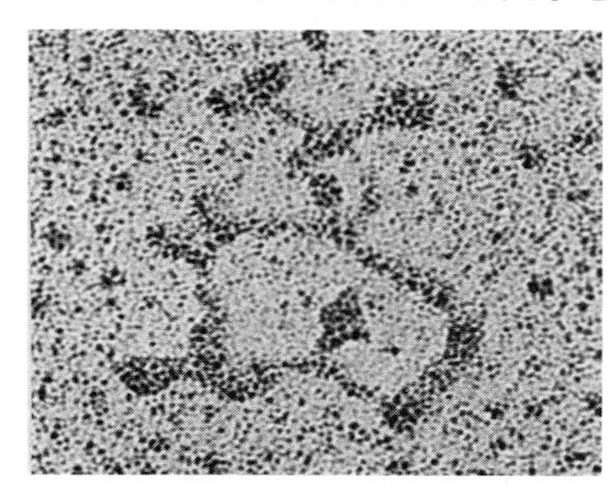

Fig. 1. Trioxsalen cross-linked secondary structures on a single-stranded SV40 DNA molecule. SV40 II was denatured and photochemically cross-linked at 16°C and 30 mM NaCl with trioxsalen. Magnification: 175,500 x.

Generation of linear SV40 DNA by restriction enzymes. SV40 I was subjected to digestion by the restriction enzymes EcoRI and HpaII, respectively, to generate full length linear SV40 molecules. Each of the two enzymes cuts SV40 DNA at one specific site, the HpaII cleavage site having been mapped at 0.74 units on the EcoRI map of SV40 (18,21,22).

Fig. 2 shows the electrophoresis patterns of various forms of SV40 DNA on an agarose gel. The molecular weights of the linear SV40 DNAs, EcoRI-SV40 I and HpaII-SV40 I, have been calculated to be 3.02×10^6 daltons (4570 base pairs) from their positions in the gel relative to the EcoRI-λ fragments, while that of the double-stranded fd was calculated to be 5750 base pairs.

Hairpins are located at specific positions on denatured EcoRI-SV40 I cross-linked by trioxsalen. EcoRI-SV40 I DNA was denatured and cross-linked at 16°C. Over 50% of the molecules cross-linked at 20 mM NaCl showed at least one hairpin near one end (Fig. 3a). As the salt was increased to 30 mM NaCl, the average number of hairpins on the denatured molecules also increased. Most of the molecules have one hairpin at one end and two at the other end, with another two to three hairpins located between the above three hairpins. Fig. 3b shows a typical molecule of denatured EcoRI-SV40 I cross-linked at 30 mM NaCl.

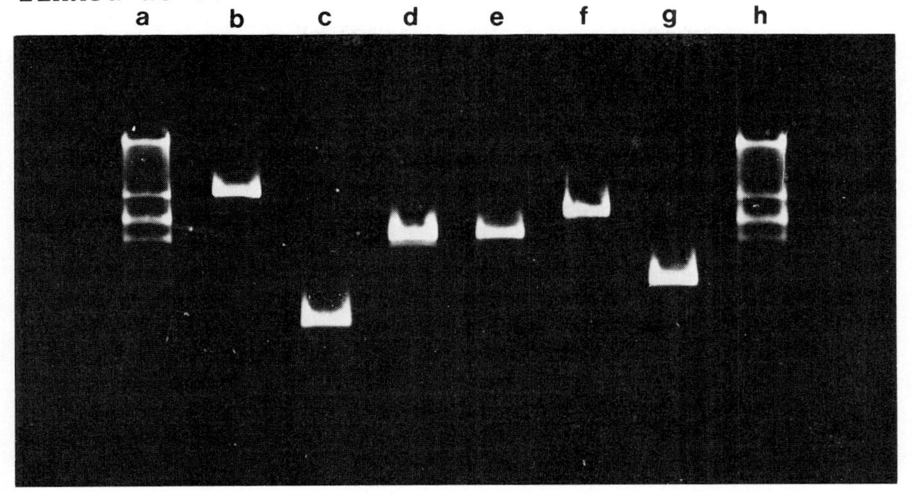

Fig. 2. Gel electrophoresis of SV40 DNAs in 1% agarose. (a) EcoRI-λ DNA fragments; (b) SV40 II; (c) SV40 I; (d) EcoRI-SV40 I; (e) HpaII-SV40 I; (f) Hind II-fd RFI; (g) fd RFI; (h) EcoRI-λDNA fragments.

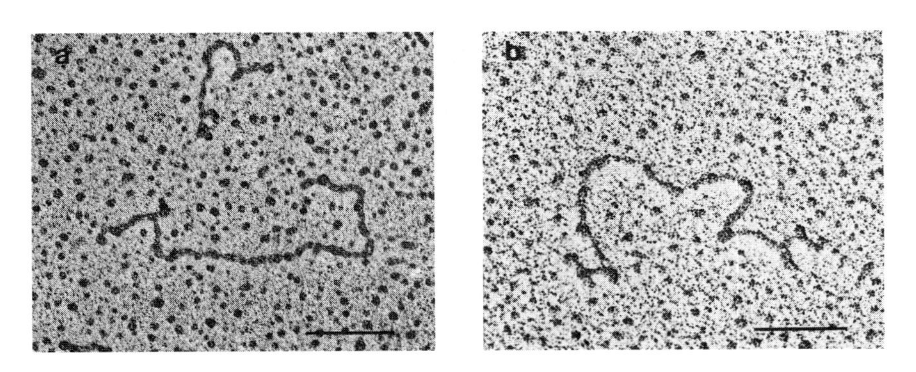

Fig. 3. Secondary structures on denatured and cross-linked EcoRI-SV40 I. (a) Denatured EcoRI-SV40 I cross-linked at 16.0 ± 0.5°C and 20 mM NaCl; (b) Denatured EcoRI-SV40 I cross-linked at 16.0 ± 0.5°C and 30 mM NaCl. The bars (—) are 500 nucleotides long.

In order to determine whether these hairpins are located at specific positions, photographs of DNA molecules were taken and the lengths of all the hairpins as well as the center to center distances of adjacent hairpins were measured and then converted to fractional lengths of SV40 DNA. The hairpin maps and histograms thus constructed are shown in Figs. 4 and 5. From Fig. 4 it can be calculated that the most stable hairpin has a length of 150 ± 60 base pairs (this is probably an overestimated value because of the shortening of the

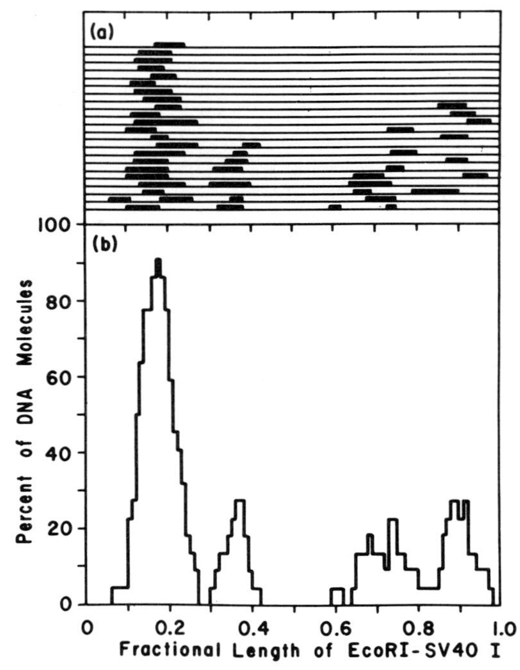

Fig. 4. Hairpin map (a) and histogram (b) of denatured EcoRI-SV40 I cross-linked at 16°C and 20 mM NaCl. In the hairpin map, the hairpins that appeared between 0.1 and 0.2 units were assigned to the left part of the molecules. The histogram shows the percentage of molecules measured that have a hairpin in a given length interval corresponding to one hundredth of the SV40 genome size.

DNA molecules by cross-linking, see Ref. 40) and its center is located 0.17 ± 0.02 units from one end of EcoRI-SV40 I. At 30 mM NaCl (Fig. 5), there are five distinct peaks (I-V) in the hairpin histogram (Fig. 5b). The 34 molecules were aligned so that the ends with two hairpins were assigned to the right part of the molecule. The centers of each of these five hairpins have been determined to be 0.15 ± 0.02 (I), 0.26 ± 0.02 (II), 0.68 ± 0.03 (III),

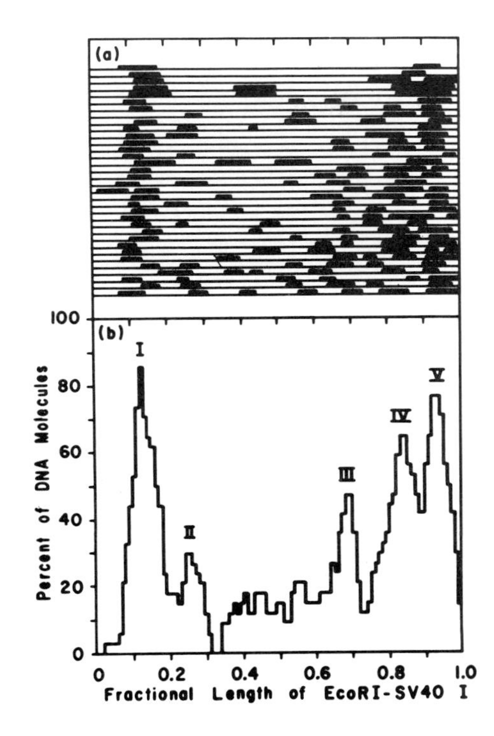

Fig. 5. Hairpin map (a) and histogram (b) of denatured EcoRI-SV40 I cross-linked at 16°C and 30 mM NaCl.

0.84 ± 0.02 (IV) and 0.94 ± 0.01 (V) units away from the left end of EcoRI-SV40 I.[†]

<u>Sequences with twofold symmetry in HpaII-SV40 I</u>. After denaturation and cross-linking, HpaII-

[†]The position of the center of the most stable hairpin was calculated by averaging the center positions of that hairpin of all the molecules shown in Fig. 4a except molecule #21.

The molecules in Fig. 5a used to calculate the center positions of hairpins I-V are listed as follows: Hairpin I, all 34 molecules; Hairpin II, #8,10,11,14,18,21,24,30,32-34; Hairpin III, #6,7, 12,13,16,18-23,25-34; Hairpin IV, #3,7,8,11-13,16, 17,19-21,23,25-27,29,31,32,34; Hairpin V, 3,6-14, 16,17,19-22,25-27,29-34.

SV40 I showed one specific hairpin at 20 mM NaCl near the middle of the molecule (Fig. 6a) while several hairpins were found in 30 mM NaCl on denatured and cross-linked HpaII-SV40 I (Fig. 6b). The hairpin histograms are shown in Fig. 7. As can be seen, the most stable hairpin is located at 0.39 ± 0.03 units from one end of denatured and cross-linked HpaII-SV40 I. This leads us to conclude that the most stable hairpin is located at 0.17 instead of 0.83 units on the EcoRI map of SV40. Similarly, an examination of the hairpin histograms of both denatured EcoRI-SV40 I and denatured HpaII-SV40 I cross-linked at 30 mM NaCl indicates that

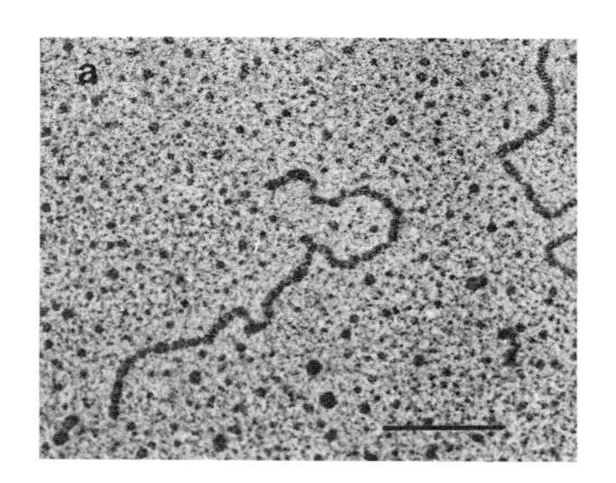

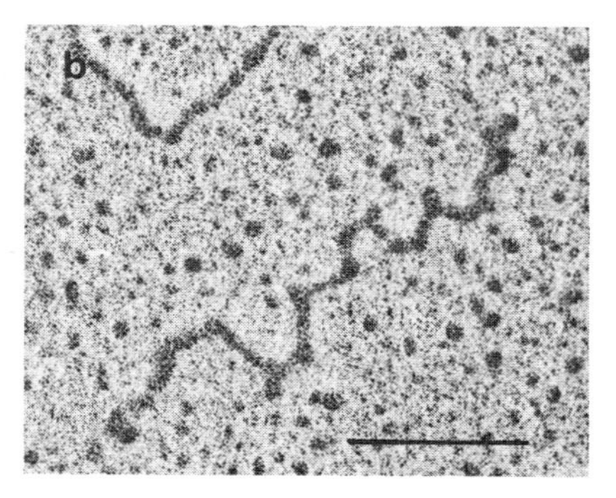

Fig. 6. Secondary structures on denatured and cross-linked HpaII-SV40 I. (a) Denatured HpaII-SV40 I cross-linked at 16.0 ± 0.5°C and 20 mM NaCl; (b) Denatured HpaII-SV40 I cross-linked at 16.0 ± 0.5°C and 30 mM NaCl. The bars (—) are 500 nucleotides long.

the positions of hairpins I-V determined in the last section are their true positions on the EcoRI-SV40 map and that hairpin I corresponds to the most stable hairpin. Hairpin VI was not observed on denatured EcoRI-SV40 I cross-linked at 30 mM NaCl, suggesting that the sequence of this hairpin contains the EcoRI cleavage site. Alternatively, this hairpin might have escaped detection on the denatured EcoRI-SV40 I molecules cross-linked in 30 mM NaCl because of its proximity to either end of the linear molecules.

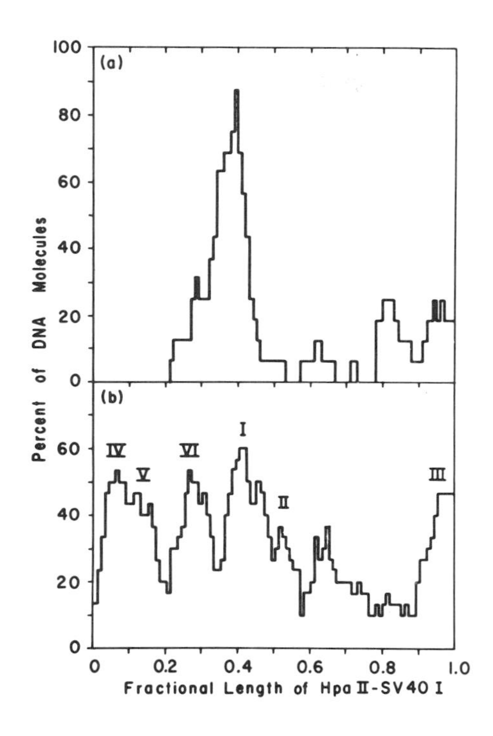

Fig. 7. Hairpin histograms of denatured HpaII-SV40 I cross-linked with trioxsalen at 16°C and two different ionic strengths: (a) 20 mM NaCl (b) 30 mM NaCl.

DISCUSSION

The evidence presented above shows that long sequences with twofold axes of symmetry are located on the SV40 genome at specific positions. Four of the six regions which are capable of forming cross-linked hairpins have been mapped to positions of biological importance.

Replication of SV40 DNA. It has been shown (24,25) that SV40 replication begins at a specific site (0.67 units on the EcoRI map of SV40) and proceeds bidirectionally, terminating about half way around the circular DNA from the initiation point. The fact that hairpins III (0.68 ± 0.03) and I (0.17 ± 0.02) are located in the initiation and termination regions of replication, respectively, suggests that these symmetric sequences may be important "start-stop" signals for SV40 replication. A 28 base pair long segment near the SV40 replication origin has been sequenced and shown to contain a high degree of twofold symmetry (9).

Transcription of SV40 DNA. The sequencing data of promoters (3-8) shows that most of them contain sequences of approximate twofold symmetry. The results of the photochemical cross-linking experiment done on fd DNA (16) are consistent with this idea.

Three messenger RNAs have been detected in and isolated from SV40 infected cells (26,27, Fig. 8). The early 19S mRNA has been mapped on the physical map of SV40 to be transcribed from the early strand with its 5' terminus located near the replication origin and the 3' terminus located near the termination site of replication (28-31). Late 19S mRNA and 16S mRNA have been detected at the late stage of SV40 infection (27) and mapped on the SV40 genome (26,28,30). The late 19S mRNA is transcribed from the late strand of SV40 from a position near the replication origin at 0.67 map units (5' mRNA end) to the termination site of replication at approximately 0.17 map units (3' mRNA end). The 5' end of the 16S mRNA is located at 0.95 map units of the EcoRI-SV40 map and also terminates at 0.17 map units. A comparison of the hairpin map we obtained with the positions of the 5' ends of these three mRNA species and the preferred E. coli RNA polymerase initiation site indicates a strong correlation between initiation sites for transcription (promoters?) and sequences with twofold symmetry axes.

SV40 mutants tsD and SV40 specific polypeptide VP3. Many temperature sensitive mutants of SV40 have been isolated and characterized (for a review, see Ref. 32). One group of the mutants, tsD mutants (33,34), have been thought to be defective in virus uncoating at high temperature. All the D mutants map between 0.85 and 0.94 map units on EcoRI-SV40. These facts suggest the portion of genome associated with mutants D may code for a virion protein (32). This point has been confirmed recently by Goff, S., Cole, C., Landers, T., Manteuil-Brutlag, S. and Berg, P. (personal communication) who found that the virus specific polypeptide, VP3, maps on SV40 genome between 0.84 and 0.95 map units. Although the polypeptide VP3 has been proposed to be generated by post-translational cleavage of a precursor polypeptide (35), the presence of hairpins (hairpins IV and V) on denatured SV40 DNA at these two positions suggests that the symmetric sequences at 0.84 and 0.94 may play important roles in the expression of VP3 at transcriptional and translational levels. They may function in the production of VP3-coding mRNA either by initiating transcription at position 0.84 and terminating transcription at 0.95, or by providing post-transcriptional cleavage sites on a precursor RNA. On the other hand, the two hairpins may appear on the late 19S mRNA and punctuate and control the translation of the mRNA segment between them to produce VP3. Baltimore et al. (36) have suggested, however, that initiation of more than one discrete polypeptide on a single eukaryotic mRNA is unlikely, an argument against translational control on a polycistronic message.

Post-transcriptional cleavage of RNA. It has been shown *in vitro* by Dunn & Studier (37) that RNase III of *E. coli* cleaves transcripts of the early region of T7 phage to small molecules identical to the phage specific mRNAs in the infected cells. Similar enzymes have been found in animal cells (38). Some evidence has been presented to show that the SV40 specific late 19S mRNA may be a precursor to the 16S mRNA (26). The presence of symmetric sequences on SV40 near the 5' ends of all three SV40 specific mRNAs suggests that these sequences may appear on the primary transcripts of SV40 (27,39) or the late 19S mRNA as hairpins and be recognized by processing enzymes.

Our results are summarized in Fig. 8 in which the positions of the six hairpin-forming sequences as well as the known maps of SV40 specific mRNAs, the SV40 specific polypeptide VP3, and regions of tsD mutants are indicated on the EcoRI-SV40 physical map. Complete sequencing data of SV40 DNA and refined mapping of the hairpins should reveal the exact locations of these sequences with twofold symmetry and facilitate the study of their biological functions.

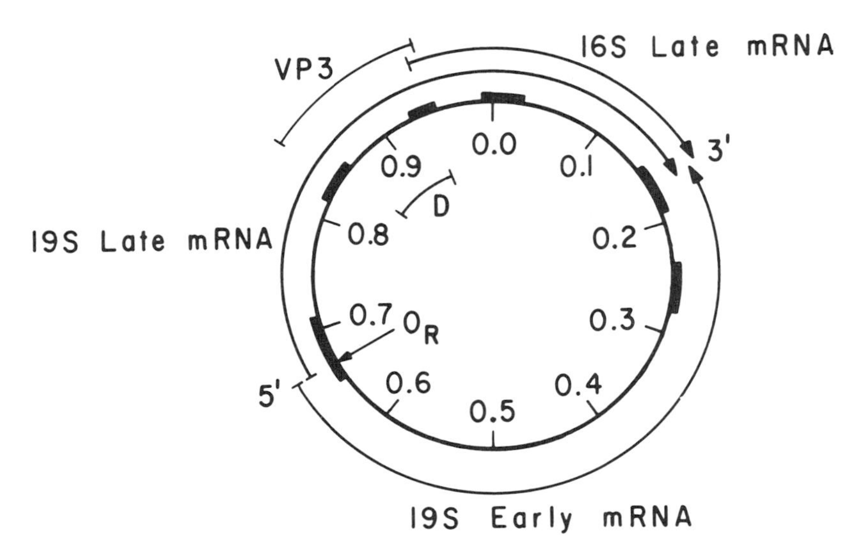

Fig. 8. Diagrammatic representation of the relationship between the positions of sequences with twofold symmetry of SV40 DNA (the blackened regions on the EcoRI-SV40 physical map) and the locations of the three SV40 specific mRNA's, the replication origin O_R, the tsD mutants, and the SV40 specific polypeptide VP3. A detailed sequence analysis of a 316 nucleotides long fragment containing the replication origin can be found in Ref. 41-44.

ACKNOWLEDGMENT

We thank Dr. L. Hallick, Dr. L. Bartholomew, Dr. P. Modrich, Dr. T. S. Hsieh, P. Rigby, H. Nagaishi, and S. Isaacs for their generosity of giving us the materials. We also appreciate the helpful comments and suggestions from Dr. M. Botchan, S. Goff, G. Wiesehahn and Dr. L. Hallick. This study was supported by the American Cancer Society Grant #NP-185, by the National Institutes of Health Grant #GM-11180, and by the National Science Foundation Grant #GB-36799. C.-K. James Shen has been supported by the Earle C. Anthony Fellowship from the University of California.

REFERENCES

1. Gilbert, W. & Maxam, A. (1973) *Proc. Nat. Acad. Sci. USA* 70, 3581-3584.

2. Maniatis, T., Ptashne, M., Barrell, B. G. & Donelson, J. E. (1974) *Nature* 250, 394-397.
3. Zain, B. S., Weissman, S. M., Dhar, R. & Pan, J. (1974) *Nucl. Acids Res.* 1, 577-594.
4. Dhar, R., Weissman, S. M., Zain, B. S., Pan, J. & Lewis, A. M., Jr. (1974) *Nucl. Acids Res.* 1, 595-614.
5. Dickson, R. C., Abelson, J., Barnes, W. M. & Renznikoff, W. S. (1975) *Science* 187, 27-35.
6. Schaller, H., Gray, C. & Herrmann, K. (1975) *Proc. Nat. Acad. Sci. USA* 72, 737-741.
7. Sekiya, T., Ormondt, H. V. & Khorana, H. G. (1975) *J. Biol. Chem.* 250, 1087-1098.
8. Pribnow, D. (1975) *J. Mol. Biol.* 99, 419-443.
9. Jay, E., Roychoudhury, R. & Wu, R. (1976) *Biochem. Biophys. Res. Comm.* 69, 678-686.
10. Davis, R. W., Simon, M. & Davison, N. (1971) in *Methods in Enzymology,* eds. Grossman, L. & Moldave, K. (Academic Press, New York) 21, 413-428.
11. Dall'Acqua, F. & Rodighiero, G. (1966) *Rc. Accad. Naz. Lincei* 40, 411-422.
12. Musajo, L., Bordin, F., Caporale, S., Marciani, S. & Rigatti, G. (1967) *Photochemistry & Photobiology* 6, 711-719.
13. Cole, R. S. (1970) Biochim. Biophys. Acta 217, 30-39.
14. Dall'Acqua, F., Marciani, S., Vedaldi, D. & Rodighiero, G. (1974) *Biochim. Biophys. Acta* 353, 267-273.
15. Cole, R. S. (1971) *Biochim. Biophys. Acta* 254, 30-39.
16. Shen, C.-K. J. & Hearst, J. E. (1976) *Proc. Nat. Acad. Sci. USA* 73, 2649-2653.
17. Isaacs, S. T., Shen, C.-K. J., Hearst, J. E. & Rapoport, H. (1977) *Biochemistry,* in press.
18. Sharp, P. A., Sugden, B. & Sambrook, J. (1973) *Biochemistry* 12, 3055-3063.
19. Shen, C.-K., J., Wiesehahn, G. & Hearst, J. E. (1976) *Nucl. Acids Res.* 3, 931-952.
20. Thomas, M. & Davis, R. W. (1975) *J. Mol. Biol.* 91, 315-328.
21. Mulder, C. & Delius, H. (1972) *Proc. Nat. Acad. Sci. USA* 69, 3215-3219.
22. Morrow, J. & Berg, P. (1972) *Proc. Nat. Acad. Sci. USA* 69, 3365-3369.
23. Acheson, N. H. (1976) *Cell* 8, 1-12.
24. Danna, K. J. & Nathans, D. (1972) *Proc. Nat. Acad. Sci. USA* 69, 3097-3100.

25. Fareed, G. C., Garon, C. F. & Salzman, N. P. (1972) *J. Virol.* 10, 484-491.
26. Weinberg, R. A., Ben-Ishai, Z. & Newbold, J. E. (1974) *J. Virol.* 13, 1263-1273.
27. Weinberg, R. A., Warnaar, S. O. & Wincour, E. (1972) *J. Virol.* 10, 193-201.
28. Khoury, G., Martin, M. A., Lee, T. N. H., Danna, K. J. & Nathans, D. (1973) *J. Mol. Biol.* 78, 377-389.
29. Khoury, G., Howley, P., Nathans, D. & Martin, M. (1975) *J. Virol.* 15, 433-437.
30. Sambrook, J., Sugden, B., Keller, W. & Sharp, P. A. (1973) *Proc. Nat. Acad. Sci. USA* 70, 3711-3715.
31. Subramanian, K. N., Dhar, R., Pan, J., Zain, B. S. & Weissman, S. M. (1976) in *Molecular Mechanisms of Gene Expression*, eds. Nierlich, D. P., Rutter, W. J. & Fox, C. F. (Academic Press, New York) 5, 367-377.
32. Lai, C.-J. & Nathans, D. (1975) *Virol.* 66, 70-81.
33. Robb, J. A. & Martin, R. G. (1972) *J. Virol.* 9, 956-968.
34. Chou, J. Y. & Martin, R. G. (1974) *J. Virol.* 13, 1101-1109.
35. Prives, C. L., Aviv, H., Gilboa, E., Revel, M. & Wincour, E. (1974) *Cold Spring Harbor Symp. Quant. Biol.* 39, 309-315.
36. Baltimore, D., Jacobson, F., Asso, G. & Juang, A. (1969) *Cold Spring Harbor Symp. Quant. Biol.* 34, 741-746.
37. Dunn, J. J. & Studier, F. W. (1974) *Proc. Nat. Acad. Sci. USA* 70, 3296-3300.
38. Büsen, W. & Hausen, P. (1975) *Eur. J. Biochem.* 52, 179-190.
39. Tonegawa, S., Walter, G., Bernardini, A. & Dulbecco, R. (1970) *Cold Spring Harbor Symp. Quant. Biol.* 35, 823-831.
40. Shen, C.-K. J. & Hearst, J. E. (1977) *Proc. Nat. Acad. Sci. USA*, in press.
41. Subramanian, K. N., Dhar, R. & Weissman, S. M. (1977) *J. Biol. Chem.* 252, 333-339.
42. Dhar, R., Subramanian, K. N., Pan, J., Weissman, S. M. & Ghosh, P. K. (1977) *J. Biol. Chem.* 252, 340-354.
43. Subramanian, K. N., Dhar, R. & Weissman, S. M. (1977) *J. Biol. Chem.* 252, 355-367.
44. Dhar, R., Subramanian, K. N., Pan, J. & Weissman, S. M. (1977) *J. Biol. Chem.* 252,368.

STUDIES ON AN SV40 DNA CARRIER STATE IN MONKEY CELLS IN CULTURE

John M. Jordan

Department of Chemistry and
the Molecular Biology Institute
The University of California, Los Angeles
Los Angeles, California 90024

ABSTRACT. We examined the capacity of several types of monkey cells in culture to produce infectious simian virus 40. All cell types studied supported the replication of SV40 DNA. Serially propagated VERO cells and clones of VERO cells failed to produce significant levels of mature virus particles. A combination of techniques showed that VERO-derived cells produced both SV40-specific "T" and "V" antigens in addition to infectious SV40 DNA. Fluorescence microscopy revealed that most of the "V" antigen in infected VERO cells is in the nucleolus and on the periphery of the nucleus. Serially propagated SV40-exposed VERO cells retain SV40 DNA and virus specific antigens in the absence of virus production. These cells demonstrate a SV40 DNA "carrier" state in which the observed limited virus production may be host cell determined.

INTRODUCTION

Although the subject of intense investigations, many of the events associated with simian virus 40 (SV40) infection of mammalian cells remain undefined (1). Interactions between the infecting virus and monkey cells are of particular importance since an understanding of these may help to clarify natural host-virus relationships. We initiated studies on non-lytic SV40-monkey cell systems with two major objectives: (i) to develop better understanding of biochemical events involved in the virus maturation cycle and (ii) the isolation and characterization of cell types capable of replicating *in vitro* produced genomes and genomes of virus deletion mutants. Such cells could prove useful in genetic analysis of animal tumor virus and in the high level production of viral gene products. We describe in this report cell types which appear to satisfactorily fulfill these objectives

RESULTS

Exposure of monkey cells in culture generally results in cell lysis which follows the production of infectious virus particles. We observed the production of mature virus particles and extensive cellular breakdown in several types of monkey kidney cells within 5-7 days after infection. As shown in Table 1, TC7, BSC-1 and AGMK cells afforded production of SV40 in high yields following infection of confluent monolayers. In contrast, VERO cells failed to demonstrate a

Table 1

Production of SV40 in virus-infected monkey cells in culture

Cell type Exposed to virus‡	No. of Determinations	Days to cytopathic effect*	Virus yield $\log(\frac{pfu}{ml})$
AGMK	3	4-6	8.5
TC7	2	5-7	8.0
BSC-1	2	5-7	7.0
VERO	3	Not observed	<3

‡Confluent monolayers of AGMK, TC7, BSC-1 or VERO cells were separately infected with SV40 at 5.0 pfu/cell, overlaid with culture medium and incubated for eight days. Cells and cellular debris were collected, subjected to three cycles of freeze-thawing, and virus present in the suspension quantitated by plaque assay using TC7 as indicator cells. During the 8-day incubation the culture medium was replaced at 3 and 5 days following virus adsorption. Plaque assays using serially diluted virus suspensions were carried out in 60mm culture dishes. Each virus dilution was assayed in triplicate.

*Time, after virus adsorption, before 50% of the cell sheet demonstrate cytopathic effect.

lytic response at times up to two weeks after infection of cells at a multiplicity of 5 or 20 pfu/cell. This last result raised the question of whether <u>all</u> cells in the VERO population behaved similarly in their response to infection by SV40. Our approach to this question involved attempted virus production in several clonal isolates of VERO. We examined ten independently isolated clones of VERO and observed these failed to produce SV40 at levels higher than the 10^3 pfu/ml of cell lysate prepared as described under Table 1. One of the clonal isolates, designated VJA, was used in the experiments described in the remainder of this report.

The limited production of simian virus 40 particles was demonstrated in two additional types of experiments. Infectious, progeny virus particles could be successfully isolated from AGMK, BSC-1 or TC7 cells using standard extraction procedures (2). SV40-infected VJA cells, on the other hand, failed to yield a substantial amount of virus particles. A mixing experiment, which will now be described, failed to isolate a significant amount of SV40 from infected VJA cells and suggested that the low yield of virus was determined by VJA cells and not the virus extraction and purification procedure. The mixing experiment utilized SV40 infected VJA cells (exposed to $^{32}P_i$ after virus adsorption) and SV40 infected African green monkey kidney secondary cultures (exposed to ^{3}H-thymidine-containing culture medium). An equal number of culture dishes containing infected VJA and African green monkey kidney cells was mixed and the mixture used to prepare simian virus 40. Additional sets of SV40 infected cells (incubated in $^{32}P_i$-containing medium) and infected AGMK cells were subjected to Hirt (3) extraction, the Hirt "supernatants" mixed and SV40 DNA analyzed following dye buoyant centrifugation(4). The distribution of radiolabel in the virus-containing extract, following buoyant centrifugation, is shown in Fig. 1, panel A. The pattern of DNA distribution in the dye buoyant gradient of the Hirt extract is illustrated in panel B of Fig. 1. The results illustrated in Fig. 1A indicate that only a small amount of virus is extractable from SV40-infected VJA cells and a correspondingly large amount of virus was isolatable from AGMK cultures. In marked contrast, panel B reveals that a significant amount of closed circular SV40 DNA could be isolated from either infected AGMK or VJA cells.

A second experiment with bearing on virus production in VJA cells involved transmission electron microscope examination of sections of SV40 infected cells following fixation, imbedding and staining. These results failed to reveal virus particles in the sections of SV40-infected VJA cells which were examined. Control experiments demonstrated a very large number of vesicle associated virus particles in SV40 infected TC7 or AGMK cells. These observations together with results from the mixing experiment just described provide substantial support for and confirmation of results of plaque assay analysis of lysates of SV40 infected VJA cells: a large quantity of SV40 virus is not produced in infected VJA and other clones of VERO cells.

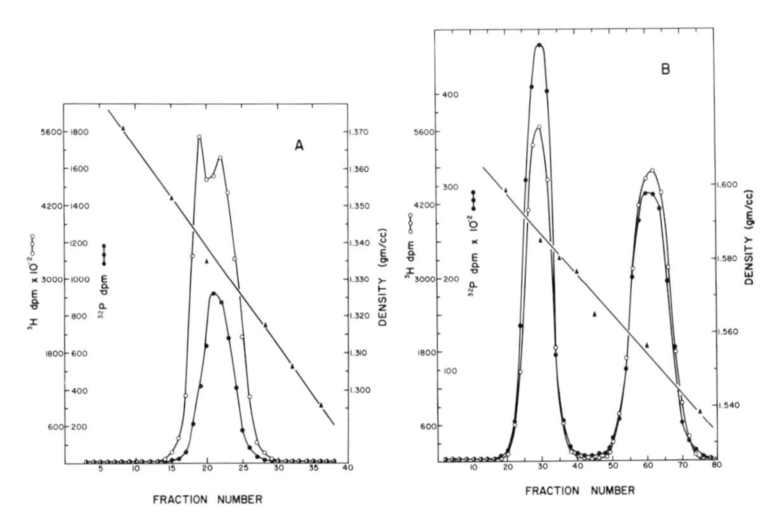

Figure 1. Within 24 hours after reaching confluence on 100mm culture dishes, 25 plates containing AGMK secondary cultures and, separately, 25 VJA cultures were infected with SV40 at a multiplicity of 5 pfu/cell. Following adsorption, the cultures were overlaid with 6ml of DME medium which contained 0.4mM inorganic orthophosphate, $^{31}P_i$, and 2% heat inactivated calf serum. Sixteen hours after infection, ^{3}H-Thymidine at 10 uc/ml was added to infected VJA cultures. At eight days after infection, cells from 22 plates of AGMK and VJA cultures were collected and combined. The mixture of ^{3}H-labeled and ^{32}P-labeled cultures was used to prepare SV40. The RNase and DNase-digested virus preparation was subjected to buoyant centrifugation producing the results shown in panel A. Three infected ^{3}H-labeled AGMK monolayers and ^{32}P-labeled VJA monolayers were Hirt extracted, the Hirt supernatants mixed and digested with DNase-free RNase A at 10 ug/ml. Following digestion and dialysis, the AGMK and VJA-cell derived DNA was dialyzed and centrifuged to equilibrium in a dye-buoyant gradient producing the result shown in B.

The persistence of SV40 DNA in VJA cells serially subcultured following exposure to simian virus 40. As described earlier, VJA cells failed to demonstrate virus-caused cytolysis at times up to two weeks following infection of confluent monolayers at multiplicities of 5 to 50 pfu/cell. Such SV40-exposed cells, designated SV-VJA, could be subcultured in Dulbecco's modified Eagle's medium containing 2%, 5% or 10% serum. Electron microscopic analysis and infectious center assay failed to demonstrate virus particles in either nucleus or cytoplasm of SV-VJA cultures. Following serial propagation for approximately 30 generations, ten confluent 100mm culture dishes were subjected to detergent extraction according to the procedure of Hirt (3). Ethidium bromide-cesium chloride buoyant centrifugation of the Hirt extract revealed a significant closed circular DNA-containing band. Electron microscopic analysis showed that more than 95% of the closed circular component was equivalent in contour length to viral SV40 DNA. Additionally, the closed circular DNA isolated from SV-VJA cells co-sedimented with viral SV40 DNA in both neutral and alkaline sucrose gradients. The identity of the closed circular DNA from SV-VJA cells was established by infectivity analysis in TC7 cells using the DEAE dextran procedure of Pagano (5). The specific infectivity—plaque forming units per μg of DNA—of circular DNA from SV-VJA cells subcultured for approximately 30 generations was comparable to the infectivity of AGMK or TC7 cell-derived SV40 DNA.

Quantitative studies showed that, five days after infection, TC7 or AGMK cells contained 14-18 μg of closed circular SV40 DNA per 100mm culture dish. In comparison, SV-VJA cells, subcultured for 30 generations, or recently infected VJA cells contained 2-5 μg of closed circular SV40 DNA per 100mm culture dish.

Whether the SV40 DNA obtained from SV-VJA cells is a result of viral DNA synthesis in a small fraction of the cultured cells was determined by cloning SV-VJA cells at approximately 10 generations after initial exposure to the virus. Several well isolated clones were selected, mass cultured and subjected to Hirt extraction. A combination of dye-buoyant centrifugation and electron microscopy was used to analyze for the presence of SV40 DNA. All of the clones of SV-VJA cells examined contained small amounts of SV40 DNA.

SV40-induced antigens in cultured SV-VJA cells. Fluorescence microscopy was used to further characterize recently infected VJA cells and SV-VJA cultures. Fluorescein conjugated antibodies directed against virus-specific proteins were employed to establish the presence and location of

SV40 "T" and "V" antigen. Although the virus-specific "T" antigen is found in both transformed and lytically infected cells, SV40 "V" antigen is generally observed only in permissive cells undergoing lytic infection following exposure to the virus. The "V" antigen is believed to correspond to viral capsid protein (6). Fluorescence microscopic analysis showed that recently infected VJA cells and serially propagated SV-VJA cells contain both "T" and "V" antigens. Fig. 2 presents results of an analysis demonstrating the presence of these antigens in SV-VJA cells after approximately 100 generations of subcultivation. Part A of Fig. 2 illustrates the distribution of "T" antigen and panel B shows the distribution of "V" antigen in the SV-VJA cultures. Our analyses demonstrated that 100% of the SV-VJA cells contain both virus-specific antigens.

VJA cells, recently infected with SV40, display virus specific antigen distribution identical to that illustrated in Fig. 2. Note is made of the presence of the "V" antigen or viral capsid protein in the nucleolus and on the periphery of the nucleus of SV-VJA cells. This distribution is determined by the infected cells and not the infecting virus particles since this virus produces the distribution of "V" antigen generally observed in cells permissive for SV40 maturation. In our experiments, SV40 infected TC7 and AGMK cells produced identical distributions of virus-specific antigens. This observation is illustrated in Fig. 3, which shows the antigen distribution in infected AGMK cells 72 hr after initial exposure to SV40.

The demonstration of SV40 "V" antigen in subcultured SV-VJA cells and in recently infected VJA cells was unexpected. The presence of "V" antigen in cells exposed to wild-type SV40 virus is a characteristic of cells which are producing mature virus particles. Studies outlined above clearly indicated limited SV40 production in infected VJA cells and additional attempts to detect SV40 in SV-VJA cells have been unsuccessful.

The basis of limited SV40 production in infected VJA cells. Experimental results outlined above show that neither recently infected VJA cells nor serially propagated SV-VJA cells support the production of substantial quantities of SV40. The demonstration of both virus-specific "T" and "V" antigens as well as SV40 DNA suggested that maturation of SV40 is blocked in these cells. Additional experiments were carried out to determine at what stage in the reproduction cycle the assembly of infectious SV40 was blocked. A critical event late in the virus maturation cycle involves associ-

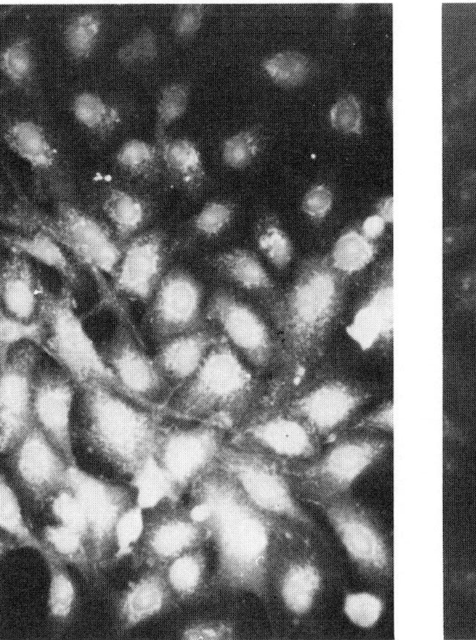

A

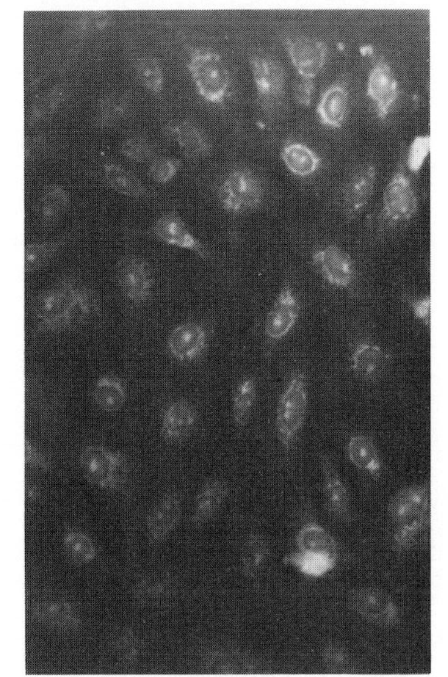

B

Figure 2. Demonstration of SV40-specific antigens in cells surviving infection by SV40. SV-VJA cells, propagated for approximately 100 generations, after initial exposure to SV40, were subcultured on 18mm micro cover slips. These cultures were prepared for fluorescence microscopic analysis as follows. The culture medium was removed by aspiration, the cells washed three times with PBS, fixed with absolute methanol and exposed to fluorescein conjugated antibodies specific for (i) SV40 induced "T" antigen or (ii) the SV40 "V" antigen. Following exposure to fluorescein conjugated antibody, the cultures were washed with PBS and mounted for microscopic examination. Panel A; cells stained with anti "T" antibody. Panel B; cells stained with anti "V" antibody. Microscope image X500.

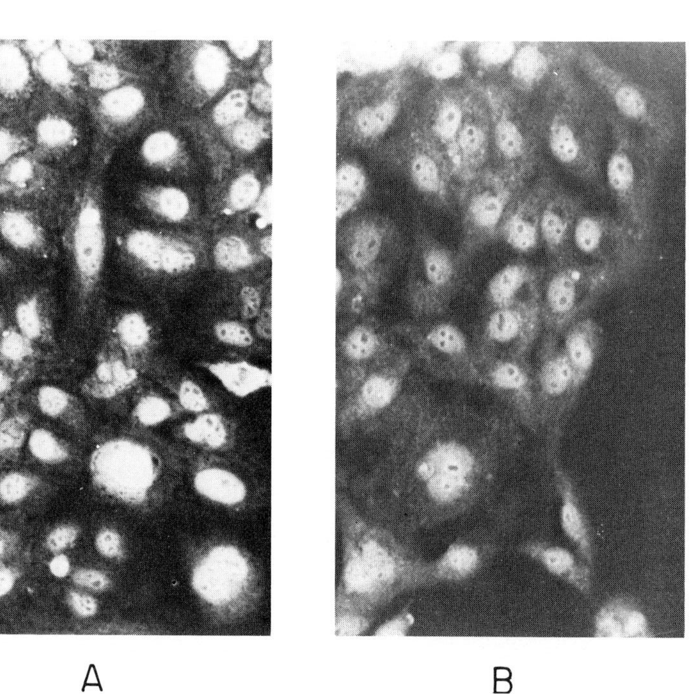

Figure 3. Demonstration of virus-specific antigens in AGMK cells surviving lytic infection by SV40. SVGMK/2 cells were subcultured for approximately 50 generations in DME medium containing 5% calf serum. The cells, subcultured on 18mm micro cover slips, were washed three times with PBS, fixed with methanol and exposed to fluorescein conjugated anti "T" antibody or fluorescein conjugated anti "V" antibody, followed by washing with PBS to remove unadsorbed antibody. The cells were mounted for microscopic examination. Panel A; SVGMK/2 cells stained with anti "T" antibody. Panel B; SVGMK/2 cells stained with anti "V" antibody. Microscope image X500.

ation of progeny closed circular duplex DNA molecules with host cell determined histones with the production of nucleoprotein complexes (7). Formation of these complexes appears to represent a step essential to "packaging" of SV40 DNA. We used procedures suggested by Green et al. (8) to determine whether SV40 nucleoprotein complexes or "minichromosomes" are formed in SV-VJA cells or recently infected VJA cultures. Fig. 4 presents results of one such analysis of infected VJA cells and TC7 cells. As shown in panel A, extraction of infected TC7 cells produces a rapidly sedimenting, 50-55S component (fractions 10-18) which contains a significant amount of ^{3}H-thymidine initially present in the culture medium. The ^{32}P-containing peak (fractions 22-26) is reference, SV40 DNA I. A distribution of incorporated ^{3}H-thymidine in extracts of infected VJA cells (panel B) is equivalent to that observed for TC7 cells. In results not presented we have demonstrated that SV40 minichromosomes or "nucleosomes" (fractions 10-18 in panel A and panel B) are also produced in serially subcultured SV-VJA cultures. These qualitative observations suggest that the block in production of large quantities of SV40 in VJA cells occurs at some as yet undefined stage beyond SV40-nucleoprotein complex formation.

Atypical localization of SV40 "V" antigen and failure of SV40 maturation has been observed in TC7 cells infected with a virus strain, tsB11, which is defective in a late gene function (9). Additional studies conducted on the VJA-SV40 system described here indicate that the failure of more significant virus production in SV40 infected VJA reflects the failure of the cells to concentrate functional "V" antigen in that region of the nucleus where viral assembly occurs.

CONCLUDING REMARKS

The studies outlined in this report show that SV40 infection of VERO cells and VERO-derived cells fails to lead to production of significant amounts of infectious virus. Although the amount of covalently closed SV40 DNA present in infected VERO cells was substantially lower than produced in infected TC7, AGMK or BSC-1 cells, the level of virus isolatable from VERO cells is probably determined by factors other than SV40 DNA I content. A key to understanding limited virus production in VERO cells is probably to be found in the unusual distribution of viral capsid protein in infected VJA cells. Although fluorescein conjugated antibodies directed against SV40 specific "V" antigen were

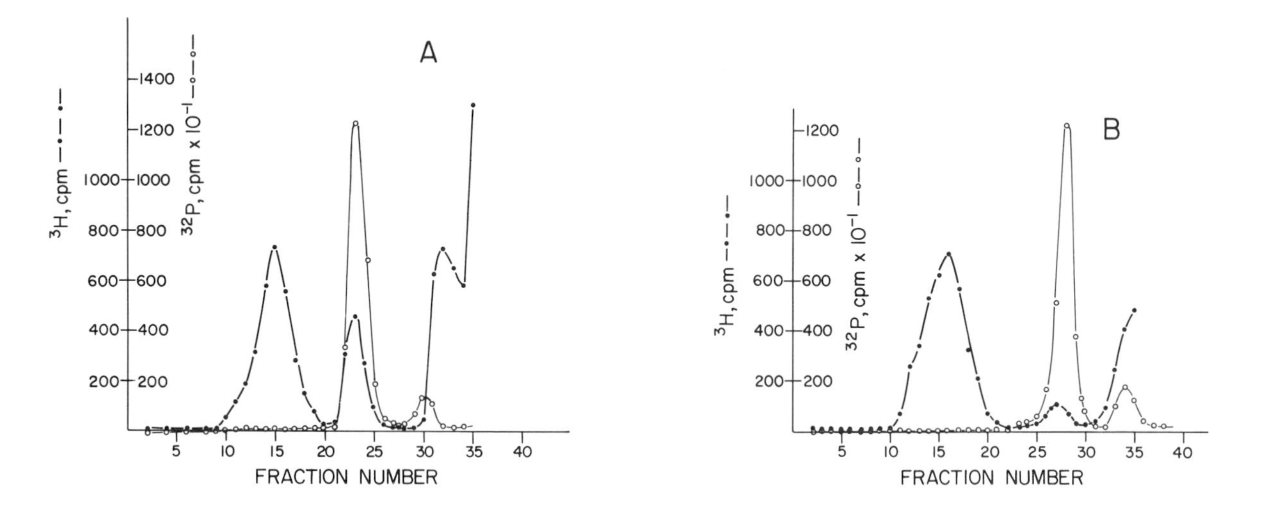

Figure 4. Analysis of SV40 "nucleosomes" isolated from infected TC7 or from infected VJA cells. Nucleosomes--nucleoprotein complexes containing SV40 DNA--were prepared from SV40-infected TC7 or VJA cells as described by Green et. al. (8). A volume of 10ul of ^{32}P-SV40 DNA I, used as a sedimentation reference, was mixed with 0.2ml of the ^{3}H-labeled "nucleosome" preparation and the mixture layered onto 4.8ml of a linear 5-20% neutral sucrose gradient which contained 200mM NaCl, 5mM EDTA and 10mM Tris·HCl, pH 7.8. The 5.0ml gradient was centrifuged for 2.0 hours in a SW 50.1 rotor operated at 6°C and 35,000 rpm in a L565 preparative ultracentrifuge. The gradient was fractionated by collection of seven drop fractions, two drops collected directly onto Whatman 3mm filter paper and five drops collected and stored for further analysis. The radiolabeled samples on filter paper were "fixed" with cold trichloroacetic acid and radioactivity was quantitated by scintillation spectrometry. The direction of sedimentation is from right to left. Panel A, "nucleosome" preparation from TC7 cells and, Panel B, "nucleosome" preparation from infected VJA cells.

observed to stain infected VJA cells the antibody preparation could not reveal structural aspects which render the "V" antigen biologically nonfunctional in participating in virus assembly. We tentatively conclude that the failure of virus assembly in SV40-infected VJA cells reflects the failure of these cells to appropriately concentrate functional viral capsid protein at the site of viral assembly. Experiments in progress are designed to further characterize "V" antigen reactive material in SV-VJA cells and features required to maintain SV40 DNA in a carrier state in these cultures.

ACKNOWLEDGEMENTS

The studies described in this report were supported by grant number 12794 from the National Institutes of Health and by a grant from the Biomedical Research Fund, UCLA.

REFERENCES

1. Tooze, J. (1973) The Molecular Biology of Tumor Viruses. J. Tooze, ed. (New York: Cold Spring Harbor Laboratory), pp. 305-349.
2. Rush, M., Eason, R. and Vinograd, J. (1971) Biochim. Biophys. Acta 228, 585.
3. Hirt, B. (1967) J. Mol. Biol. 26, 365.
4. Radloff, R., Bauer, W. and Vinograd, J. (1967) Proc. Nat. Acad. Sci. USA 57, 1514.
5. Pagano, J.S. (1969) "Assay of infectious DNA," in Fundamental Techniques of Virology. K. Habel and N. Salzman, eds. pp. 184-197.
6. Widmer, C. and Robb, J. (1974) J. Virol. 14, 1530.
7. White, M. and Eason, R. (1973) Nature New Biol. 241, 46.
8. Green, M.D., Miller, H.I. and Hendler, S. (1971) Proc. Nat. Acad. Sci. USA 68, 1032.
9. Tegtmeyer, P., Robb, J., Widmer, C. and Ozer, H. (1974) "Altered protein metabolism in infection by the late tsB11 mutant of simian virus 40," J. Virol. 14, 997-1007.

THE HIS4 FUNGAL GENE CLUSTER IS NOT POLYCISTRONIC

Ramunas Bigelis, Joseph Keesey, and G. R. Fink

Department of Genetics, Development, and Physiology
Cornell University, Ithaca, New York 14853

ABSTRACT. The his4 region of yeast encodes a single 95,000 MW protein which catalyzes the 3rd, 2nd and 10th steps of histidine biosynthesis. Active fragments of the protein result from nonsense or frameshift mutations or from proteolytic cleavage of the 95,000 protein. These results show that his4 is not polycistronic like bacterial operons.

INTRODUCTION

It is not known whether polycistronic messenger RNA's, characteristic of bacteria operons, exist in eucaryotes. Studies of fungal gene clusters, which show several features typical of bacterial operons, including the clustering of genetic information for several steps in a biosynthetic or degradative pathway, and polarity of nonsense and frameshift mutations shed little light on this question. By definition, a polycistronic region encodes a single messenger RNA from which discrete polypeptides are translated (13). The products of fungal gene clusters, however, tend to remain associated after extensive purification (8, 10, 14, 17), although separation of these activities has been reported in some cases (3, 7, 18, 20). Because of these contradictory findings it has remained unclear whether the fungal gene clusters code for: (1) an aggregate of proteins forming a multi-enzyme complex (2) a single protein with several catalytic activities (3) a single protein which is cleaved to form several distinct proteins which remain active. The inability to determine which of these possibilities is correct has caused difficulties in deciding whether the fungal gene cluster is monocistronic or polycistronic.

In this report we show that his4, a typical fungal gene cluster, is monocistronic, specifying a single polypeptide chain of 95,000 molecular weight. This polypeptide is multifunctional and catalyzes the 2nd, 3rd and tenth steps in the pathway of histidine biosynthesis. Proteins smaller than 95k, capable of carrying out one or the other of the reactions result from two features common to the products of many fungal gene clusters: (1) fragments of the protein produced by proteolysis or premature termination in nonsense mutants may be active in the individual reactions; (2) proteolysis occurs even in highly purified preparations. When care is taken to prevent degradation of the protein, it can be shown that the 95,000 protein is the only species present.

RESULTS

The his4 region. The image of the his4 region created by a combined genetic and complementation analysis of nonsense, frameshift, and deletion mutations is consistent with the idea that there are three distinct cistrons, his4A, B, and C (4, 5, 6). Missense mutations defective in the same biochemical reaction map together in the same portion of the his4 region. All mutations in his4A are defective in step 3, all mutations in his4B are defective in step 2, and all mutations in his4C are defective in step 10. Missense mutations in any one of these regions complement missense mutations in either of the other two regions. These genetic and physiological data suggest that the his4 region is comprised of three independently functioning units (Figure 1).

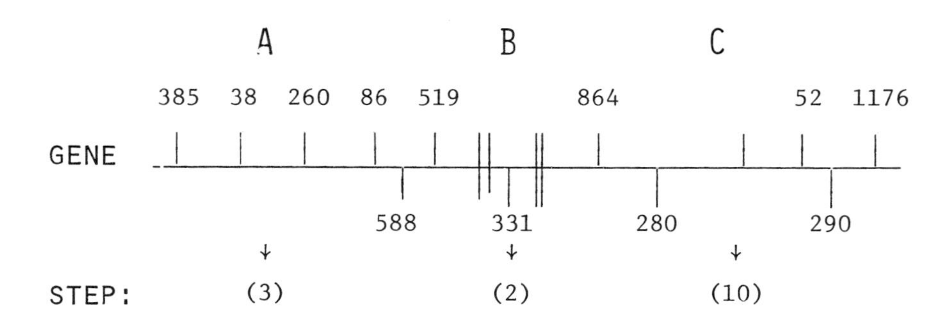

Fig. 1. Mutations at the his4 region. The line after "GENE" represents the his4 region and the numbers above and below represent a few of the sites of mutation which have been ordered within this region by genetic analysis. The genetic map has been sub-divided into the A, B, and C regions which encode respectively the third, second and tenth steps in the pathway of histidine biosynthesis (as shown under "STEP"). The mutations below the line are missense and those above the line are frameshifts (-38, -519) and nonsense: (UAG-385, UGA-260, UAA-86, -864, -52, -1176).

This picture of his4 is supported by studies on nonsense and frameshift mutations. These types of mutations show polarity from his4A → his4C (4, 6, 18): nonsense mutations in his4A are missing his4A, -B, and -C, those in his4B are missing his4B and -C and those in his4C are missing his4C. The fact that a nonsense or frameshift in his4C complements a his4A or -B mutation emphasizes the independent functioning of these regions. The complete polarity of the his4A nonsense and frameshift mutations suggests that the whole region is a single unit of transcription from A to C.

Recent work in our laboratory shows that the his4 gene specifies a single polypeptide chain of 95,000 MW (95k) capable of carrying out all three enzymatic activities (his4A, his4B, and his4C) encoded by his4. Two key tools which allowed us to discover the structure of the his4 gene product were affinity chromatography on AMP-Sepharose and a highly specific antibody for the his4 gene product. Experiments leading to the conclusion that the 95k protein is the sole protein responsible for all three his4 activities are described below:

(1) His4A, -B, and -C activities are associated with a single 95k protein after chromatography first on AMP-Sepharose and then on DEAE cellulose. These two steps effect purification of all three activities in high yield.

(2) The 95k protein is missing in strains carrying nonsense and frameshift mutations. Extracts of strains carrying UAG, UAA, and UGA mutations have been examined on SDS-polyacrylamide gels after chromatography on AMP-Sepharose. Many different strains carrying nonsense and frameshift mutations in various portions of the his4 gene have been studied and none has the 95k protein. Several missense mutations in his4 were studied in parallel and all were found to have the 95k protein (Figure 2).

(3) All three his4 activities in crude extracts are precipitated by antibody against the 95k protein. When the immunoprecipitate is redissolved in SDS and subjected to electrophoresis on 10% SDS-polyacrylamide gels, a single 95k protein is found (Figure 2).

(4) Immunoprecipitation of crude extracts reveals a single 95k protein in wild type and strains carrying missense mutations in his4A, his4B, and his4C. Extracts of strains carrying nonsense or frameshift mutations in his4A do not form a detectable immunoprecipitate nor do they give any reaction in Ochterlony double-diffusion (Figure 3).

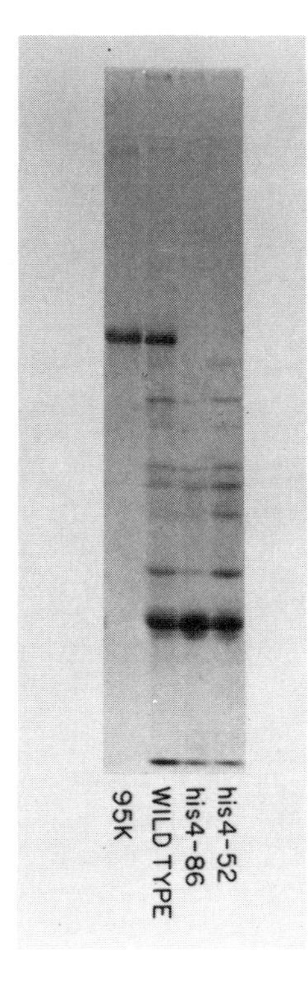

Fig. 2. A comparison of proteins from wild type and nonsense mutants on SDS gels. Crude extracts prepared by breaking yeast in the Braun homogenizer were concentrated by the addition of ammonium sulfate to 70% saturation. The concentrated protein was resuspended, dialyzed overnight, and then added to a column of N^6-aminohexyl AMP-Sepharose. All three his4 activities eluted together in high yield from the column in the 5mM AMP wash (9). Active fractions from the nonsense mutants (which are devoid of all three activities) were combined, concentrated, and prepared for electrophoresis on SDS gels. Shown here is a 10% SDS-polyacrylamide slab gel of the proteins eluted from 3 AMP-Sepharose columns. Extracts were from wild type, his4-86 (UAA), and his4-52 (UAA). In addition, pure 95k protein (prepared by chromatography of the wild type on DEAE-cellulose after the AMP-Sepharose chromatography) was run on the same gel to show the position of the 95k protein specified by his4.

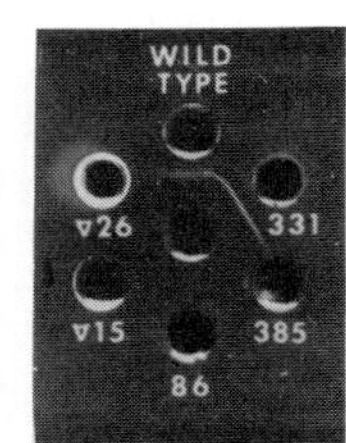

Fig. 3. Antigenic properties of wild type and altered his4 proteins. The immunological cross-reactivity of crude extracts from a variety of strains was tested by Ochterlony double diffusion. On the top, antibody prepared against pure 95k protein can be seen to interact with protein from a crude extract of wild type and a his4B missense mutant (331) but not with nonsense mutants 385 (UAG), 86 (UAA) or deletions 15 and 26. On the bottom the interaction of purified 60k and 95k proteins with antibody prepared against the 95k protein is shown.

(5) The 95k protein is the only immunologically active protein in crude extracts. Proteins from wild type and a strain carrying *his4A* nonsense mutation (*his4-385*) were labeled with ^{35}S. Extracts of each strain were prepared by breaking the cells in a Braun homogenizer. The extracts were then centrifuged and pure 95k protein was added to each extract. Antibody was added to the extracts and the immunoprecipitate from each was solublized and subjected to SDS-polyacrylamide electrophoresis. A single radioactive protein of 95k was found in wild type and none was found in nonsense mutant *his4-385*.

(6) Nonsense mutant *his4-864* is missing the normal 95k protein found in wild type and has instead a protein of 45k. *His4-864* is a UAA mutation which maps in the middle of the *his4* gene. Sucrose gradient centrifugation shows that the residual *his4A* and *his4B* activities in strains carrying *his4-864* sediment with a molecular weight of 45k (18, 19). The protein formed in this nonsense mutant is immunologically active and cross-reacts with antibody to the 95k protein (9). Analysis of the immunoprecipitate on SDS gels shows a single 45k protein. The presence of the 45k protein and the absence of the 95k protein in this nonsense mutant rules out the possibility that the region specifies three different 95k proteins.

Altered forms of the protein. Previous attempts to purify the three activities by conventional procedures resulted in the loss of *his4A* (cyclohydrolase) and *his4B* (pyrophosphorylase) activities. Concomitant with the loss of the *his4A* and *his4B* activities, the molecular weight of the *his4C* (histidinol dehydrogenase) activity shifted to 60,000. Had we separated one of the polypeptides of a *his4* multi-enzyme complex or had we isolated a degraded but still active fragment of a larger polypeptide chain? From the experiments described below we conclude that the 60k protein is an active fragment produced by proteolytic degradation of the 95k protein.

We were able to visualize the two forms of the enzyme by utilizing an activity stain specific for histidinol dehydrogenase (*his4C*) after electrophoresis of partially purified extracts on a non-dissociating polyacrylamide gel (11). When the protein was purified by chromatography on valine-Sepharose and DEAE-cellulose we observed a single, slow-moving band of *his4C* activity in this gel system (Form I). When the extract was incubated for a week at 4°C, a faster migrating form (Form II) begins to appear (Figure 4). The generation of Form II and the disappearance of Form I with time suggested that proteolysis might be responsible for the conversion.

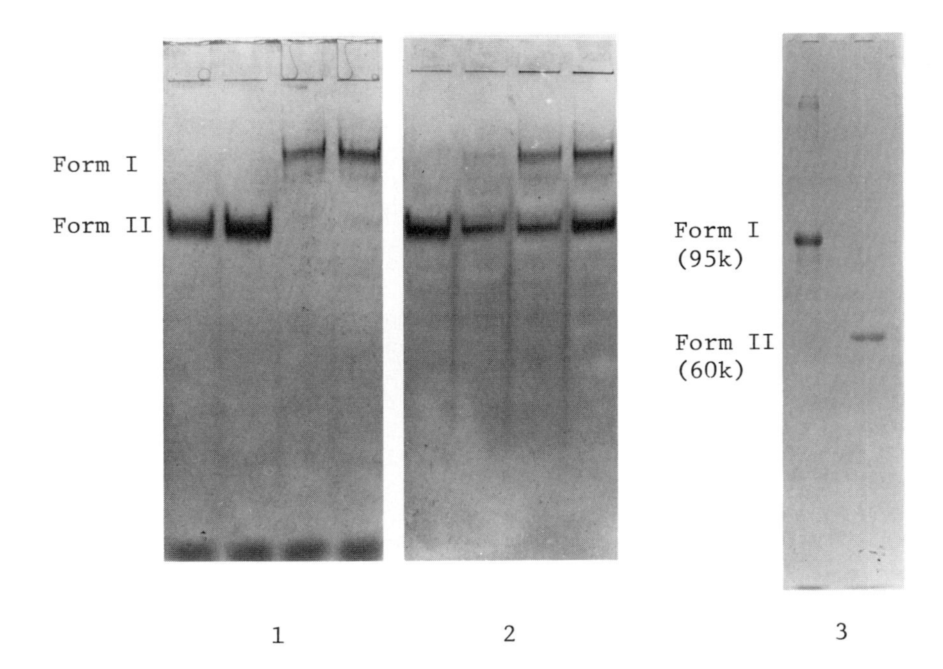

Fig. 4. Two forms of histidinol dehydrogenase. The two gels on the left (1 and 2) are non-dissociating 7½% polyacrylamide gels run in a Tris-borate buffer system (12). They were stained to display histidinol dehydrogenase activity (11). Shown are the four products of a meiotic tetrad resulting from the cross: high proteinase mutant X wild type. Extracts were purified by chromatography first on valine-Sepharose and then on DEAE-cellulose prior to electrophoresis. Gel 1 shows the mobility of histidinol dehydrogenase from the four ascosporal clones of a tetrad one week after purification. The first two tracks on the left show histidinol dehydrogenase from the spores with high proteinase activity and the next two show histidinol dehydrogenase from the two sister, wild type spores. On gel 2, extracts of the same four spores have been run after aging the extracts for an additional week at 4°C. The two forms of histidinol dehydrogenase were sliced out and eluted from the activity stained gels. The proteins resulting from this treatment were then subjected to electrophoresis on the 10% SDS acrylamide gel, labeled 3. Calibration of gel 3 with standards of known molecular weight revealed that Form I is 95,000 MW and Form II is 60,000 MW.

Analysis of a yeast mutant with high levels of proteinases provided important insights into the conversion of Form I to Form II. Recently, a mutation was found (1) which causes high levels of proteinases A and B (16). The abnormally high levels of proteinases appear to result from the absence of the proteinase A inhibitor. Extracts of strains carrying this mutation are remarkable in that most of the histidinol dehydrogenase is in Form II (Figure 3). Moreover, when protein from the high proteinase strain is added to a freshly purified sample of the histidinol dehydrogenase from wild type, Form I from the wild type extract is rapidly converted to Form II. These experiments show that Form II is associated with high proteolytic activity. Moreover, large quantities of Form II can be obtained by addition of high proteinase extracts to the protein from wild type.

Both Form I and Form II were purified to homogeneity by electrophoretic elution of the protein from the activity-stained gels. After elution of the proteins from SDS gels, both Form I and Form II were re-purified to electrophoretic homogeneity on SDS polyacrylamide gels. Using marker proteins of known molecular weight we could show that Form I had a molecular weight of 95,000 and Form II had a molecular weight of 60,000. Form I and the 95k *his4* protein (containing all three activities) purified by AMP-Sepharose chromatography have the same migration upon co-electrophoresis on an SDS gel. Antibody made against the AMP-Sepharose enzyme immunoprecipitates Form I. Moreover, the purified AMP-Sepharose enzyme and Form I show a line of identity in Ochterlony double diffusion. Form II also interacts with the antibody to the 95k protein (Figure 4) showing that the two proteins are related. Thus, the 60k protein appears to be a degraded fragment of the 95k protein.

DISCUSSION

Our results show that the *his4* region of yeast encodes a single protein of 95,000 molecular weight which carries out three enzymatic activities. Fragments of this protein produced by mutation or by proteolysis can carry out some of the reactions normally carried out by the intact 95k polypeptide. Nonsense mutant *his4-864* produces a 45k fragment with *his4A* and *his4B* enzymatic activities. Proteolysis produces a 60k fragment with *his4C* activity. These enzymatically active fragments are immunologically active and can be immunoprecipitated. Yet, none of these lower molecular weight species are present in crude extracts of log-phase wild type cells. All three catalytic activities for the his4 reactions reside solely in the 95k protein.

The his4 region is not polycistronic like a bacterial operon. The region encodes a single, multifunctional protein. Many fungal gene clusters have properties similar to his4. Though the evidence is by no means complete, indications so far are that each of these systems, like his4, specifies a single, multifunctional polypeptide chain. The absence of polycistronic regions in fungi could be related to the inability of eucaryotic translation systems to initiate at internal start (AUG) signals. In agreement with these kinetic experiments (15) show that most yeast messages are monocistronic. Work on the CYC1 gene (21) has shown that the yeast translation system is incapable of reinitiating after the sequence AUG...UAA AUG. In bacteria this same sequence permits reinitiation after the second AUG. Although the exact mechanism responsible for this difference between eucaryotes and procaryotes is unclear at the present time, one possibility is that eucaryotic ribosomes actually dissociate from the message when they encounter termination signals and can only re-enter at the 5' end. Perhaps the modified 5' end is a requirement for effective entry of the ribosome onto the messenger. If the ribosome encounters a termination signal, it may have to return to the modified 5' end in order to reinitiate even though there are potential AUG start signals downstream in the message.

ACKNOWLEDGEMENTS

These studies were supported by grants GM15408 and GM01035 to Gerald R. Fink. We wish to thank Debi Ferguson and Jim Hicks for help in preparation of the manuscript.

REFERENCES

1. Beck, I., Fink, G. R., and Wolf, D. H. (1976) *Proc. 10th Int. Congress Biochem., Hamburg*, p. 420.
2. Carsiotis, M., Appella, E., Provost, P., Germershausen, J., and Suskind, S. (1965) *Biochem. Biophys. Res. Comm.* 18, 877.
3. Case, M. E. and Giles, N. (1971) *Proc. Nat. Acad. Sci.* 68, 58.
4. Culbertson, M., Charnas, L., Johnson, M. T., and Fink, G. R. (1977) *In preparation.*
5. Fink, G. R. (1965) *Genetics* 53, 445.
6. Fink, G. R. and Styles, C. (1974) *Genetics* 77, 231.
7. Gaertner, F. and DeMoss, J. (1969) *J. Biol. Chem.* 244, 2716.
8. Hulett, F. M. and DeMoss, J. (1975) *J. Biol. Chem.* 250, 6648.

9. Keesey, J., Bigelis, R., and Fink, G. R. (1977) *In preparation.*
10. Kirschner, K. and Bisswanger, H. (1976) *Ann. Rev. Biochem.* 45, 143.
11. Loper, J. and Adams, E. (1965) *J. Biol. Chem.* 240, 788.
12. MacIntyre, R. J. (1971) *Biochemical Genetics* 5, 45.
13. Martin, R. G. (1969) *Ann. Rev. Genet.* 3, 181.
14. Matchett, W. H. and DeMoss, J. (1975) *J. Biol. Chem.* 256, 2941.
15. Peterson, N. S. and McLaughlin, C. (1973) *J. Mol. Biol.* 81, 33.
16. Saheki, T. and Holzer, H. (1974) *Eur. J. Biochem.* 42, 621.
17. Schweitzer, E., Kniep, B., Castorph, H., and Holtzner, U. (1973) *Eur. J. Biochem.* 39, 353.
18. Shaffer, B., Rytka, J. and Fink, G. R. (1969) *Proc. Nat. Acad. Sci.* 63, 1198.
19. Shaffer, B., Brearly, I., Littlewood, R., and Fink, G. R. (1971) *Genetics* 67, 483.
20. Shaffer, B., Edelstein, S. and Fink, G. R. (1972) *Genetics* 67, 483.
21. Sherman, F. and Stewart, J. W. (1975) *Proc. 10th FEBS Meeting, North-Holland/American Elsevier, N.Y.* 38, 175.

The Yeast Workshop

E. P. Geiduschek
Department of Biology
University of California at San Diego
La Jolla, California

The workshop was too brief to sample more than a fraction of all the aspects of eucaryotic molecular biology to which current work with saccharamyces yeasts contributes importantly. Nevertheless, this brief session held at least its share of new ideas, of intriguing observations and of detailed analysis of significant features of genome organization.

The five talks at the session were presented (in order of appearance) by I. Herskowitz, J. R. Warner, J. S. Beckmann, J. H. Cramer and T. Petes. P. Philippsen, who also presented work in this area in other workshops, contributed briefly.

Herskowitz talked about the remarkable genetics of cell (mating) type in saccharomyces. Each mating type (*a* or *α*) is determined by a small number of regulator genes (at least two for *α*) inserted at the mating type locus, on chromosome III of *S. cerevisiae*, which can be in one of two states, *a or α*. Haploid cells of either mating type can switch to the other type at characteristic, low frequencies in heterothallic (ho) strains. In homothallic (HO) strains these transitions occur at high frequencies, following rules that have been inferred from pedigree analysis. A novel mechanism for mating type transition was proposed. The new hypothesis rests on genetic analyses of mating type switching in homothallic strains, the results of which are themselves most striking. The hypothesis, promulgated as the "cassette" model, involves *silent* copies of both the *a* and *α* regulatory genes that are "stored" at loci HM_a and HM_α near the opposite ends of chromosome III. The stored copies of the regulatory genes can be activated by insertion at the mating type locus. When such insertion replaces one set of regulatory genes with the other set, the mating type also changes. In heterothallic (ho) strains, replacement of regulatory genes at the mating type locus occurs at a low frequency and the mating type is relatively stable. Thus, the "cassette" model of sex in yeast has these important implications: 1) that mating types are determined by alternate sets of regulatory genes; 2) that these must be inserted at a specific chromosomal site to be active;

3) that their rate of transposition is genetically determined, presumably through the activities of transpositional enzymes.

Warner discussed the coordinate regulation of the synthesis of ribosomal proteins in S. cerevisiae. The transcriptional basis of this coordinacy has been established by use of an appropriate cell-free protein synthesizing system. Particular interest attaches to this coordinate regulation because it involves a large number of widely dispersed genes. The production of mRNA for the ribosomal proteins is transiently and selectively sensitive to temperature shift-up even in wild type S. cerevisiae. The regulatory mechanisms coordinating the expression of these genes should be of great interest.

The remaining talks dealt with cloning as a means of analyzing the disposition and structure of genes for the stable RNAs of S. cerevisiae.

Beckmann discussed the yeast tRNAs cloned into the tet^r gene of the E. coli plasmid pBR313. Two lines of evidence suggest that the tRNA genes are widely dispersed (as the suppressor tRNA genes are already known to be): 1) the frequency of isolation of tRNA gene clones relative to all yeast DNA-containing clones; 2) the testing of tRNA-containing clones with individual purified yeast tRNAS. The presence of small clusters of tRNA genes is not excluded. If the transcription units for tRNA genes are small, then the yeast tRNA clones are likely to provide several avenues for analyzing regulation and discrimination in the synthesis of eucaryotic RNA.

Cramer, Petes and Philippsen described restriction enzyme maps of ribosomal transcription units. The stable transcripts (25, 17, 5.8 and 5S RNA) have now been accurately mapped onto these units, which are evidently homogeneous in length and highly clustered (see below). Neither S. cerevisiae nor the closely related S. carlsbergensis has heterogeneous spacers between ribosomal genes. Despite strong implications of homogeneity in the ribosomal genes of individual strains of S. cerevisiae, different haploid strains of S. cerevisiae evidently differ somewhat in the nucleotide sequences of their untranscribed "spacer" regions. Petes has utilized this polymorphism to analyze the segregation of ribosomal cistrons in sporulation. He concludes that all the ribosomal cistrons in normal haploids are located in one linkage group. More tentatively, there is a suggestion that the segregation and propagation of recombinant arrays of ribosomal genes in haploids is strongly depressed (for example, because of reduced combination frequency).

The sequencing of the ribosomal transcription units is already well under way. In one of the principal talks, Gilbert briefly presented the sequence of approximately 160 nucleotide pairs adjacent to both ends of the 5S RNA gene. The nucleotide sequences of other regions of the ribosomal transcription units are being analyzed in several laboratories.

MATING TYPE INTERCONVERSION IN YEAST AND ITS RELATIONSHIP TO DEVELOPMENT IN HIGHER EUCARYOTES

Ira Herskowitz, Jeffrey N. Strathern, James B. Hicks,* and Jasper Rine

Department of Biology and Institute of Molecular Biology, University of Oregon, Eugene, Oregon 97403

ABSTRACT. In this paper we discuss studies on the regulation and interconversion of cell type (mating type) in the lower eucaryote, *Saccharomyces cerevisiae*, which lead us to propose the cassette hypothesis, in which determination of cell type involves genetic rearrangement. Our studies with yeast strains which rapidly interconvert between the two cell types also reveal a specific pattern of interconversion. We discuss this pattern in the context of clonal development and, with an extension of the cassette hypothesis, generalize it to account for some aspects of development in multicellular organisms.

INTRODUCTION

Development of a multicellular organism involves the generation of diverse cell types in a clone of cells. A focus of molecular studies of development is to provide mechanisms responsible for cellular differentiation and for the timing of this differentiation. At least three different molecular elements can be invoked to play a role in production of stably differentiated cells--regulatory proteins, DNA modification, and DNA rearrangement. Circuits involving repressors and activators have ample biological precedent in procaryotes (1,2,3,4). The hypothesis that cell type is stably altered by methylation or other modifications of DNA is discussed by Holliday and Pugh (5). In this paper we discuss studies on the regulation and interconversion of cell type in the lower eucaryote, *Saccharomyces cerevisiae*, which lead us to propose the cassette hypothesis, in which determination of cell type involves genetic rearrangement. Our studies with yeast strains which rapidly interconvert between the two cell types also reveal a specific pattern of interconversion. We discuss this pattern in the context of clonal development and, with an extension of the cassette hypothesis, generalize it to account for some aspects of development in multicellular organisms.

*Present address: Department of Genetics, Development and Physiology, Cornell University, Ithaca, New York 14850

BACKGROUND AND DEFINITIONS

The yeast S. cerevisiae exists as either a haploid or a diploid cell (summarized in ref. 6). Haploid cells are of two types, mating type a or mating type α, determined by alleles of the mating type locus, MATa and MATα, located on chromosome III. Cells of one mating type mate efficiently with cells of the other mating type to form diploids. In heterothallic yeast the mating type of a cell is stable through many generations, although matings between cells of like mating type do occur at low frequency (7,8). a/α diploids can be induced to undergo meiosis and sporulation, producing an ascus with four haploid spores, two of mating type a and two of mating type α. MATa and MATα are codominant, as a/α diploids have properties not associated with either single mating type--they sporulate and do not mate, unlike a/a and α/α cells which behave as their respective haploids. We view the MATa and MATα alleles as being non-homologous blocs of DNA. Support for this view comes from analysis of non-mating mutants with defects in the MATα locus (9,10,11) which indicates that MATα codes for at least two functions not present in MATa (12).

Some strains of yeast have a life cycle with a distinct difference from that of heterothallic cells. In these homothallic strains, a single haploid spore of either mating type can give rise to a colony containing diploid cells capable of sporulation. The "standard" homothallic S. cerevisiae strain (13) differs from standard heterothallic strains by a single determinant--homothallic strains carry the dominant allele HO and heterothallic strains the recessive allele ho (14,15). Two other genes involved in homothallism, HMa and HMα, are discussed below. Studies from a number of laboratories (16, 17,8) provide an explanation for ability of homothallic cells to produce sporulating diploids: a homothallic cell is able to switch mating type at high frequency; diploid cells are produced by mating between siblings of opposite mating type. The HO gene is necessary for the high frequency of switching between mating types but its presence is not required for the maintenance of the new mating type (17,8). Thus, when an HO cell switches from α to a mating type, the HO gene can be removed by genetic crosses with an α ho strain to produce a heterothallic (ho) a cell. The determinant for mating type of this a ho cell is allelic with the mating type locus of ordinary heterothallic strains and is indistinguishable from the a mating type locus in heterothallic strains.

The ability of yeast strains to interconvert mating types, either at low frequency in heterothallic strains or at high frequency in homothallic strains, indicates that all cells contain the information to be a and α, but that only one or the other mating type is expressed. The HO gene promotes a change at the mating type locus which is stable after removal of HO.

HOMOTHALLIC INTERCONVERSION OF YEAST MATING TYPES

One can follow the process by which mating type interconversion occurs by taking advantage of two characteristics of yeast cells: (i) a cells can be distinguished from α cells by their sensitivity to the pheromone α-factor--in the presence of α-factor, a cells but not α cells are arrested in the G1 phase of the cell division cycle and undergo a characteristic shape change (18,19); (ii) reproduction occurs by budding, allowing a parent cell to be distinguished from its daughter. Using these characteristics, one can start with a single cell, separate mother from daughter cell by micromanipulation, and produce a pedigree of the fates of the progeny derived from the original cell (8,15). (See Figure 1.)

Homothallic asci contain two spores sensitive to α-factor (a cells) and two spores insensitive to α-factor (α cells). After one cell division cycle, α cells give rise to two cells, S (the original spore cell) and D1 (its daughter), both of which are always insensitive to α-factor and which can mate readily with a cells. When S produces its second daughter, D2, an interesting change is often observed: both S and D2 are now able to respond to α-factor and are capable of mating with α cells. These cells have switched to the a mating type. In contrast, the D1 cell and its first daughter, D1-1, invariably remain as α cells. In clones at the four-cell stage, a switch is observed in 86% of the cases for the strain we have studied most extensively (8,12). The mating type switch always occurs in pairs of cells and always in cells from the older half of the pedigree, i.e., in S and D2 as opposed to D1 and D1-1.

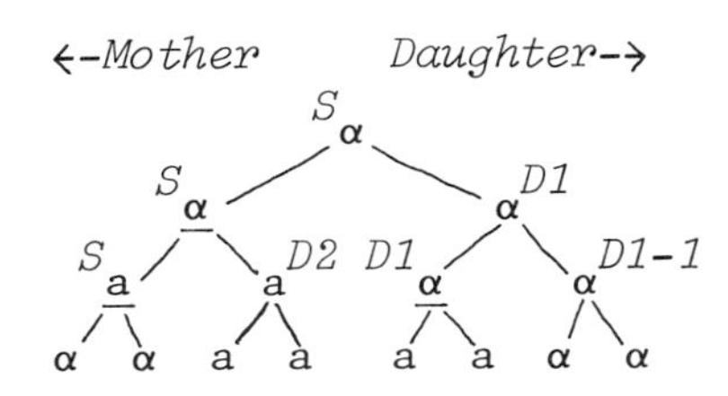

Underlined cells are competent to switch.
Figure 1.

Cells which have sustained a change in mating type are capable of subsequent changes in mating type. Indeed, when the S cell at the four-cell stage (now of a mating type) buds again, most of the time the two resulting cells are now of the α mating type (12). The rules which govern the pattern of switching are derived from noting the ability of cells at the four-cell stage to give rise to cells of switched mating type in their next cell division cycle: the S cell, as already noted, and D1 often give rise to switched cells; D2 and D1-1 never give rise to switched cells. From this analysis, we can form two general rules:

1. Mating type switches result in pairs of switched cells. The older cell at any given stage in the pedigree need not switch mating type every cell division but when a switch occurs it is always seen in both cells from a given cell division. Since our assay for a change in mating type (a change in sensitivity to $\underline{\alpha}$-factor) requires that a cell be in G1, the event responsible for changing mating type must occur sometime after G1 of one cell division cycle and before the next G1. Because the switch causes a stable alteration of the mating type locus and is observed in pairs of cells, we propose that the switching event affects the DNA at or before the time of DNA replication of the mating type locus.

2. Cells differ in their capability to switch: Inexperienced cells, those which have not previously undergone a cell division cycle, are not capable of giving rise to switched cells. Experienced cells, those which have completed at least one cell division cycle, are capable of giving rise to switched cells and do so at a frequency of greater than 50%. Consideration of cases in which switching has not occurred at the four-cell stage raises an important, but as yet unresolved, question. Failure of the S cell at the two-cell stage to produce cells of different mating type in its next cell division cycle may be due either to "chastity or impotence" (20). We cannot yet distinguish between the possibility that the S cell is competent to switch but does not, or that the S cell is not competent to switch. Whatever the reason, it is clear that the products of one cell division cycle, the mother and daughter cells, differ from each other in their "developmental potential", that is, in their ability to switch mating types. The asymmetry may be caused by any of a number of differences between mother and daughter cells--for example, differences in a cytoplasmic component which is localized in mother cells or perhaps differences in modification in the DNA strands of mother and daughter cells (see ref. 5).

THE CASSETTE HYPOTHESIS

In this section, we shall briefly describe the evidence which leads us to propose that determination of cell type, i.e., mating type, occurs by insertion of information into the mating type locus. We propose that yeast cells contain unexpressed copies of MATa and MATα information and that this information is activated by insertion of this information (or a copy) into the mating type locus.

We are led to propose this hypothesis from the following considerations (additional discussion can be found in refs. 8 and 21):

1. As noted earlier, the existence of mating type interconversion indicates that cells contain the information to be both a and α.

2. The mating type of a strain is stable through growth and through meiosis.

3. The mating type locus of a strain with a defective MATα locus (matα⁻) can be converted to a functional MATa locus and subsequently to a functional MATα locus. The recovery of functional MATα has been observed in strains carrying four independent matα⁻ mutants, one of which carries an ochre mutation in MATα (21,22,12).

How can MATα function be restored to a strain with a defect in the α mating type locus? In earlier work with phage λ, we had shown that λ mutants doubly defective in genes S and R could acquire these functions by recombination from a previously unrecognized defective prophage (23). We thus interpreted the recovery of MATα function as indicating the existence of silent MATα information elsewhere in the yeast genome.

The elegant studies of Harashima, Nogi, and Oshima (24) and of Naumov and Tolstorukov (25) on the genes involved in homothallism led us to propose that genes HMa and HMα correspond to non-expressed copies of MATα and MATa respectively. By a series of crosses involving "cerevisioid" strains such as S. norbensis (26), it was shown that two genes in addition to HO are required for mating type interconversion. HMa is required for conversion from a to α, and HMα is required for conversion from α to a. Thus, strains a HO HMa hmα and α HO hma HMα switch mating type, but a HO hma HMα and α HO HMa hmα do not. The observation that both HO HMa HMα and HO hma hmα switch mating types regardless of the original mating type suggests that hma is not simply the absence of HMa function and that hmα is not simply the absence of HMα function. Instead, we propose (21,22) that the HMa and hmα loci are silent copies of MATα [denoted as "(MATα)"] and that HMα and hma loci are silent copies of MATa ["(MATa)"]. Thus both HO HMa HMα and HO hma hmα contain both (MATa) and (MATα). Strains such as α HO HMa hmα, which do not switch mating type, correspond to α HO (MATα) (MATα): We propose that these strains appear not to switch mating types because they do not carry the MATa information.

In summary, we propose that the HO gene promotes mating type interconversion by stimulating the insertion of copies of (MATa) or (MATα) into the mating type locus. When these cassettes are inserted into the mating type locus, they are expressed because they are adjacent to a functional promoter (see Figure 2). We note that the insertion process must be a non-reciprocal exchange, as conversion from α to a does not affect the HMa and HMα loci.

Figure 2

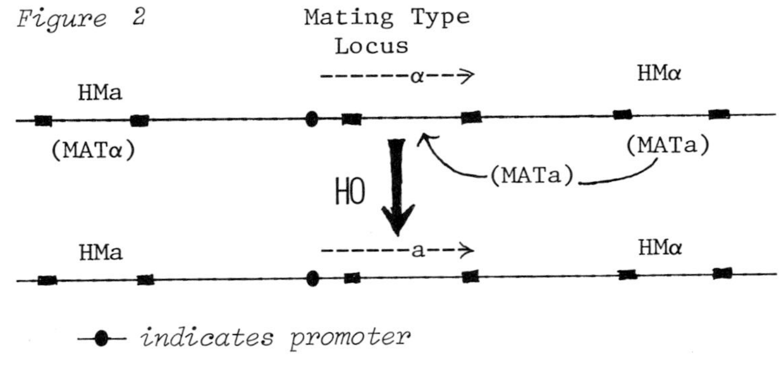

indicates promoter

PREDICTIONS OF THE MODEL

The role attributed to HMa and HMα genes provides an explanation for a puzzling observation by Hawthorne (7). He noted that heterothallic strains are capable of switching from α to a by a deletion extending from the mating type locus rightwards past THR4 and ending before the MAL2 gene. It has recently been shown (27) that the HMα locus resides on chromosome III, distal to THR4 and in the vicinity of MAL2. We propose that Hawthorne's deletion links an essential site (for example, a promoter) of the mating type locus with the silent MATa information at HMα. This explanation requires that HMα be located just distal to the endpoint of the deletion.

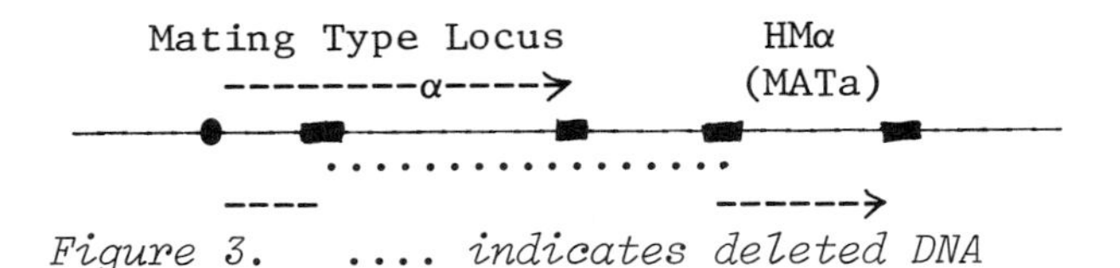

Figure 3. .... indicates deleted DNA

Mapping experiments confirm this prediction (12). A further prediction from our hypothesis for the nature of HMa and HMα is that heterothallic strains which are α ho HMa hmα cannot switch to mating type a by a "Hawthorne deletion", but that an a ho HMa hmα strain might be able to switch to α by a deletion analogous to the Hawthorne deletion. These tests are in progress. A decisive test of the cassette model requires a mutation, preferably a unique mutation such as an ochre mutation, in HMa. One could then ask whether in the presence of HO the ochre mutation is inserted into the mating type locus to produce a strain which is matα-ochre. Finally, recent advances in isolation of yeast DNA sequences should allow a direct test of the identity of HMa and MATα, etc. by DNA heteroduplex analysis.

WHY ARE THE SILENT COPIES SILENT?

We proposed above that adjacent to the mating type locus is a promoter which controls transcription of the genetic information next to it. There are, of course, other mechanisms by which transposition of information to the mating type locus can lead to its expression. For example, both MATα and MATa blocs may contain their own promoter as part of the translocatable element but be turned off by a repressor acting on an operator adjacent to the HMa and HMα loci. In this case, movement of information into the mating type locus would separate the cassette from its regulatory site. This model predicts the existence of mutations in the repressor which allow expression of the information at the silent loci. Such mutations could allow the expression of α functions in an a cell and the expression of a functions in an α cell. Furthermore, such mutations might suppress the defects associated with mutations in the α mating type locus.

A candidate for a mutation allowing at least partial expression of information at the silent loci has been found. The mutation (ssp515) was originally isolated as a suppressor of the mating defect of strains carrying a mutation in MATα (28). The suppressor is unlinked to the mating type locus, is recessive, and suppresses the defects of all known matα$^-$ mutants (28,12). Furthermore, a/a diploids homozygous for the suppressor are able to sporulate efficiently to produce four a spores (11). Our interpretation is that the suppressor in some manner supplies the functions required for sporulation that are normally supplied by MATα. matα$^-$/matα$^-$ diploids homozygous for this suppressor also sporulate efficiently, and each ascus contains four α cells. The suppressor thus may also be providing the functions required for sporulation normally provided by MATa. These observations are consistent with the suppressor being a mutated repressor of the silent loci. A strong prediction of such a model is that matα$^-$ mutations should not be suppressed in hma HMα strains since there should be no silent MATα information.

DEVELOPMENTAL ANALOGIES

Cell pedigrees. The rules for homothallic interconversion of mating type described above assure that cells of the original phenotype will exist in each generation of the growing clone. These features are precisely the requirements necessary to establish and maintain a stem cell population. Namely, division of the original cell results in one cell with the same potential as in the previous generation (the stem cell) and one cell with an altered developmental potential. It is interesting to note that daughter cells form the stem cell line rather than mother cells, in that yeast cells are capable of only a fixed number of cell divisions due to accumulation of bud scars (29). The yeast pedigrees take on more familiar stem cell line characteristics under conditions where homothallic switching is not reversible; for example, (i) $\underline{\alpha}$ to $\underline{a}$ conversions in the presence of $\underline{\alpha}$-factor, so that the $\underline{a}$ phenotype is terminal (see Fig. 1, cells S, D2, D1, and D1-2), or (ii) $\underline{\alpha}$ to $\underline{a}$ conversions in *HO hma* $\underline{\alpha}$ *HMα* cells, so that the $\underline{a}$ cells produced can proliferate but have a stable phenotype.

The generation of more than one cell type from a single cell requires divisions which are asymmetric either in terms of the environment into which the daughter cells are segregated or in terms of restrictions placed upon the potential of the cell. *S. cerevisiae* does not experience the variety of cell contacts and micro-environments which occur during development of multicellular organisms and which have been postulated to account in part for differentiation. The asymmetry demonstrated between experienced and inexperienced cells must be a reflection of an intrinsic difference between these cells, for example, in a cytoplasmic component, a nuclear regulator, or a modification of the genome. The pattern of divisions observed in yeast is not sufficient to account for all cellular pedigrees. In order to generate clones of determined size the stem cell must lose its potential to divide. This restriction of the stem cell could be accomplished by an alteration of the cellular environment or an autonomous cellular clock such as exhibited by "mortal" yeast cells or by a genome modification mechanism of the kind proposed by Holliday and Pugh (5).

Sequential cassette insertion. Development in multicellular organisms involves a series of cellular transformations occurring during the production of a clone of cells. In contrast, a clone of homothallic yeast cells contains only two different types of cells. We shall present an extension of the cassette hypothesis to account for the production of multiple cell types.

An interesting result of the quantitative analysis of mating type interconversion by homothallic cells is that the mating type switch is not random--that is, switches to the opposite mating type have been observed more than 50% of the time. In terms of the cassette model, in a cell with an $\underline{\alpha}$ mating type locus the *MATa* cassette is preferentially inserted. (We do not know yet whether the failure to observe a switch reflects the absence of insertion or whether the cassette of the same mating type is inserted.) The significance of this observation is that the homothallic cell has a mechanism for inserting the cassette different from the one at the mating type locus. One can imagine two kinds of explanations for this bias: (i) The active mating type locus, e.g., *MATα* produces a protein which specifically activates transposition of the silent *MATa* cassette. (ii) The active mating type locus contains recognition regions (analogous to prophage attachment sites) which can recombine preferentially with the recognition regions of the *MATa* cassette.

The directional nature of cassette insertion leads to a natural extension of the cassette hypothesis to produce cell lineages of increasing complexity. For example, consider a hypothetical stem cell, S, which can give rise to four kinds of differentiated cells--type A, type B, type C, and type D. We propose that this cell contains four cassettes, *casA*, *casB*, *casC*, and *casD*, each coding for regulators for the particular differentiated cell phenotype. In addition, each cassette codes for a protein which triggers the insertion of one other cassette. For example, *casA* codes for A regulatory proteins and an enzyme catalyzing insertion of the *casB* element, *casB* codes for B regulatory proteins and a protein catalyzing insertion of the *casC* element, etc. Thus by a process of sequential cassette insertion, a pedigree of the following form can be generated.

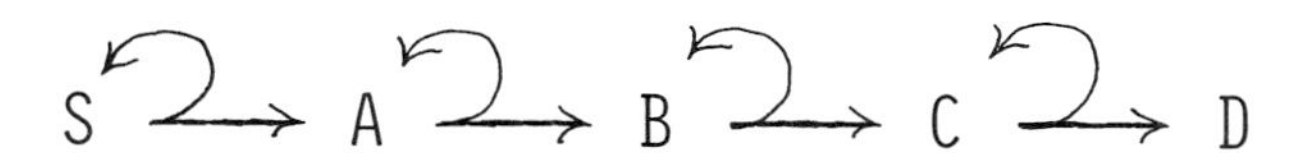

We note that generation of differentiated cells by genetic rearrangement need not require extensive reorganization of the genome. Furthermore, such rearrangements would not be reflected in the genetic behavior of an organism if they occur only in somatic cells or if they are reversible. In the model proposed here, activation of a small number of cassettes coding for regulatory proteins is sufficient to produce profound changes in cell type.

ACKNOWLEDGMENTS

We would like to thank Robert Horvitz and all the members of our laboratory for discussion, Douglass Forbes for comments on the manuscript, and Kathleen Teichman for preparation of the manuscript. This work was supported by a Training Grant to the Institute of Molecular Biology and by a Research Grant and a Research Career Development Award from the Public Health Service.

REFERENCES

1. Monod, J. and F. Jacob (1961). Cold Spring Harbor Quant. Biol. 26: 389.
2. Thomas, R. (1971). In The Bacteriophage Lambda, A.D. Hershey (Editor), Cold Spring Harbor Laboratory, p. 211.
3. Echols, H. (1972). Ann. Rev. Genetics 6: 157.
4. Herskowitz, I. (1973). Ann. Rev. Genetics 7: 289.
5. Holliday, R. and J.E. Pugh (1975). Science 187: 226.
6. Mortimer, R.K. and D.C. Hawthorne (1969). In Yeast Genetics, A.H. Rose and J.S. Harrison (Editors), Academic Press, Inc., New York, p. 385.
7. Hawthorne, D.C. (1963). Genetics 48: 1727.
8. Hicks, J.B. and I. Herskowitz (1976). Genetics 83: 245.
9. MacKay, V. and T.R. Manney (1974). Genetics 76: 255.
10. Strathern, J.N., J.B. Hicks, and I. Herskowitz (1977), in preparation.
11. J. Rine, unpublished observations.
12. Strathern, J.N. (1977). Ph.D. Thesis, University of Oregon.
13. Winge, Ö. and C. Roberts (1949). Compt. Rend. Trav. Lab. Carlsberg, Ser. Physiol. 24: 341.
14. Hopper, A.K. and B. D. Hall (1975). Genetics 80: 77.
15. Hicks, J.B., J.N. Strathern, and I. Herskowitz (1977). Genetics 87: in press.
16. Hawthorne, D.C. (1963). (Abstr.) Proc. 11th Intern. Congr. Genet. 1: 34.
17. Takano, I. and Y. Oshima (1970). Genetics 65: 421.
18. Duntze, W., V. MacKay, and T.R. Manney (1976). Science 168: 1472.
19. Bucking-Throm, E., W. Duntze, L.H. Hartwell, and T.R. Manney (1973). Exptl. Cell Res. 76: 99.
20. S. Brenner, personal communication.
21. Hicks, J.B. and I. Herskowitz (1977). Genetics 87: in press.
22. D. Hawthorne, personal communication.
23. Strathern, A. and I. Herskowitz (1975). Virology 67: 136.
24. Harashima, S., Y. Nogi, and Y. Oshima (1974). Genetics 77: 639.
25. Naumov, G.I. and I.I. Tolstorukov (1973). Genetika 9: 82.
26. Santa Maria, J. and D. Vidal (1970). I.N. Invest. Agron. (Madrid) 30: 1.
27. Harashima, S. and Y. Oshima (1976). Genetics 84: 437.
28. Hicks, J.B. (1975). Ph.D. Thesis, University of Oregon.
29. Mortimer, R.K. and J.R. Johnston (1959). Nature 183: 1751.

CO-ORDINATE REGULATION OF THE SYNTHESIS OF YEAST RIBOSOMAL PROTEINS

Charles Gorenstein and Jonathan R. Warner

Departments of Biochemistry and Cell Biology
Albert Einstein College of Medicine
Bronx, New York 10461

ABSTRACT. When wild type cells of *Saccharomyces cerevisiae* are subjected to a temperature shift-up, there is a co-ordinate repression of the synthesis of more than 50 of their ribosomal proteins. Translation *in vitro* of the RNA isolated from such cells suggests that the repression occurs at the level of transcription of mRNA for the ribosomal proteins (mRNA(R.P.)). On the other hand the transcription of ribosomal precursor RNA (rpreRNA) continues normally after a temperature shift, but the RNA matures slowly and is unstable. After cells have been maintained at the elevated temperature for twenty to thirty minutes, the synthesis of mRNA(R.P.) resumes, and by sixty minutes the balance between the syntheses of rRNA, ribosomal proteins, and total proteins has been reestablished. A mutation in any one of a number of genes prevents the resumption of synthesis of mRNA(R.P.).

On the other hand, when ribosomal precursor RNA synthesis is selectively inhibited, either by actinomycin D, by UV irradiation, or by starving the cells for an essential amino acid, there is a selective inhibition of the synthesis of ribosomal proteins.

INTRODUCTION

The biosynthesis of the ribosome is intriguing not only because of its role in protein synthesis, but also because it is the cell's smallest and best understood organelle. The way in which the cell co-ordinates the synthesis of the seventy or more components and assembles them into a finished ribosome may serve as a model for the synthesis of more complex organelles. In E. coli there is a co-ordinate regulation of the transcription of rpreRNA and of mRNA(R.P.), all under the control of the *rel* gene, (2,3). We have been attempting to determine if such regulation exists in a eukaryotic organism, where transcription and translation are physically separate.

ANALYSIS OF RIBOSOMAL PROTEINS IN WHOLE CELL EXTRACTS

By means of an adaptation of a method used for E. coli, (1), we have been able to extract and purify ribosomal proteins from yeast spheroplasts irrespective of whether they are assembled into ribosomes (5). Figure 1 is a two dimensional polyacrylamide gel of total yeast protein, and for comparison a similar gel of proteins isolated from purified ribosomes. In order to measure the rate of synthesis of any protein, a culture of cells is labelled for several generations with ^{14}C leucine and then pulsed with ^{3}H leucine. The proteins are extracted, displayed on a gel, each spot excised, dissolved, and its $^{3}H/^{14}C$ ratio determined (5). The rate of synthesis of the I th spot can be expressed as:

$$A_i = \frac{(^{3}H/^{14}C)_i}{(^{3}H/^{14}C)\ \text{TOTAL PROTEIN}}$$

CO-ORDINATE REGULATION OF RIBOSOMAL PROTEINS

In Vivo: When the temperature of a culture of yeast is suddenly raised from 23° to 36°, there is a sharp decline in the synthesis of ribosomal proteins (5). Representative data is shown in Figure 2B. Under these conditions the synthesis of fifty ribosomal proteins is repressed, while the synthesis of non-ribosomal proteins is unaffected. The rate of decline in the synthesis of ribosomal proteins is about the same as the rate of decay of mRNA in yeast at 36° (9,10). After about twenty minutes the rate of synthesis of ribosomal proteins recovers, and reaches normal levels by 60 to 70 minutes. These data suggest that one result of the temperature shift-up is a specific but transient repression of the synthesis of mRNA for ribosomal proteins. There is, however, little, if any, repression of the transcription of rpreRNA (19).

In Vitro: In an attempt to confirm the suggestion that during the period immediately following a shift-up the cells are deficient in mRNA(R.P.), we have measured the translation of yeast mRNA in a wheat germ extract (18,15). The ability of such a system to synthesize the ribosomal proteins of yeast is demonstrated in Figure 3. A number of points are clear from Fig. 3 and analogous experiments. The location and shape of the radioactive spots representing in vitro products are identical to the stained spots of authentic yeast ribosomal proteins. Therefore it appears that ribosomal proteins are not synthesized in the form of a precursor, and their migration to the nucleus (20) does not depend on a "signal sequence" (4). Secondly, if polyA(+) RNA is used as a template, all the ribosomal proteins are synthesized (18), indicating that, unlike the case of the histones, the mRNA for ribosomal proteins contains poly A.

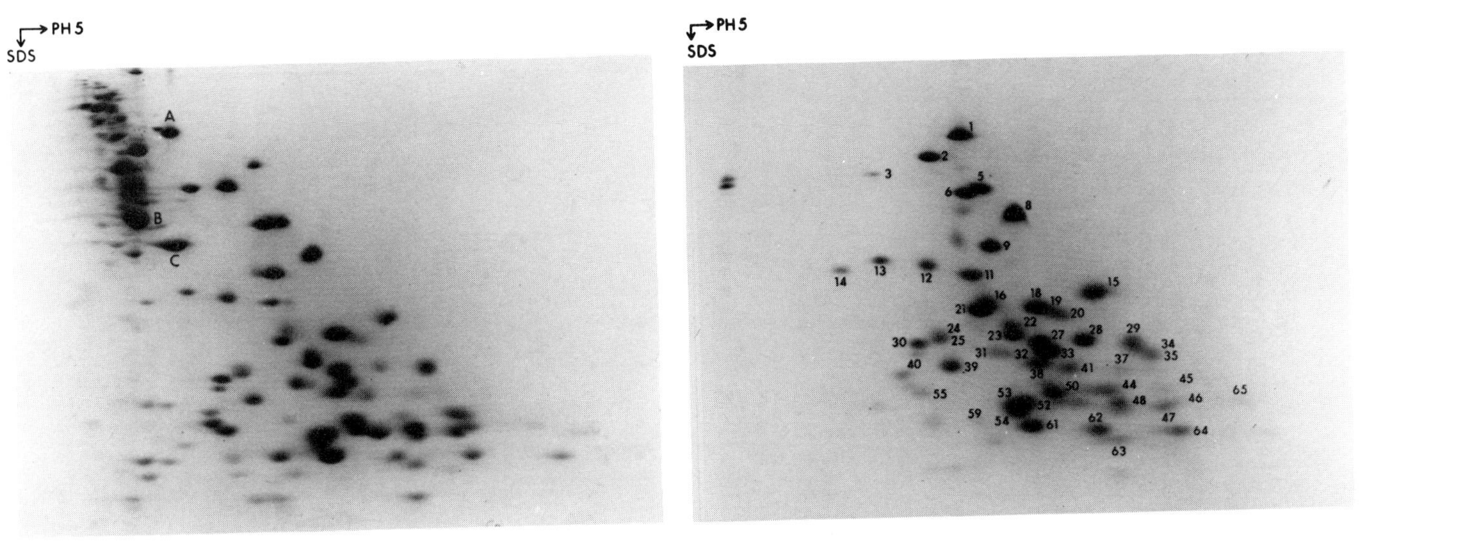

Fig. 1. Polyacrylamide Gel Analysis of Total Protein and Ribosomal Protein of Yeast. LEFT: Total protein was extracted from yeast spheroplasts using 67% acetic acid as described (3). 400 μg was run on a two dimensional polyacrylamide gel (pH 5 x SDS (5,12)). RIGHT: Protein was extracted from yeast 60S and 40S ribosomal subunits using 67% acetic acid. 200 μg was run on an identical gel. The gels were stained with Coomassie Blue. Ribosomal proteins are numbered; some non-ribosomal proteins are denoted by letters. In the whole cell extract nearly all the ribosomal proteins are clearly distinct from other cell proteins.

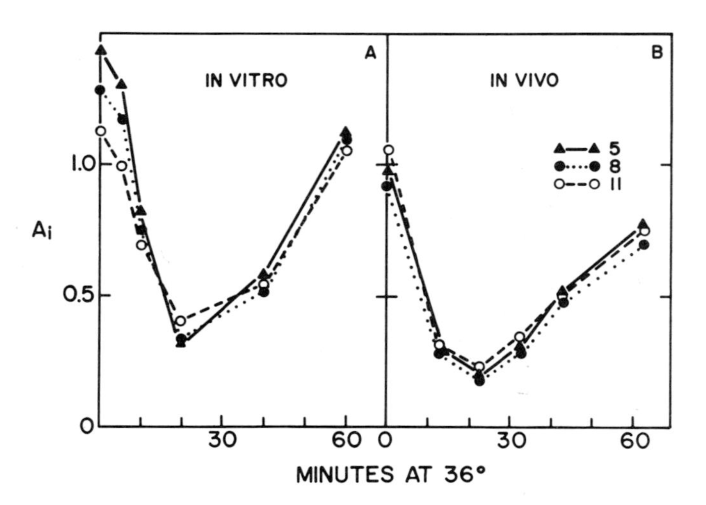

Fig. 2. Synthesis of Ribosomal Proteins *In Vitro* and *In Vivo*. *A*: The template activity, for three ribosomal proteins, of total RNA isolated from a culture shifted from 23° to 36°. In this case protein of cells labelled uniformly with ^{3}H methionine was added to the sample and

$$A_i = \frac{(^{35}S/^3H)_i}{(^{35}S/^3H)_{TOTAL}}$$

B: The rate of synthesis, A_i, of three ribosomal proteins as a function of time after the culture was shifted from 23° to 36°, measured as described in the text and in (5). This figure represents only three of more than 40 proteins examined both *in vivo* and *in vitro*. (From ref. 18).

When total RNA is isolated from a culture at various times after a temperature shift-up, translated at limiting concentration in a wheat germ extract, and the products analyzed on a two dimensional gel, results like those in Fig. 2A are obtained. After the shift-up, the template activity for nearly all the ribosomal proteins falls dramatically during the first twenty minutes and then begins to recover, reaching a normal value by 60 minutes, paralleling the results obtained *in vivo*. The template activity for non-ribosomal proteins remains constant (18). These results confirm the suggestion that the synthesis of mRNA(R.P.) is selectively repressed for a period of time after a temperature shift-up.

Figure 4 is a schematic representation of the kinetics of ribosomal protein synthesis and the inferred kinetics of mRNA(R.P.) synthesis after a temperature shift-up. Such a

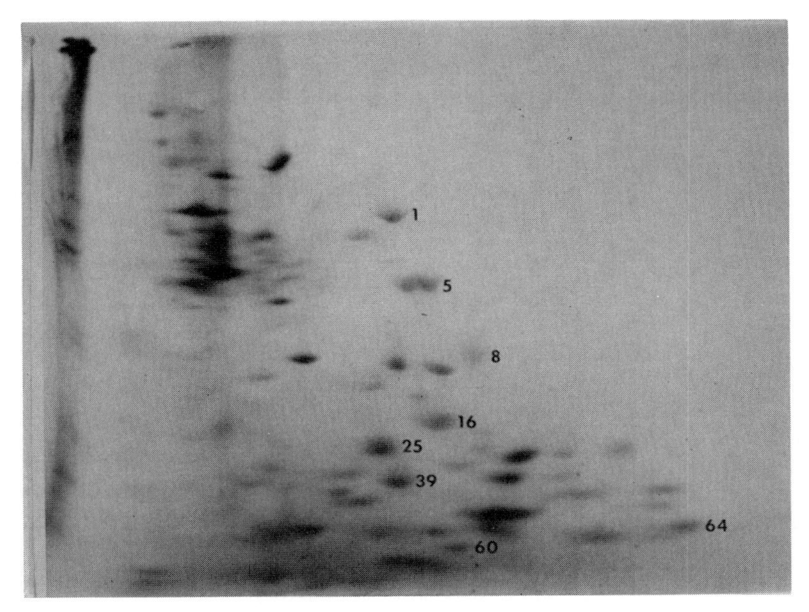

Fig. 3. Detection of Yeast Ribosomal Proteins Synthesized in a Wheat Germ Extract. Total yeast RNA was added to wheat germ extract in the presence of ^{35}S methionine (18,15). The product was isolated with 67% acetic acid, displayed on a pH 5 X SDS gel as in Fig. 1, and autoradiographed. Several ribosomal proteins have been numbered for comparison with Fig. 1B. In the absence of yeast RNA there were no spots in the region of the ribosomal proteins.

repression of mRNA(R.P.) synthesis occurs in all wild type strains tested, in rich and in minimal media, and under conditions, e.g. in the presence of tetracycline, where the concentration of ppGpp does not change (14). A temperature down-shift has no detectable effect on ribosomal protein synthesis.

Mutants of Ribosomal Protein Synthesis: We previously isolated a number of *ts* mutants defective in ribosome synthesis at the restrictive temperature (19,7). By comparing these mutants with the wild type using both *in vivo* and *in vitro* experiments of the kind described in figure 2, we have concluded that the mutants respond initially to a temperature shift just as the wild type does, but that they are unable to recover the ability to synthesize mRNA(R.P.), which

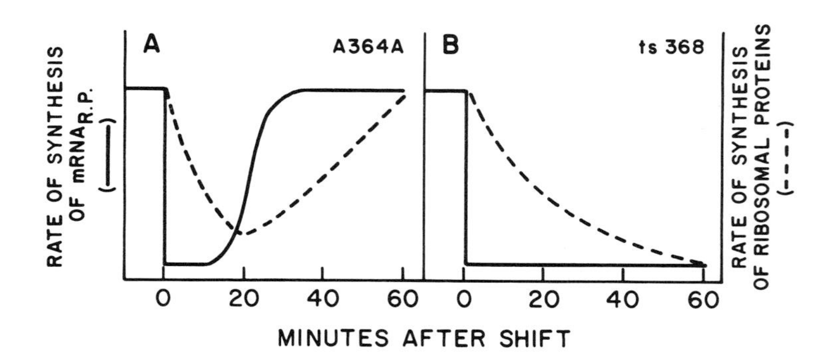

Fig. 4. Schematic diagram of the kinetics of the rate of synthesis of ribosomal proteins and of mRNA (R.P.) as a function of time after a temperature shift from 23° to 36°. A364A is the wild type strain, ts 368 is the temperature sensitive strain (7).

eventually declines to a basal level of 5 to 10%. (See Fig. 4.) That this is due to a deficiency in specific mRNA's is supported by studies of mRNA abundance under permissive and restrictive conditions (8). These results suggest that the lesion in the mutant is an inability to restore metabolic equilibrium after the shift up. The importance of this function is attested by the number of genes involved, ten or more, and the frequency of such mutations, 23 out of 400 unselected _ts_ mutants (7).

THE SYNTHESIS OF RIBOSOMAL PROTEINS DEPENDS ON THE SYNTHESIS OF rpreRNA

Although the previous experiments have described situations in which the synthesis of ribosomal RNA and ribosomal proteins are uncoupled, such is not always the case. We have shown that when cells are deprived of an essential amino acid, tyrosine, the transcription of rpreRNA falls to less than 20% of normal while synthesis of mRNA and tRNA are relatively unaffected (16). Translation _in vitro_ of RNA prepared from cells deprived of tyrosine reveals that such cells have a relative deficiency of mRNA for ribosomal proteins. (Table 1) The kinetics of the inhibition of rpreRNA and mRNA(R.P.) synthesis after cells are deprived of tyrosine, suggest that the order of events is: inhibition of protein synthesis ⟶ inhibition of rpreRNA synthesis ⟶ inhibition of mRNA(R.P.) synthesis.

TABLE 1

RIBOSOMAL PROTEIN SYNTHESIS IS DEPENDENT ON rpreRNA SYNTHESIS

	Control	-Tyr	Ai Control	+Act.D	Control	+UV
Ribosomal Protein						
5	1.19	.40	1.34	0.34	1.44	0.63
8	1.23	.51	1.08	0.33	1.56	0.46
11	1.06	.69	0.95	0.34	1.55	0.59
Non-Ribosomal Protein						
B	1.57	1.66	1.29	0.71	1.28	0.94
C	1.08	1.04	1.66	1.95	1.83	2.54

Minus tyrosine: A culture of A364A was washed and the cells suspended in medium lacking tyrosine. Immediately and five hours later, RNA was isolated from the cells and translated in a wheat germ extract (18).

Actinomycin D: Spheroplasts of strain J98, a strain sensitive to actinomycin D, were treated with 10 μg/ml of actinomycin D for 60 minutes. Under these conditions the synthesis of rpreRNA was inhibited by more than 90%.

UV: Spheroplasts of strain JM1510, which is unable to repair thymine dimers, were irradiated with a germicidal lamp at 200 μwatts/cm^2 for 45 seconds. The synthesis of rpreRNA was inhibited by more than 90%.

In all cases the synthesis of ribosomal and non-ribosomal proteins was measured as described in figure 2. Data is presented for only three of more than thirty ribosomal proteins examined, all of which behaved similarly.

The apparent dependance of mRNA(R.P.) synthesis on rpreRNA synthesis led us to look for other conditions in which rpreRNA synthesis could be preferentially inhibited. Two such conditions are described in Table 1. Using actinomycin D or UV irradiation, ribosomal RNA synthesis can be inhibited by more than 90% while total protein synthesis is inhibited by only 50-60%. In both cases the synthesis of ribosomal proteins is specifically inhibited three fold more than total protein synthesis. (Table 1).

These experiments suggest that the synthesis of mRNA for ribosomal proteins depends on continued synthesis of rpreRNA. On the other hand, the synthesis of rpreRNA, while dependant on total protein synthesis, is not dependant on the synthesis of ribosomal proteins specifically.

HOW DOES IT WORK?

In *E. coli* most of the ribosomal protein genes map within two or three discrete areas of the genome, and appear to be transcribed as several large polycistronic mRNAs (11). On the other hand, the genetics of the yeast ribosomal proteins is primitive. Three antibiotic resistant markers, presumed to be ribosomal proteins, have been mapped at widely scattered locations (6,13,17), as is generally the case for related proteins in eukaryotic organisms. By extrapolation the postulated metabolic signals activated by the temperature shift-up or by the deficiency of rpreRNA must be able to identify and control fifty or more regulatory sites on the genome. The intriguing implication is that all these regulatory sites have some common feature.

ACKNOWLEDGEMENTS

This work has been supported by grants from the NSF, the ACS, and the NCI. Jonathan R. Warner is a Faculty Research Awardee of the ACS.

REFERENCES

1. Dennis, P. (1974) J. Mol. Biol. 88,25.
2. Dennis, P., and Nomura, M. (1974) Proc. Nat. Acad. Sci., U.S. 72,5016.
3. Dennis, P., and Nomura, M. (1975) J. Mol. Biol. 97,61.
4. Devillers-Thierry, A., Kind, T.T., Scheele, G., and Blobel, G. (1975) Proc. Nat. Acad. Sci., U.S. 72,5016.
5. Gorenstein, C., and Warner, J.R. (1976) Proc. Nat. Acad. Sci., U.S. 73,1547.
6. Grant, P., Schindler, D., and Davies, J. (1976) Genetics 83,667.
7. Hartwell, L.H., McLaughlin, C.S., and Warner, J.R. (1970) Molec. Gen. Genetics 109,42.
8. Hereford, L.M., and Rosbash, M. (1977) Cell (in Press).
9. Hutchison, H.T., Hartwell, L.H., and McLaughlin, C.S. (1969) J. Bacteriol. 99,807.
10. Hynes, N.E., and Phillips, S.L. (1976) J. Bacteriol. 125, 595.
11. Jaskunas, S.R., Lindahl, L., and Nomura, M. (1975) Nature 256,183.
12. Metz, L.J., and Bogorad, L. (1974) Analit. Biochem. 57, 200.
13. Mortimer, R., and Hawthorne, D. (1966) Genetics 53,165.
14. Pao, C.C., Paietta, J., and Gallant, J.A. (1977) Biochem. Biophys. Res. Comm. 74,314.
15. Roberts, B.E., and Paterson, B.M. (1973) Proc. Nat. Acad. Sci. U.S. 70, 2330.
16. Shulman, R.W., Sripati, C.E., and Warner, J.R. (1977) J. Biol. Chem. 252, (in Press).
17. Skogerson, L., McLaughlin, C., and Wakatama, E. (1973) J. Bacteriol. 116,818.
18. Warner, J.R., and Gorenstein, C. (1977) Cell (in Press).
19. Warner, J.R., and Udem, S.A. (1972) J. Mol. Biol. 65, 243.
20. Wu, R.S., and Warner, J.R. (1971) J. Cell Biol. 51,643.

ISOLATION AND CHARACTERIZATION OF *ESCHERICHIA COLI* CLONES CONTAINING GENES FOR THE STABLE YEAST RNA SPECIES

Jacques S. Beckmann, Peter F. Johnson, John Abelson and Shella A. Fuhrman

Department of Chemistry and Biology
University of California, San Diego, La Jolla, Ca. 92093

ABSTRACT: Recombinant plasmids were constructed by insertion and ligation of yeast DNA within the tetracycline resistance gene of the *E. coli* plasmid pBR313. The plasmid tet^r gene was split by endonucleases R· *Bam*H I or *Hind* III. Yeast DNA cut with endo R· *Bgl* II was recombined with the former and endo R· *Hind* III cleaved DNA with the latter. Some 4000 amp^r tet^s clones were isolated from 33 independent transformations, all of which were then tested for the presence of yeast specific genes for 4S, 5S or 5.8S RNAs. Of these, 127 and 123 clones were identified as containing the genes for the yeast 5S or 5.8S ribosomal RNAs, respectively. Another 187 clones were recognized as carrying yeast transfer RNA genes. We describe here the distribution of individual yeast tRNA genes among these plasmids. The data indicate that yeast tRNA genes are, in general, not highly clustered. In addition it was observed that the reiteration number of different tRNA genes may vary extensively.

INTRODUCTION

Recombinant DNA technology provides an approach to study the fine structure and organization of eukaryotic genomes. The yeast *Saccharomyces cerevisiae* has been chosen by many laboratories for study because it is a eukaryotic system well suited for detailed analysis of its genetic organization. For example, the ribosomal RNA genes, reiterated some 140 times per haploid genome, (*1,2,3*) appear to be tandemly repeated (*4*) and it was suggested that they might be present on only one chromosome (*5*). The transfer RNA genes, of which there seem to be about 360 copies (*1,2*), appear to be located on several chromosomes (*6-9*). The genes coding for $tRNA^{Tyr}$ have been mapped (*7*) and they are located at eight different loci on several chromosomes (*10*).

We report here the construction of a bank of *Escherichia coli* clones containing yeast DNA inserted into the plasmid pBR313 (*11*), and the subsequent identification of clones containing rRNA or tRNA genes. We present data, described in part in a previous report (*12*), regarding the distribution of several specific tRNA genes among these clones. The data suggest that the yeast tRNA genes are not generally present in tandemly repeated units nor significantly clustered, although limited clustering does exist.

MATERIALS AND METHODS

Strains and Materials: All yeast and *E. coli* strains were described elsewhere (*12*). The endonuclease restriction enzymes endo R· *Bam*H I, *Bgl* II and *Sma* I were purified according to Roberts (personal communication). Endo R· *Hind* III was prepared as described by Smith (*13*) and Endo R· *Eco* RI was a gift from D. Helinski. Bacteriophage T4 DNA ligase and polynucleotide kinase were obtained respectively from Miles Laboratory, Inc. and Biogenics Research Corporation. γ-labeled ATP was prepared as described (*14*). Purified tRNA species were gifts from several generous colleagues as indicated in Table 2.

DNA preparation. Yeast cells were grown and then collected by centrifugation. Spheroplasts were obtained by treatment with β-mercaptoethanol and glusulase (Endo Laboratories, Inc.). These were then lysed (*15*), gently extracted with phenol chloroform (6:1) and precipitated with ethanol. The DNA pellet was dissolved in 0.015 M NaCl, 0.0015 M NaCitrate, pH 7.0 (0.1 x SSC)/2mM EDTA, pH 8.0/1% Sodium Dodecyl Sulfate (SDS) and treated with pancreatic RNAase followed by digestion with preincubated Pronase and finally purified by equilibrium sedimentation in CsCl. The procedures for the restriction, annealing and ligation of the DNA and for the transformations were described previously (*16*). This work was done under P2 conditions according to the guidelines as defined by the National Institute of Health.

Labeling of yeast RNA. Yeast cells were grown in rich medium depleted of phosphate (*17*). ^{32}P orthophosphate was present in the medium at 50-100 μCi/ml. Cells were labeled for three hours followed by a chase step of 30 minutes by adding Potassium Phosphate (pH 7.0) to a final concentration of 10 mM. Low molecular weight RNA, extracted from these cells as described by Rubin (*17*), was purified by passage through a small DEAE-cellulose column and finally separated into tRNA and rRNA components by electrophoresis at 20°C on a 40 cm long 10% polyacrylamide slab gel in 4 M urea (*7*). After autoradiography, the desired RNA bands were cut out and the RNA eluted in 0.3 M NaCl containing 0.2% SDS. The RNA was finally purified by Millipore filtration, passed through a DEAE-cellulose column and precipitated with two volumes of ethanol. Individual *in vivo* labeled tRNAs were purified by two dimensional polyacrylamide gel electrophoresis according to the method of Piper, *et al.* (*7*). After electrophoresis the tRNA positions were revealed by autoradiography and the tRNAs eluted and recovered as described above. In some cases the crude tRNA preparation was first fractionated with benzoylated DEAE-cellulose (*7*) prior to polyacrylamide gel electrophoresis. Purified species of tRNA (gifts from several laboratories, Table 2) were labeled *in vitro* by incubation with γ-labeled ATP and bacteriophage T4 polynucleotide kinase (*14*). The labeled RNAs were purified through a short DEAE-cellulose column.

Colony Hybridization Method. Since it was our intention to screen a large number of clones with various probes we have modified the basic colony hybridization technique developed by Grunstein and Hogness (*18*) as follows. After transformation, all ampicillin resistant tetracycline sensitive clones (tetracycline sensitive clones result from a DNA insertion within the gene governing resistance to tetracycline (*11*)) were transferred to wells of a microtiter dish (Falcon #3040 Microtest II tissue culture plate, 96 wells/plate) containing 0.2 ml L-broth and ampicillin (80 μg/ml) per well. After growth the cells were replica-plated onto 15 cm diameter petri dishes containing L-agar and ampicillin using an aluminum plate containing 96 screws spaced so as to fit the exact matrix of the wells in the dish (*19*). After overnight incubation at 37°C, a rectangular piece of dry Whatman 540 filter paper (P.C. Wensink (personal communication) had previously shown that nitrocellulose filters could be replaced by Whatman 541 or 3MM filter papers in the original technique of Grunstein and Hogness (*18*)) was pressed lightly onto the agar and lifted off the plate. Most of the bacteria remained on the filters. These were air dried and treated with 0.5 N NaOH for 10 minutes. The NaOH was applied from the bottom of the filters through sheets of Whatman paper in order to prevent the colonies from washing off the filter. This step was followed by two rinses in 0.5 M Tris-HCl, pH 7.4, then one 1 x SSC rinse and finally one 95% ethanol wash. All those steps were carried out at room temperature for five minutes each. After drying, the filters were ready for hybridization (baking was not found necessary).

Hybridization was performed batchwise in a minimal volume, i.e., just enough to wet all filters, in 50% deionized formamide/4 x SSC/0.4% SDS for 16 hours at 37°C. 100,000 cpm of a pure RNA species were utilized per filter. For total 4S RNA, about 10-20 x 10^6 cpm were utilized. After incubation, the filters were then washed twice with 50% formamide/4 x SSC for 20 minutes each at room temperature, then three to four times with 2 x SSC, again for 20 minutes each. No RNAase treatment was found necessary. The filters were finally dried and autoradiographed. It is worth mentioning that it is possible under these hybridization conditions to perform competition experiments between different labeled and unlabeled RNA preparations. In addition, upon completion of an experiment, these filters can be recycled for further use.

Normally, the location of the bacterial colony is easily recognizable on these filters. In case of doubt the DNA can be visualized by exposing the dry filter to ultraviolet light or by presoaking the filters in an ethidium bromide solution, washing off excess ethidium bromide with water and then visualization with either ultraviolet or visible light (Fig. 2A).

RESULTS

Cloning experiments. We established a large collection of *E. coli* clones carrying yeast DNA fragments which could subsequently be screened for those clones containing sequences coding for the 4S, 5S and 5.8S RNAs. The vehicle used for these experiments was the plasmid pBR313 (*11*) which carries antibiotic resistance genes for tetracycline (*tet*^{r}) and ampicillin (*amp*^{r}). Foreign DNA can be inserted by ligation into this plasmid at a number of single restriction sites. In our experiments we used Endo R· *Hind* III and *Bam*H I restriction endonucleases to cleave the plasmid DNA. Both these enzymes, recognizing respectively the DNA sequences A*AGCTT and GGATCC (*20*), cleave the circular DNA at a single site in *tet*^{r} gene, yielding linear molecules with 5' single stranded ends, 4 nucleotides in length. Yeast DNA was purified from *Saccharomyces cerevisiae* strain X2180-1A (*21*) and cleaved with endo R· *Hind* III or *Bgl* II. The latter enzyme recognizes and splits an AGATCT sequence and produces the same single stranded sticky ends as endo R· *Bam*H I (*20*). The yeast DNA fragments were mixed in three-fold molar excess with cleaved plasmid DNA and the mixture was ligated (*16*) to produce a random collection of hybrid DNA molecules. These recombinant DNA plasmids were introduced into *E. coli* strain C600 SF8 (*22*) by transformation (*16*). After addition of the DNA the transformation mixture was divided into a number of separate growth tubes in order to minimize the chances of obtaining sibling clones. The transformation mixtures were plated on ampicillin containing medium to select for clones carrying the plasmid. From 33 independent transformations about 12,000 ampicillin resistant clones were isolated. Four thousands of these clones were tetracycline sensitive, indicating that foreign DNA had been inserted into the *tet*^{r} gene.

Screening for yeast rRNA and tRNA genes. The collection of four thousand clones containing yeast DNA was screened by the colony hybridization technique to detect those clones with DNA sequences complementary to yeast 4S, 5S and 5.8S RNA. In these experiments the RNA probes were labeled *in vivo* by growing yeast cells in medium containing ^{32}P orthophosphate (*17*). The labeled RNA was purified by polyacrylamide gel electrophoresis (*7*). One hundred and eighty seven clones hybridized specifically with 4S RNA; 127 hybridized with 5S RNA and 123 with 5.8S RNA. Only two of these clones seemed to contain both 5 and 5.8S RNA genes. The 5 and 5.8S clones also hybridized with 4S RNA, but this hybridization was shown to be due to the presence of contaminating ribosomal RNA sequences in our 4S RNA and was not attributable to tRNA. Figure 1 illustrates the detection of clones hybridizing with 4S RNA.

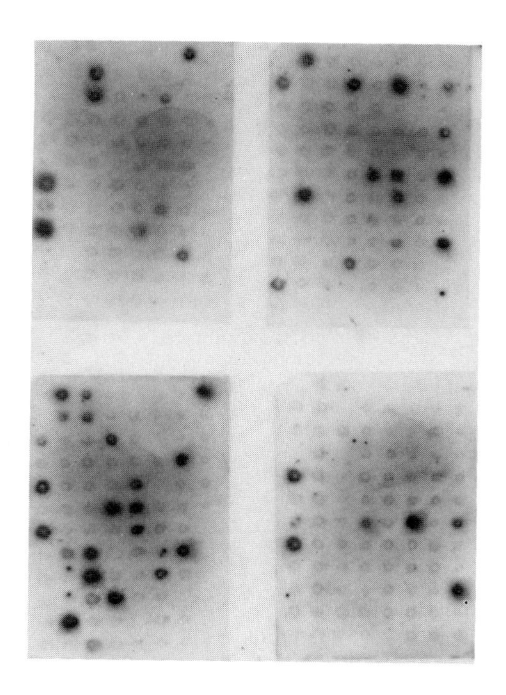

Fig. 1. Screening for recombinant clones carrying yeast tRNA genes. Three hundred and eighty clones carrying fragments of the yeast genome inserted into pBR313 are shown after having been assayed for the presence of yeast tRNA genes by hybridization with *in vivo* ^{32}P-labeled 4S RNA.

Preliminary analysis of one of the 5S clones. One of the recombinant clones produced by ligation of endo R· *Bgl* II restricted yeast DNA with endo R· *Bam*H I cleaved pBR313 DNA and carrying the yeast 5S RNA cistron was chosen for analysis. This clone, JB151, carries a plasmid molecule of 8.4 x 10^6 daltons. Whereas the parent pBR313 vehicle had only one endo R· *Bam*H I site and no endo R· *Bgl* II target, the recombinant plasmid is insensitive to both enzymes. Therefore, it has lost its only GGATCC sequence within the *tet*^{r} gene, has acquired no new one within the inserted yeast DNA fragment. This latter piece contains no AGATCT sequence indicating that the inserted segment is neither a partial digestion product nor the result of a multimolecular ligation event, but represents a unique piece of terminally digested DNA.

The DNA of this plasmid was analyzed by various restriction enzymes and shown to contain the endo R· *Sma* I target site and one of the two endo R· *Hind* III sites found within the yeast ribosomal repeats (*4*,*23*). Upon digestion with endo R· *Eco* RI the plasmid yielded five fragments, the largest of which contained most of the parental pBR313 DNA; the second largest encoded the 5S cistron. In addition, three smaller fragments were observed. We are presently attempting to align this ribosomal DNA segment with the restriction maps proposed by other laboratories (*4*,*5*) though the observed differences between these maps may indicate a degree of freedom for variation in the ribosomal DNA from strain to strain (*5*).

Screening for individual tRNA genes. Our collection of 187 clones hybridizing to 4S RNA were screened with purified tRNA species. Prior to the final screening process each clone was purified and retested for hybridization to 4S RNA. Then two grids were formed containing the entire 4S collection. In a preceding report we describe the results obtained after screening these clones with 18 different tRNA species (*12*). We have extended this study and are now reporting the data obtained with a total of 25 non-overlapping tRNAs.

Pure labeled tRNAs were obtained in two ways as described in Materials and Methods. The purity and in some cases the identity of the individual *in vivo* labeled tRNAs were ascertained by studying the T1 ribonuclease products of each species. Twenty-two pure species of yeast tRNAs have been isolated in this way. In addition we were fortunate enough to obtain a number of pure yeast tRNAs from our colleagues (Table 2). These tRNAs were made radioactive by incubation with γ-labeled ^{32}P ATP and bacteriophage T4 polynucleotide kinase (*14*). In this way, we have obtained eleven pure tRNA species. In several cases these proved to be identical to members of the first set so that in all we have screened our set of 4S clones with 25 different species of tRNA. The acceptor activity of 12 of these is unknown. Figure 2 gives an example of the detection of 4S clones hybridizing to pure species of tRNA.

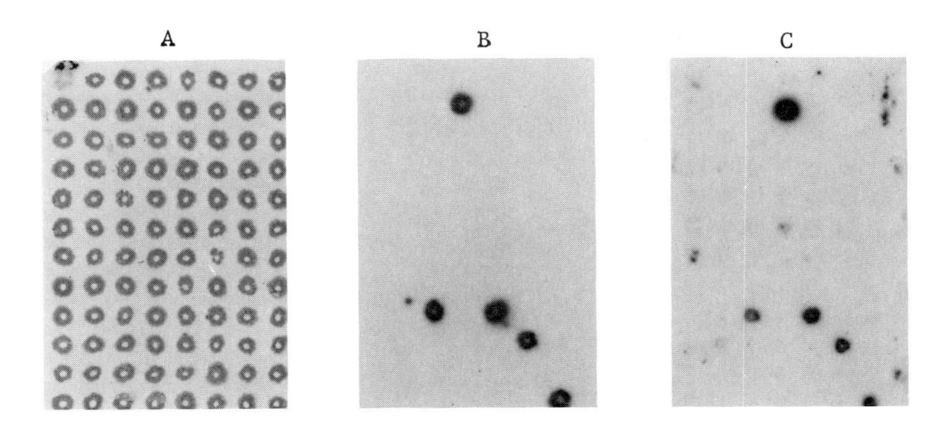

Fig. 2. Screening for specific tRNA genes among 4S containing clones. The 187 clones that were shown to hybridize with ^{32}P-labeled 4S RNA (see Figure 1) were tested for their ability to anneal different purified tRNA species. Panel A shows 95 bacterial colonies, lysed *in situ* on the Whatman 540 filters, after staining with ethidium bromide. Panel B shows the hybridization to a tRNA sample labeled *in vivo* and purified by two dimensional electrophoresis. This tRNA was later identified by its ribonuclease T1 fingerprint as $tRNA_3^{Leu}$. Panel C shows the hybridization of another identical filter to purified $tRNA_3^{Leu}$ (supplied to us by S. H. Chang, and by G. Pixa and G. Keith) labeled *in vitro* as described (*14*).

Figure 2B illustrates the hybridization to a tRNA purified by two dimensional polyacrylamide gel electrophoresis (*7*). This tRNA was identified by analysis of its T1 ribonuclease products as $tRNA_3^{Leu}$ (*24*). Five clones have hybridized with this tRNA. Panel C of figure 2 shows the hybridization of the same clones with purified $tRNA_3^{Leu}$, labeled *in vitro* as described earlier. As expected, the same five clones are found to hybridize with this $tRNA_3^{Leu}$.

Legend to Table 1: All 187 clones containing 4S genes obtained by a series of 33 independent transformations are listed. The identity of the various purified tRNA species found to hybridize with each

TABLE

Clones	tRNA species	Clones	tRNA species	Clones	tRNA species
1-i	Thr	6-a	Leu$_3$(R5)	10-m	Arg$_3$(R19),Asp,R9
2-b	Val$_2$(R12),R18	c		p	Arg (R19),R9
c	Asp	g		q	T2(Leu$_{UUA}$)
e	Thr	h	Tyr(R10)	11-d	Thr
f	Thr	l	Gly	f	Leu$_3$(R5)
j		7-j	T2(Leu$_{UUA}$)	g	Ser$_2$(T3)
k	Leu$_3$(R5)	m	Gly	h	R7
m	Leu$_3$(R5)	8-a	Thr	j	
p	Thr,R9	c	Thr	k	R18
r	Val$_2$(R12)	e	T2(Leu$_{UUA}$),T9	l	
3-b		f	Asp	n	R18
d	Gly,T9	j		o	Ser$_2$(T3)
h		k	R18	p	
i	Val$_2$(R12)	l	R13	q	T8
p	R8(Ala)	n		s	Phe(R11),R8(Ala)
q	T15	p	Asp	12-a	R17
4-a	Leu$_3$(R5),R1	q		d	Leu$_3$(R5),R1
d	R7	s	T8	e	Asp
e		u	Leu$_3$(R5)	o	
g		9-d	T9	p	
i		e	T9	q	Asp
j		g	Ser$_2$(T3)	13-a	
k	R8(Ala)	k	T15	14-c	T9,R15
l	Ser$_2$(T3)	l	R18	d	
o	R13	n		e	T15,R17
5-a	Asp	10-b	R13	g	R4(Ser$_{UCG}$)
e	Phe(R11)	e		k	R8(Ala)
f	Gly	f		15-a	
i	R8(Ala),T16	g		c	
j		h	Tyr(R10)	d	
		i		g	T2(Leu$_{UUA}$)
		l	R7	h	T9

Catalog of 4S Clones

Clones	tRNA species
16-h	
17-a	
c	Leu_3(R5)
d	R15
f	Asp
h	T9
p	T9
18-c	
f	R15
g	R4(Ser_{UCG})
k	R9
l	R7
n	R18
q	Val_2(R12)
s	
u	Arg_3(R19),Tyr (R10),Asp
19-b	
f	Arg_3(R19),Asp, R9
g	T9
19-i	T15
k	T15
m	Leu_3(R5),R1
n	T8
r	R18
t	
20-b	Trp
f	T9
g	
m	Trp
n	Ser_2(T3)
o	

Clones	tRNA species
21-d	Arg_3(R19),Asp, R9
k	R7
n	Arg_3(R19),Asp, R9
22-c	
f	
g	
h	
i	R7
23-a	T9
f	Leu_3(R5)
g	
h	Gly
k	Tyr(R10)
l	R4(Ser_{UCG})
24-b	Asp
c	T15,R17
d	
k	
o	
25-l	R18
p	Val_2(R12)
26-d	R8(Ala)
f	Val_2(R12)
h	T9
27-a	
b	Ser_2(T3)
d	R9
g	
j	
28-b	R13
c	T9

Clones	tRNA species
28-e	
i	Asp
29-e	
i	T15,R17
l	Val_2(R12),T16
n	Arg_3(R19),R9
p	R7
q	T2(Leu_{UUA})
r	Arg_3(R19),T9, R9
30-e	
g	T15
h	Gly
i	R6
l	T9
31-l	Asp
32-e	
f	T16
g	T16
h	T16
141	
142	
143	Ser_2(T3)
144	Tyr(R10)
241	Arg_3(R19)
242	Leu_3(R5)
243	Val_2(R12)
244	
341	Ser_2(T3)
342	
541	R6
542	Leu_3(R5)

clone is indicated. Screening for the various tRNA species was done using either individual 4S spots obtained by 2-dimensional separation of *in vivo* labeled tRNAs (T1-T16;R1-R19) or by incubating purified tRNA species with bacteriophage T4 polynucleotide kinase and γ-labeled ^{32}P ATP (*14*). Spots T2,T3,R4,R5,R8,R10 and R11 were respectively shown by fingerprint analyses to be tRNAs specific for Leu_{UUA}, Ser_2, Ser_{UCG}, Leu_3, Ala, Tyr and Phe; T9 was not fingerprinted but did cohybridize with the same clones as $tRNA^{Lys}$. Abbreviations for the tRNA species are as follows: Asp, aspartic acid; Arg, arginine; Gly, glycine; Leu, leucine; Lys, lysine; Phe, phenylalanine; Ser, serine; Thr, threonine; Trp, tryptophan; Tyr, tyrosine.

The results of hybridization with all 25 pure tRNA probes are given in Tables 1, 2 and 3. One hundred and twenty nine of the 187 clones in the 4S collection hybridize to at least one of the tRNAs (66% of our collection). Only 21 of the 4S clones hybridize to more than one tRNA species. Six of these clones are found to hybridize to three different tRNAs and can be subdivided into three groups according to the tRNAs to which they hybridize (Table 3). It is interesting to note that a possible cluster of Arg_3, Asp and R9 tRNA genes occurs four times in our collection.

DISCUSSION

The distribution of various yeast tRNA genes among the recombinant clones suggest some tentative conclusions regarding the organization of these genes in the yeast genome. The average molecular weight of an endo R· *Hind* III or *Bgl* II fragment is 2.5 x 10^6 daltons. Thus, each fragment represents one four thousandth of the yeast haploid genome (ca. 10^{10} daltons) (*25*). We have isolated 4000 clones at random. If the 360 tRNA genes contained in a haploid genome (*1,5*) were widely spaced we would have expected to isolate by chance about 360 clones, each containing a gene for only one tRNA species (although in our collection the chance of obtaining a particular tRNA gene as determined by Poisson distribution is $1-e^{-1}$ or 0.63). On the other hand, if tRNA genes are tightly clustered we would have detected far fewer 4S clones. The size of our collection (187 4S clones) suggests that the tRNA genes are not tightly clustered. These results are in good agreement with data from Craig Chinault and John Carbon (personal communication). If our collection of 187 clones contains 360 tRNA genes, on the average there will be two genes per clone. The possibility that yeast tRNA genes are widely spaced in the genome is consistent with the genetic mapping of tyrosine-

TABLE 2

Frequency of Occurrence of the Different tRNA Clones

tRNA species	Number of clones found	Minimum independent number
Ala(R8) (G. Keith) *(27)*	6	6
Asp (G. Keith; P. Bolton and D. Kearns) *(28)*	15	12
Arg_3(R19) (J. Weissenbach and G. Keith) *(29)*	9	8
Gly (P. Bolton and D. Kearns)	6	6
Leu_{UUA}(T2) *(30)*	5	5
Leu_3(R5) (S. H. Chang; G. Pixa and G. Keith) *(24)*	12	10
Phe(R11) (B. Reid) *(31)*	2	2
Ser_2(T3) (H. Zachau) *(32)*	8*	6
Ser_{UUG}(R4)	3	3
Thr (J. Weissenbach and G. Keith)	7	4
Trp (G. Keith; B. Reid) *(33)*	2	1
Tyr(R10) (B. Reid; P. Bolton and D. Kearns) *(34)*	5	5
Val_2(R12) (S. Montasser and G. Keith) *(35)*	8	7
T8	3	3
T9	14	12
T15	8	7
T16	5	3
R1	3	3
R6	2	2
R7	7	7
R9	10	7
R13	4	4
R15	3	3
R17	4	4
R18	8	7

The data reported under Table 1 are summarized as follows: Column 1 lists the total number of clones hybridizing to each tRNA. Column 2 represents the minimum number of independently obtained clones for each tRNA, computed by substracting any possible siblings. We gratefully acknowledge the various donors of the different purified tRNA species as indicated for each tRNA.

*In a previous report (12) we described the screening of the 4S clones with a $tRNA_2^{Ser}$ preparation obtained by two dimensional separation of ^{32}P-labeled total tRNAs. These data allowed us to recognize five clones as carrying the Ser_2 specific gene. Six other clones hybridized weakly to this tRNA preparation and therefore their designation as Ser_2 clones remained uncertain. Hybridization of these clones to purified $tRNA_2^{Ser}$ (a gift from Dr. H. Zachau), kinased *in vitro*, enabled us to recognize three more Ser_2 specific clones. In total, we now have eight clones carrying $tRNA_2^{Ser}$ coding sequences.

inserting suppressors. These suppressors are unlinked and often map on separate chromosomes (*10*). Clarkson and collaborators (*26*) have studied the organization of tRNA genes in *Xenopus*. Their data suggest a model in which identical tRNA genes are arranged in tandem, interspersed by spacer DNA. The results presented here, though not excluding the possibility of tandem repetition of genes for identical tRNA species, do argue in favor of a different general principle or organization in yeast.

Table 3

Possible Clusters of Two or More tRNA Genes

tRNA species	Number of clones found	Minimum independent number
Gly, T9	1	1
Ala(R8), T16	1	1
Leu_{UUA}(T2), T9	1	1
Ala(R8), Phe	1	1
Arg_3, Asp, R9	4	3
Arg_3, Asp, Tyr	1	1
Arg_3, T9, R9	1	1
Arg_3, R9	2	2
Val_2, R18	1	1
Thr, R9	1	1
Leu_3, R1	3	3
T9, R15	1	1
T15, R17	3	3

The clones which react with more than one of the tRNAs tested and their frequency of occurence are listed as described under Table 2.

Our data also suggest that the frequency of occurrence of the different tRNA genes on the yeast genome may vary extensively from one tRNA species to another. Table 2 shows that whereas ten clones hybridize with $tRNA_3^{Leu}$, only two hybridize with $tRNA^{Phe}$. It is obviously of importance to analyze and compare the DNA from various of our clones. In doing so we hope to answer the following questions: for a set of clones carrying genes for a particular tRNA, what is the conserved sequence--is it confined to the structural gene or does it contain auxiliary information? Do iso- or hetero- clusters of tRNA genes exist and if so, what is their organization and mode of transcription? Can one by comparison of the sequences adjacent to the structural genes recognize any special features which could serve as regulatory signals for transcription?

ACKNOWLEDGMENTS

We would like to thank our colleagues listed in Table 2 for sending us numerous pure species of tRNA. Drs. P. Piper and H. Feldman supplied us with valuable information used in identifying yeast tRNA fingerprints. We thank Dr. Peter Geiduschek for his continous interest in our work. This research was supported by a grant from the National Cancer Institute, CA 10984.

REFERENCES

1. Schweizer, E., MacKechnie, C., and Halvorson, H. O. (1969) *J. Mol. Biol.* 40, 261.
2. Rubin, G. M. and Stulston, J. E. (1973) *J. Mol. Biol.* 79, 521.
3. Feldman, H. (1976) *Nucleic Acids Research* 3, 2379.
4. Cramer, J. H., Farrelly, F. W., Barnitz, J. T., and Round, R. H. (1977) This Symposium.
5. Petes, T. (1977) This Symposium.
6. Capecchi, M. R., Hughes, S. H., and Wahl, G. M. (1975) *Cell* 6, 269.
7. Piper, P. W., Wasserstein, M., Engbaek, F., Kaltoft, K., Celis, J. E., Zeuthen, J., Liebman, S., and Sherman, F. (1976) *Nature* 262, 757.
8. Brandiss, M. C., Stewart, J. W., Sherman, F., and Botstein, D. (1976) *J. Mol. Biol.* 102, 467.
9. Gesteland, R. F., Wolfner, M., Grisafi, P., Fink, G., Botstein, D., and Roth, J. R. (1976) *Cell* 7, 381.
10. Gilmore, R. A., Stewart, J. W., and Sherman, F. (1971) *J. Mol. Biol.* 61, 157.
11. Bolivar, F., Rodriguez, R. L., Betlach, M. C., and Boyer, H. W. (submitted to *Gene*).
12. Beckmann, J. S., Johnson, P. F., and Abelson, J. (1977) *Science* 196, 205.
13. Smith, H. O. (1975) *Methods in Molecular Biology* 7, 71 (Ed. R. B. Wickner).
14. Maxam, A. M. and Gilbert, W. (1977) *Proc. Nat. Acad. Sci. U.S.A.* 74, 560.
15. Shalitin, C. and Fisher, I. (1975) *Biochim. Biophys. Acta* 414, 263.

16. Velten, J., Fukada, K., and Abelson, J. (1976) *Gene* 1, 93.
17. Rubin, G. M. (1975) *Methods in Cell Biology* 12, 45 (Ed. D. Prescott).
18. Grunstein, M. and Hogness, D. S. (1975) *Proc. Nat. Acad. Sci. U.S.A.* 72, 3961.
19. Brenner, M., Tisdale, D., and Loomis, W. F., Jr. (1975) *Exptl. Cell Res.* 90, 249.
20. Roberts, R. J. (1976) *Crit. Rev. Biochemistry* 3, 123.
21. This strain, originally from Dr. Mortimer was described by Duntze, S., Mackay, V., and Manney, T. R. (1970) *Science* 168, 1472.
22. Struhl, K., Cameron, J. R., and Davis, R. W. (1976) *Proc. Nat. Acad. Sci.* U.S.A. 73, 1471.
23. Meyerink, J. H., Retel, J., Planta, R. J. and Heidekamp, F. (1976) *Molecular Biology Reports* 2, 393.
24. Chang, S. H., Kuo, S., Hawkins, E., and Miller, N. R. (1973) *Biochem. Biophys. Res. Comm.* 51, 951.
25. Bhargava, M. M. and Halverson, H. O. (1971) *J. Cell Biol.* 49, 423.
26. Clarkson, S. G. and Birnstiel, M. L. (1973) *Cold Spring Harb. Symp. Quant. Biol.* 38, 451.
27. Holley, R. W., Apgar, J., Everett, G. A., Madison, J. T., Marquisee, M., Merrill, S. H., Penswick, J. R., and Zamir, A. (1965) *Science* 147, 1462.
28. Gangloff, J., Keith, G., Ebel, J. P., and Dirheimer, G. (1972) *Biochem. Biophys. Acta* 259, 198.
29. Kuntzel, B., Weissenbach, J., and Dirheimer, G. (1972) *FEBS Letters* 25, 189.
30. Piper, P. W. and Wasserstein, M. (Submitted to *Europ. J. Biochem.*).
31. RajBhandary, U. L., Chang, S. H., Stuart, A., Faulkner, R. D., Hoskinson, R. M., and Korana, H. G. (1967) *Proc. Nat. Acad. Sci. U.S.A.* 57, 751.
32. Zachau, H. G., Dutting, D., and Feldmann, H. (1966) *Angew, Chem.* (Int. Ed. Engl.) 5, 422.
33. Keith, G., Roy, A., Ebel, J. P., and Dirheimer, G. (1971) *FEBS Letters* 17, 306.
34. Madison, J. T., Everett, G. A., and Kung, H. (1966) *Science* 153, 531.
35. Axelrod, V. D., Kryukov, V. M., Isaenko, S. N., and Bayev, A. A. (1974) *FEBS Letters* 45, 333.

ORGANIZATION OF RIBOSOMAL DNA IN YEAST

Jane Harris Cramer, Frances W. Farrelly,
Joy T. Barnitz and Robert H. Rownd

Laboratory of Molecular Biology and
Department of Biochemistry
University of Wisconsin, Madison, Wisconsin 53706

ABSTRACT. We have constructed hybrid plasmids by ligating EcoRI partial digestion fragments of purified *Saccharomyces cerevisiae* ribosomal DNA to the bacterial plasmid RSF2124. These plasmids have been used to map the SmaI, HindII, HindIII, and EcoRI restriction sites in the rDNA repetitive units. Analyses of a number of the hybrid plasmids have shown that the repetitive units are homogeneous in size and composition and that they are not separated by large heterogeneous spacer regions. We have also prepared a denaturation map of the *S. cerevisiae* rDNA and aligned it with the restriction map. A comparison of the two maps has enabled us to locate the ribosomal RNA coding regions on the denaturation map and has demonstrated that the spacer regions have a higher A·T content than the RNA coding regions.

The SmaI, HindIII and EcoRI restriction fragments of rDNA from the closely related yeast *Saccharomyces carlsbergensis* coelectrophorese with those from *S. cerevisiae* rDNA.

INTRODUCTION

S. cerevisiae contains approximately 140 copies of the 18S and the 25S ribosomal RNA (rRNA) genes (1). These two rRNA species as well as the 5.8S rRNA are synthesized as part of a 35S rRNA precursor which has a molecular weight of 2.5×10^6 (2). Approximately 20% of this precursor is removed and degraded during subsequent processing (2). The 5S RNA genes are present in the same number of copies as the rRNA precursor genes and are closely linked to them in an alternating arrangement (3,4); however, they are transcribed separately from the rRNA precursor genes (2,5). Thus, the individual repeating units of ribosomal DNA (rDNA) contain a 35S rRNA precursor gene, a 5S RNA gene and possibly some non-transcribed spacer DNA. Some experiments have suggested that the repeating units also contain tRNA genes (6); however, this has not been demonstrated conclusively. The DNA which contains the rDNA repeating units has a density of 1.705 g/cc in neutral CsCl density gradients and can be separated from the rest of the nuclear DNA—called α DNA—which has a density of 1.699 g/cc (7). Approximately 60-70% of the rDNA repeating units are located on chromosome I (8,9); the chromosomal location of the remainder is unknown. The units on any one chromosome are not all contiguous but appear to be arranged in small clusters of tandem repeating units which are interspersed with regions of nonribosomal DNA (7).

We have used restriction endonuclease digestion and electron microscopic techniques to examine the size and degree of homogeneity of the individual repeating units in purified _S. cerevisiae_ rDNA. The restriction fragments produced by digestion with SmaI, HindIII, HindII + III and EcoRI have been characterized (Table 1). In each case the pattern of digestion is reproducible, the fragments are present in stoichiometric amounts, and the sum of the fragment molecular weights is 5.6×10^6 (10; Table 1). SmaI is a particularly useful enzyme since it cleaves each repeating unit at only one site and the resulting fragment represents a monomer unit of rDNA. To determine if the repeating units were homogeneous in composition as well as size, we dissociated the complementary strands of rDNA SmaI fragments and allowed them to reanneal. Of the duplex molecules formed, all were homoduplexes. We observed no molecules with substitution or deletion loops (10). These experiments suggest that the repeating units in yeast rDNA have a molecular weight of 5.6×10^6 and are homogeneous in size and composition.

METHODS

Procedures for restriction endonuclease digestion and gel electrophoresis have been described (10). The methods for constructing and selecting the yeast rDNA-RSF2124 hybrid plasmids will be reported in detail elsewhere (12). Denaturation mapping was done by the method of Inman and Schnös (13).

RESULTS

Construction of hybrid plasmids containing S. cerevisiae rDNA. We have constructed hybrid plasmids by ligating EcoRI-digested purified yeast rDNA to the bacterial plasmid RSF2124, a ColE1 derivative which carries ampicillin resistance (14). Because EcoRI cleaves each rDNA unit seven times, we used partial digestion fragments in the ligation reaction to obtain plasmids which contain larger segments of yeast DNA. The resulting plasmids were transformed into _E. coli_ and screened for the presence of rDNA fragments. We have obtained a number of plasmids which contain different rDNA partial digestion fragments. Several of the plasmids which contain segments larger than one rDNA repeating unit have

TABLE 1: MOLECULAR WEIGHTS AND GC COMPOSITION OF *S. CEREVISIAE* rDNA RESTRICTION FRAGMENTS

RESTRICTION ENZYME	SmaI	HindIII	HindII + III	EcoRI
Molecular Weights of Restriction Fragments ($\times 10^6$)				
A	5.62[a](44.9%)[b]	4.00(45.9%)	1.60	1.79[a](47.2%)
B		1.60(42.9%)	1.60	1.46[a](41.8%)
C			1.03	1.19[a](44.9%)
D			.90	.41[a]
E			.42	.34[a]
F			.09	.22[a]
G				.17[a]
Total Molecular Weight ($\times 10^6$)	5.62	5.60	5.64	5.58

[a]The molecular weights of these fragments were determined by measuring their contour lengths in the electron microscope. The molecular weights of the remaining fragments were determined by coelectrophoresis with DNA markers of known molecular weight.

[b]Numbers in parentheses show the GC composition of the restriction fragments. These values have been calculated from the fragment buoyant densities in neutral CsCl by the equation of Schildkraut, *et al.* (11).

been used to determine the location of the EcoRI, HindII, HindIII, and SmaI restriction sites in yeast rDNA (12, see below). In addition, we have found several plasmids which contain EcoRI fragments not present in rDNA. We are currently characterizing the EcoRI fragments in these plasmids to determine if they represent fragments which are located at the interface between α DNA and rDNA in the yeast chromosomes.

Restriction mapping of *S. cerevisiae* rDNA. The restriction mapping experiments were done primarily with pJHC11, which contains a 9.6 x 10^6 dalton yeast rDNA segment and yields a complete monomer rDNA fragment upon digestion with SmaI. Our mapping strategy involved the following approaches. We determined the molecular weights and the stoichiometry of the fragments of pJHC11 generated by digestion with SmaI and HindIII. The orientations of these fragments with respect to one another and with respect to the locations of the EcoRI cleavage sites were determined by digesting the plasmid with combinations of SmaI + HindIII, SmaI + EcoRI, and HindIII + EcoRI, and by redigestion of purified SmaI-HindIII restriction fragments with EcoRI (12). The location of the HindII sites on pJHC11 was determined by the same approach—an analysis of the digestion products from HindII alone and in combination with other enzymes (12). This information is summarized in Figure 1 which shows the restriction map of pJHC11.

We have characterized a number of other rDNA hybrid plasmids, some containing less than a complete rDNA unit, others containing more. In the different plasmids the rDNA EcoRI fragment order is circularly permuted and in cases where the rDNA segment is larger than 5.6 x 10^6 daltons, different EcoRI fragments are duplicated. We have found plasmids with rDNA inserted in the same or in the opposite orientation to the rDNA in pJHC11. In all the hybrid plasmids studied to date, the restriction fragment map is the same. The analysis of these plasmids has confirmed our earlier observations with purified rDNA from yeast. The rDNA repeating units are uniform in size and composition, they are not separated by heterogeneous spacer regions, and they are arranged in head to tail array.

Figure 2 shows a restriction map of two tandem rDNA repeating units based on mapping information from the hybrid plasmids. Because there is only one SmaI restriction site per unit, we have used this site to orient the map. However, we do not know the location of this site with respect to the ends of the repeating units as they exist in the yeast chromosomes. Our EcoRI map for *S. cerevisiae* rDNA agrees with the results of others (Kramer, personal communication, Petes, personal communication) but differs slightly from the map reported for *S. carlsbergensis* rDNA (15).

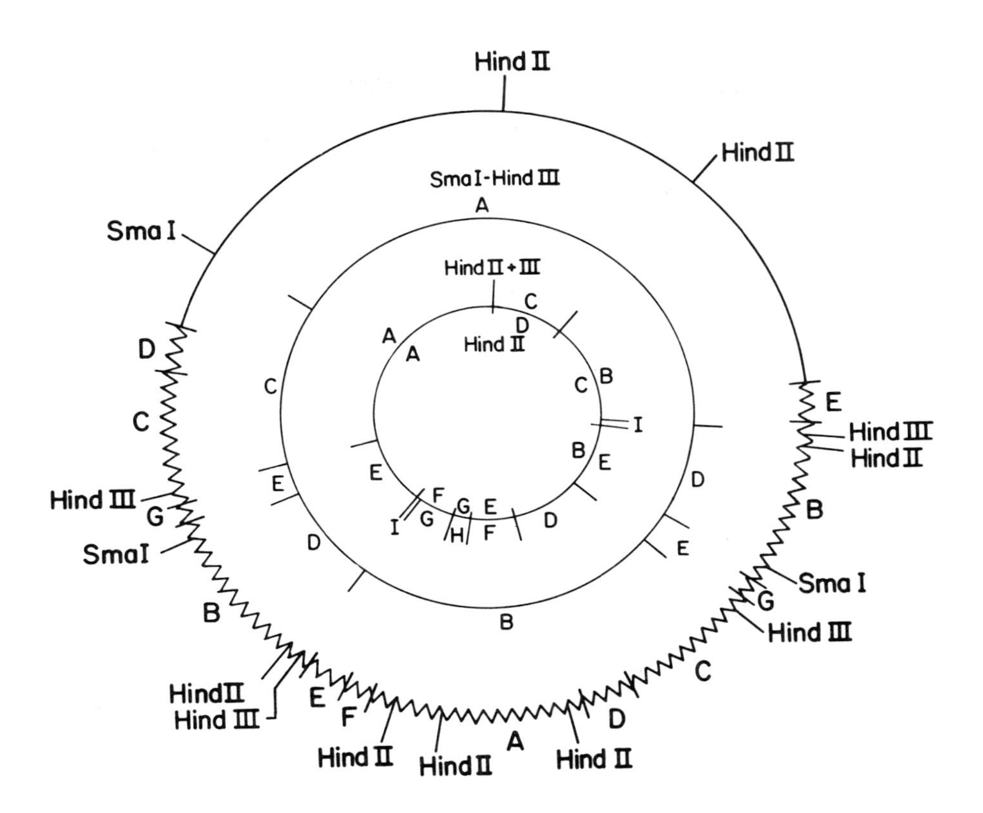

Fig. 1. Restriction map of hybrid plasmid pJHC11 DNA. In the outermost circle the RSF2124 portion of the plasmid is represented by a solid line (——) and the yeast rDNA portion is represented by a sawtooth (⋀⋀⋀). The SmaI, HindII and HindIII cleavage sites are labeled. The EcoRI cleavage sites are indicated by the short divider lines and the yeast rDNA EcoRI fragments are labeled with the appropriate letters. In the inner circles the entire plasmid is represented by a solid line (——). The middle circle shows the pJHC11 SmaI-HindIII fragments; the cleavage sites for these two enzymes are designated. The innermost circle shows the pJHC11 HindII + III fragments on the outside and the pJHC11 HindII fragments on the inside. The HindII and the HindIII cleavage sites are indicated.

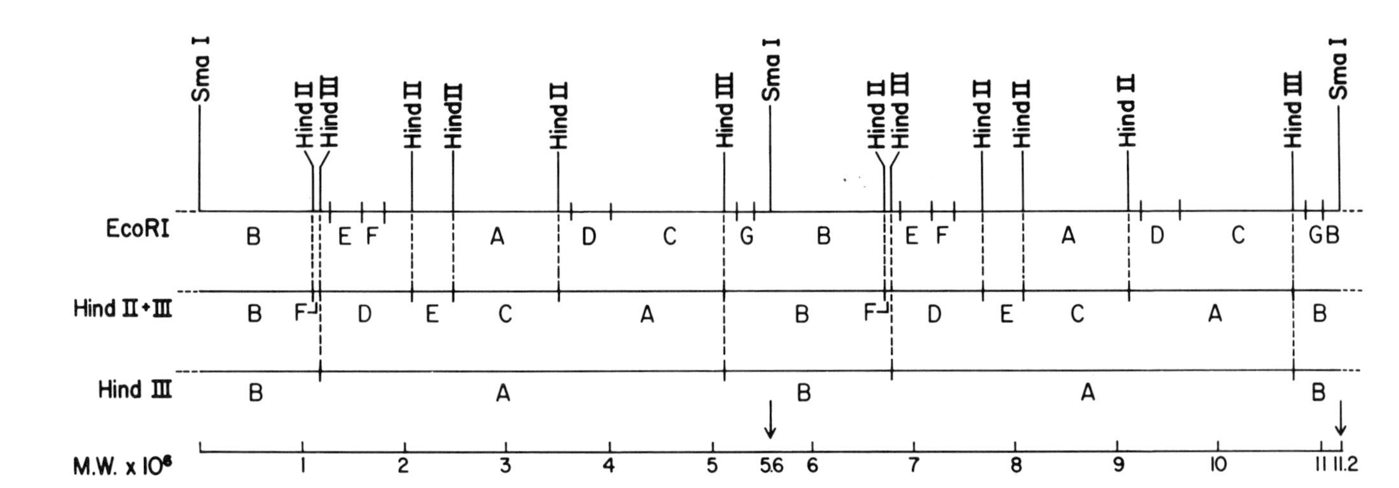

Fig. 2. Restriction map of S. cerevisiae rDNA. Each horizontal line represents two complete repeat lengths of yeast rDNA beginning at the SmaI cleavage site. Each repeat has a molecular weight of 5.6×10^6. Short unmarked divider lines intersecting the top line represent EcoRI cleavage sites; cleavage sites of SmaI, HindII and HindIII are labeled. The letters under each line mark the restriction fragments produced by digestion with the enzyme or enzymes indicated to the left. Those restriction fragments which are cleaved by SmaI are shown only in part at the ends of the map.

Denaturation mapping of *S. cerevisiae* rDNA. The thermal denaturation curves of *S. cerevisiae* rDNA are biphasic (Cramer, unpublished results), indicating that this DNA contains regions which differ widely in their G·C composition. The midpoints of the first and second phases of the rDNA temperature transition curve are representative of DNA regions with G·C compositions of approximately 28% and 44%, respectively (Cramer, unpublished results). Since the G·C composition of yeast rRNA is 46-47% (16,17), the rRNA coding regions must be located in the DNA with the higher melting transition. In order to locate the regions of different base composition in the rDNA, we have prepared a denaturation map of the repeating units. Purified rDNA was digested with SmaI to yield uniform fragments of monomer length, the fragments were partially denatured at pH 10.6, and a histogram of the denatured sites was prepared (Figure 3a). As in the case of the restriction map, we have used the SmaI restriction site for orienting the denaturation map of the repeating units. Therefore, a SmaI cleavage site marks both the beginning and the end of the histogram. To align the denaturation map with the restriction map, we have prepared denaturation maps of the three rDNA fragments produced by a double digestion with SmaI and HindIII. In Figure 3b, these three denaturation maps have been juxtaposed and aligned with the map of the SmaI fragments. By locating the two HindIII restriction sites on the denaturation map, we can determine the positions of the other restriction sites which we have mapped (Figure 3).

From the work of others (4,18) we know that EcoRI fragments A, F and E hybridize to 25S rRNA, while EcoRI fragment C and possibly part of fragments D and G hybridize to 18S rRNA. The regions of the denaturation map which correspond to the location of the 18S and 25S rRNA genes show very few denatured sites. In contrast, the more highly denatured, A·T rich regions in EcoRI fragments B and D appear to be located in transcribed or nontranscribed spacer regions. More detailed information on the location of the 18S rRNA coding region will be necessary before we can determine if the denaturation site in EcoRI fragment C lies within the 18S rRNA gene or a spacer region.

We have calculated the G·C percentages of some of the purified restriction fragments (Table 1). The relative values correlate very well with what would be expected for these fragments based on their positions with respect to the denaturation map.

Restriction endonuclease analysis of *S. carlsbergensis* rDNA. Retel and Van Keulen (19) have suggested that in *S. carlsbergensis* the rRNA precursor genes are interspersed with heterogeneous, A·T rich spacer sequences which account for about 40% of the rDNA. Such a pattern of organization

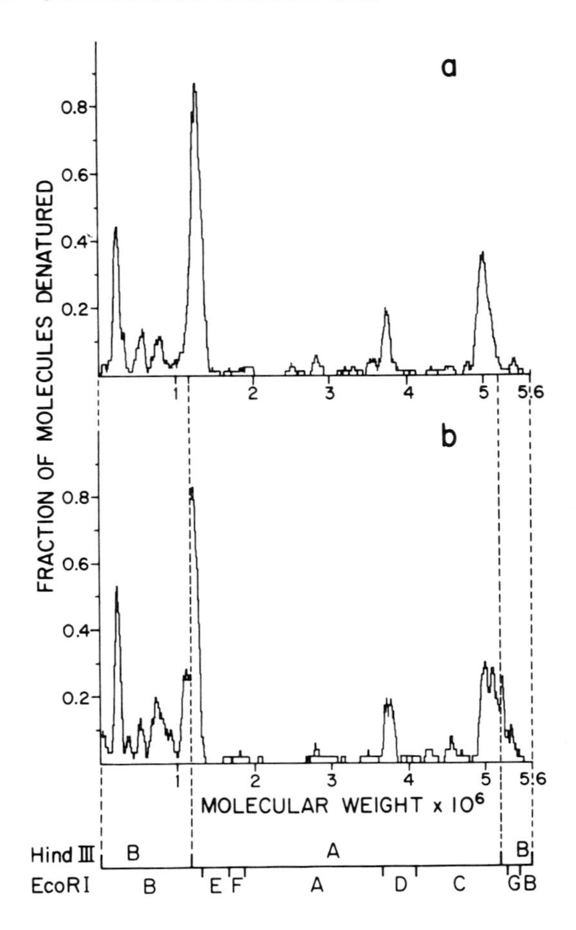

Fig. 3. Denaturation map of S. cerevisiae rDNA restriction fragments. Purified S. cerevisiae rDNA was digested with SmaI alone or SmaI + HindIII, partially denatured, and examined by electron microscopy. Histograms of the denatured sites in the SmaI (a) and the three SmaI + HindIII fragments (b) were prepared. In (b) the maps of the three fragments have been juxtaposed and aligned with the SmaI fragment denaturation map in (a) to locate the HindIII restriction sites on the denaturation map.

The SmaI restriction site marks the beginning and the end of the histogram. The vertical dashed lines represent the SmaI and HindIII restriction sites and show the boundaries of the fragments in each map. The restriction map at the bottom of the figure shows the location of the HindIII and EcoRI restriction fragments with respect to the denaturation map.

would be quite different from what we observe for the rDNA of S. cerevisiae. In order to examine the degree of similarity in the rDNAs of these two yeast species, we have compared their restriction endonuclease patterns. Figure 4 shows the SmaI, HindIII and EcoRI restriction patterns of purified S. carlsbergensis rDNA compared with the patterns of the hybrid plasmid pJHC11. Each set of gels in Figure 4 shows the digestion pattern of pJHC11, which contains S. cerevisiae rDNA restriction fragments, on the left, S. carlsbergensis rDNA on the right and a coelectrophoresis of the two in the center. The rDNA fragments are labeled in each gel set. The additional restriction fragments in the pJHC11 digests result from the RSF2124 portion of the plasmid (12, see Figure 1). It is evident from these results that S. carlsbergensis and S. cerevisiae rDNA are the same, at least with respect to their repeat size and the location of the restriction sites for SmaI, HindIII and EcoRI. It is possible that minor differences observable only by more sensitive techniques, such as heteroduplex mapping, exist between the two rDNAs.

DISCUSSION

We have constructed a number of different plasmids which contain varying amounts of S. cerevisiae rDNA. These plasmids will be very useful in detailed mapping and structural studies of RNA coding regions, spacer sequences, RNA polymerase binding sites and other important functional regions in rDNA. In addition to their importance in examining the organization of the rDNA repeating units, the hybrid plasmids should be valuable in studies of the transcription and replication of yeast rDNA both in vivo and in vitro.

Our studies have shown that the individual yeast rDNA repeating units contain regions which differ widely in their G·C composition. The spacer sequences, both transcribed and nontranscribed, are A·T rich in comparison with the rRNA coding regions. Perhaps this difference in base composition is related to the different functional roles of the spacer and the RNA coding regions.

The structure of the rDNA repeating units is highly conserved in two closely related yeast species, S. cerevisiae and S. carlsbergensis. Studies are currently in progress to determine the degree of similarity in the rDNAs of a number of different yeast species.

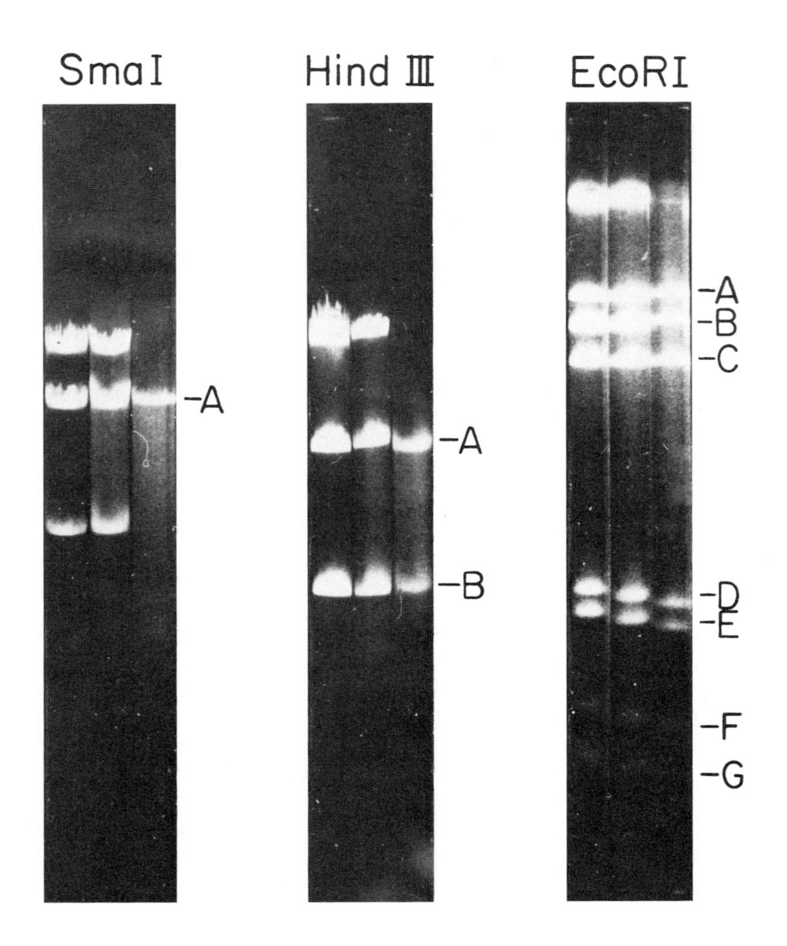

Fig. 4. Restriction patterns of S. carlsbergensis rDNA. Purified rDNA from S. carlsbergensis was digested with SmaI, HindIII or EcoRI. pJHC11 was digested with the same enzymes and the fragments from the rDNA portion of the plasmid were used as electrophoretic markers for S. cerevisiae rDNA restriction fragments.

Each set of three gels includes a pJHC11 DNA digest on the left, S. carlsbergensis rDNA digest on the right, and a coelectrophoresis of the two in the center. The positions of the rDNA restriction fragment(s) for each particular enzyme are indicated to the right of each gel set. The additional restriction fragments in the pJHC11 digests result from the RSF2124 portion of the plasmid.

ACKNOWLEDGEMENTS

This work has been supported by National Science Foundation grants GB41551 and 144J278 and U.S. Public Health Service grant GM14398.

REFERENCES

1. Schweizer, E., MacKechnie, C. and Halvorson, H.O. (1969) J. Mol. Biol. 40, 261.
2. Udem, S.A. and Warner, J.R. (1972) J. Mol. Biol. 65, 227.
3. Rubin, G.M. and Sulston, J.E. (1973) J.Mol. Biol. 79,521.
4. Kramer, R., Philippsen, P., Cameron, J. and Davis, R. (1976) in Molecular Mechanisms in the Control of Gene Expression, eds. Nierlich, D.P., Rutter, W.J. and Fox, C.F. (Academic Press: New York) p. 581.
5. McLaughlin, C.S. (1974) in Ribosomes, eds. Nomura, M. Tissières, A. and Lengyel, P. (Cold Spring Harbor Press) p. 815.
6. Aarstad, K. and Øyen, T.B. (1975) FEBS Letters 51, 227.
7. Cramer, J.H., Bhargava, M.M. and Halvorson, H.O. (1972) J. Mol. Biol. 71, 11.
8. Finkelstein, D.B., Blamire, J. and Marmur, J. (1972) Nature New Biol. 240, 279.
9. Kaback, D.B., Bhargava, M.M. and Halvorson, H.O. (1973) J. Mol. Biol. 79, 735.
10. Cramer, J.H., Farrelly, F.W. and Rownd, R.H. (1976) Molec. Gen. Genet. 148, 233.
11. Schildkraut, C.L., Marmur, J. and Doty, P. (1962) J. Mol. Biol. 4,430.
12. Cramer, J.H.,Farrelly, F.W.,Barnitz, J.T. and Rownd, R.H. (1977) Molec. Gen. Genet. in press.
13. Inman, R.B. and Schnös, M. (1970) J. Mol. Biol. 49, 93.
14. So, M.,Gill, R. and Falkow, S.F. (1975) Molec. Gen. Genet 142, 239.
15. Meyerink, J.H. and Retel, J. (1976) Nucleic Acids Res. 3, 2697.
16. Fauman, M.,Rabiwitz, M. and Getz, G.S. (1969) Biochim. Biophys. Acta 182, 355.
17. Sogin, S.J., Haber, J.E. and Halvorson, H.O. (1972) J. Bact. 112, 806.
18. Nath, K. and Bollon, A.P. (1976) Molec. Gen. Genet. 147, 153.
19. Retel, J. and Van Keulen, H. (1975) Eur. J. Biochem. 58, 51.

ORGANIZATION OF YEAST RIBOSOMAL DNA

Thomas D. Petes, Lynna M. Hereford* and David Botstein

Department of Biology
Massachusetts Institute of Technology
Cambridge, Massachusetts 02139
*Rosensteil Basic Medical Sciences Research Center
and Department of Biology
Brandeis University
Waltham, Massachusetts 02154

ABSTRACT. The intracistronic and intercistronic organization of yeast ribosomal DNA has been studied. A yeast strain containing two types of ribosomal cistrons (distinguishable by their EcoR1 restriction patterns) is described. Restriction maps of both types of ribosomal cistrons have been constructed.

The meiotic segregation pattern of ribosomal DNA has been examined in a yeast strain containing the two types of rDNA. The heterogeneity (for seven of the eight tetrads examined) segregated as a single Mendelian unit, indicating that all of the information for the yeast ribosomal DNA is located on a single chromosome. The data also suggest that meiotic recombination among the ribosomal DNA cistrons may be much less frequent than recombination in the remainder of the yeast genome.

INTRODUCTION

The yeast Saccharomyces cerevisiae, in common with other eucaryotes, contains many copies of the sequences which code for ribosomal RNA. A haploid yeast cell has about 100 to 140 copies of the ribosomal DNA cistrons (1). In this report, both the intracistronic and intercistronic arrangement of these repeating units has been studied.

RESTRICTION MAPPING OF YEAST RIBOSOMAL DNA

Ribosomal DNA from the diploid yeast strain +D4 was isolated and digested with EcoR1 restriction enzyme. The resulting fragments were examined on agarose gels. As shown in Figure 1a, eight bands were observed. Seven of these bands were similar in electrophoretic mobility to those observed by other workers (2,3,4). Fragments A, B, C, D,

E, F and G had molecular weights of 1.7, 1.45, 1.15, 0.4, 0.35, 0.22 and 0.17 x 10^6 daltons respectively. Fragment X' had a molecular weight of 1.85 x 10^6 daltons.

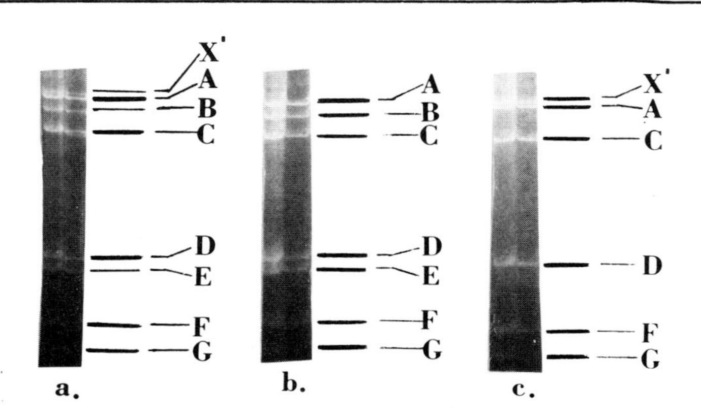

Figure 1. EcoRl restriction analysis of yeast ribosomal DNA from a diploid yeast strain (+D4) and its haploid parental strains (A364a and 2262). Yeast ribosomal DNA was isolated by dye-density gradient centrifugation as described by Williamson (5) substituting the fluorescent dye Hoechst 33258 for DAPI (Williamson, personal communication). Ribosomal DNA from each strain was digested with EcoRl restriction enzyme and electrophoresed on a 1.4% agarose gel. The bands were photographed under short wavelength ultraviolet illumination (6). Next to each photograph of a gel, a schematic drawing of the bands is given, since the smaller fragments (D,E,F and G) are not well visualized.

Figure 1a. Restriction digest of rDNA from the diploid strain +D4. The thin lines in the schematic drawing for bands X', B and E indicates that the number of copies per cell for these fragments is half that for fragments A,C,D,F and G.

Figure 1b. Restriction digest of rDNA from the haploid strain A364a.

Figure 1c. Restriction digest of rDNA from the haploid yeast strain 2262.

When ribosomal DNA from the haploid parents of +D4 was analyzed by a similar procedure, the restriction patterns shown in Figures 1b and 1c were found. The haploid parent strain A364a contains seven EcoRl fragments, A,B,C,D,E,F and G. This seven fragment pattern will be referred to in this paper as the Type I ribosomal cistron. The second haploid parent, 2262, had ribosomal DNA which produced six fragments after digestion with EcoR1, X',A,C,D,F and G (Type II ribosomal cistron). The diploid +D4, therefore, shows the pattern expected if each haploid strain contributed an equal number of ribosomal cistrons to the diploid.

The order of the EcoR1 fragments in the two types of ribosomal cistrons was examined by analyzing recombinant DNA molecules (constructed in vitro) which contained insertions of yeast ribosomal DNA from the strain +D4 (Petes, Broach, Wensink, Hereford, Fink and Botstein, in preparation). Recombinant molecules with yeast ribosomal DNA were analyzed by digesting isolated plasmid DNA from individual clones with the EcoR1 restriction enzyme (Petes, Hereford and Skryabin, in preparation). Since the clones were constructed by insertion of sheared yeast DNA into the single EcoR1 restriction site of the bacterial plasmid pMB9 (provided by H.W. Boyer) using the poly-d(A-T) connector method (7), all the sites cleaved by EcoR1 are within the yeast DNA segment of the recombinant molecule.

As expected from the analysis of the haploid parents of +D4, no recombinant clones were observed which contained both the X' fragment and the B or E fragments, although X' was found in clones containing the A,C,D,F and G fragments. This confirms the idea that two types of ribosomal cistrons were present in the diploid +D4. Maps of these two types of cistrons were constructed using two different techniques.

Of the seventy-five clones analyzed, twenty-four contained either fragment B or fragment E plus other fragments (Type I ribosomal cistron). As shown in Table 1, by determining which fragments are found together in clones containing a small number of fragments, we constructed an unambiguous map of the Type I cistron. Although only six clones were used to construct the maps shown in Table 1 and Figure 2a, the structure of all clones with fragments B and E conforms to these maps.

Only five of the clones analyzed contained the X' fragment. Since there were not enough clones to construct an unambiguous map by the same technique used for the Type I cistrons, a different approach was used. The recombinant clone pY1rA12 which contained all six fragments found in Type II cistrons (X', A,C,D,F and G) was analyzed with the restriction enzymes HindIII and Bgl II. This analysis allowed the construction of the restriction map shown in Figure 2b. The structures of all clones containing X' are consistent with this map.

A summary of the two restriction maps is given in Figure 2. The restriction maps are circular because the clones which were analyzed were random fragments of a tandem

TABLE 1

MAPPING OF ECORl RESTRICTION FRAGMENTS
OF TYPE I RIBOSOMAL CISTRONS

Clone Designation	Fragments Produced After EcoRl Treatment*	Derived Map**
pYlrF2	BG	GB
pYlrF10	ABEF	GB(AEF)
pYlrG12	BCEG	CGBE(AF)
pYlrB3	BEFG	CGBEFA
pYlrFl	ABDEFG	CGBEFAD
pYlrBl	ACDEF	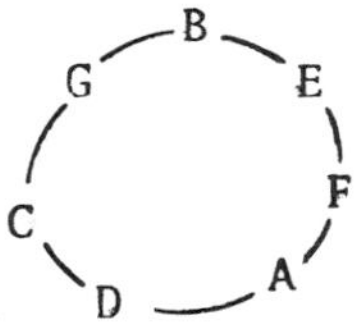

*excluding fragments containing pMB9 DNA.
**order not determined for fragments within brackets.

array of ribosomal cistrons (Petes, Hereford and Skryabin, in preparation). Each ribosomal cistron is not a small circular DNA molecule (8). As expected from the restriction patterns of the haploid parental strains, the fragment X' of the Type II cistron replaces the fragments B and E of the Type I cistron. Additional evidence also suggests that X' and, B and E, are alternative possibilities. First, the size of X' is approximately the sum of the sizes of B and E. Second, hybridization studies (Petes and Hereford, unpublished data) showed that the X' fragment hybridized to both B and E. The X' fragment also has several restriction sites (for enzymes other than EcoRl) at the positions expected if X' were a fusion of B and E. Finally, the stochiometry of the fragments produced by EcoRl digestion of ribosomal DNA isolated from the diploid +D4 indicated that the fragments A,C,D,F and G were present in about the same number of copies per cell; fragments X', B and E were also present in approximately equal molar ratios but there only half the number of fragments per cell as for A,C,D,F and G. All these experiments support the conclusion that the yeast strain +D4 contains two types of ribosomal DNA cistrons in roughly equal amounts. Although the only difference between Type I and Type II ribosomal cistrons which has been detected thus far is a change at a single EcoRl site, much more extensive heterogeneity between these different types of cistrons may exist.

The analysis of the recombinant DNA molecules, in addition to allowing restriction mapping of the ribosomal DNA, yielded information on the homogeneity and clustering of the ribosomal cistrons. Seventy-four of seventy-five clones contained only EcoR1 fragments with the mobility of X',A,B, C,D,E,F or G (in addition to the pMB9 fragment). This result suggests that yeast ribosomal cistrons, including the non-transcribed spacer, are homogenous in size. It also indicates that the ribosomal cistrons are highly clustered since junctions of unique DNA with ribosomal DNA would have been detected as recombinant clones with EcoR1 fragments of unique mobilities.

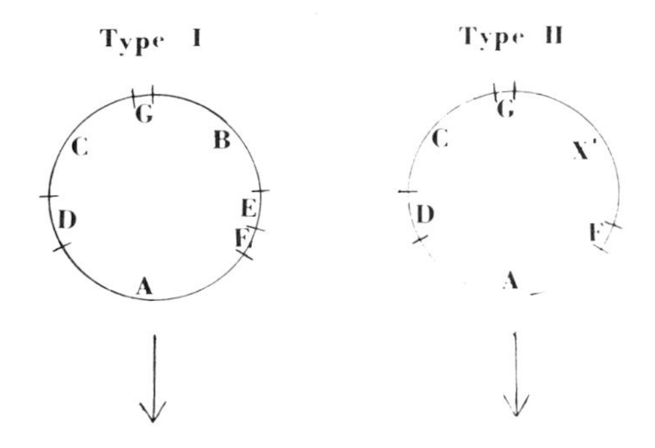

Figure 2. EcoR1 restriction maps of the two types of ribosomal cistrons present in the diploid yeast strain +D4. As described in the text of this paper, the restriction maps were assembled by analysis of recombinant DNA molecules (constructed in vitro) which contained insertions of yeast ribosomal DNA. The restriction maps are circular since the ribosomal cistrons are tandomly arranged on the chromosomes (8,9).

GENETIC ANALYSIS OF YEAST RIBOSOMAL DNA

The interchromosomal arrangement of ribosomal DNA was investigated by genetic techniques. Since the yeast strain (+D4) which has both Type I and Type II rDNA cistrons is a diploid, the strain can be induced to undergo meiosis. The EcoR1 restriction pattern of the ribosomal DNA from each of the four spores produced by a single meiosis can be analyzed. If the ribosomal cistrons are located on a single chromosome and the rDNA cistrons are not mixed by recombination, the different restriction patterns for the ribosomal DNA should segregate 2:2 (two spores having only the Type I pattern and two spores having only the Type II pattern). If the ribosomal cistrons are on two chromosomes, half of the meiotic events will result in a 2:2 segregation for the restriction patterns; the other half of the meiotic events will result in four spores, each containing a mixture of Type I and Type II rDNA. If the ribosomal cistrons are distributed over many yeast chromosomes, since non-homologous chromosomes segregate independently at meiosis, almost all the spores will contain mixtures of Type I and Type II rDNA.

Eight tetrads have been dissected and the restriction patterns of rDNA isolated from the 32 spores have been analyzed. Seven of the eight tetrads showed a 2:2 segregation of the ribosomal cistrons. An example of this segregation is shown in Figure 3.

One of the eight tetrads did not show a 2:2 segregation for the different rDNA patterns. In this tetrad, two of the spores had the Type I rDNA pattern and two of the spores had mostly the Type I rDNA pattern (about 80%) but also contained some rDNA of the Type II pattern. The possibility that this tetrad is an artifact has not been ruled out although the 2:2 segregation of other genetic markers in this tetrad indicate that it is unlikely to be a false tetrad. An alternative explanation for this tetrad is that it represents segregation of ribosomal cistrons in a cell in which a mitotic cross-over occured before induction of meiosis.

The observation that in seven of the eight tetrads examined the two types of ribosomal cistrons segregate 2:2 strongly suggests that all the information for the yeast ribosomal cistrons is located on a single chromosome. The result also indicates that none of the information for the rDNA is extra-chromosomal. Experiments are in progress to determine whether these ribosomal cistrons are located on chromosome I of yeast, as indicated by some previous studies (10,11).

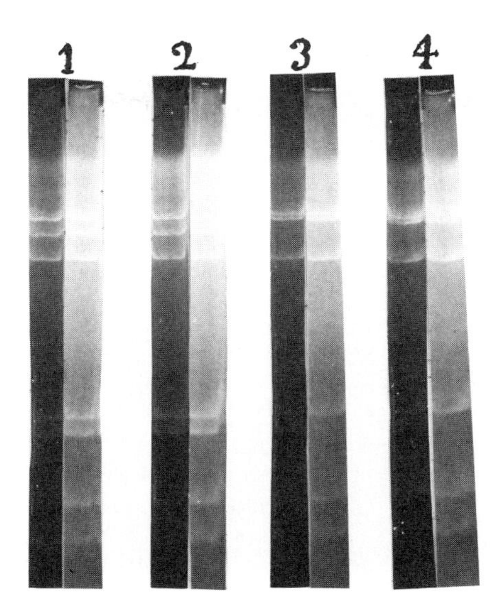

Figure 3. 2:2 segregation pattern of ribosomal DNA in the four spores of a yeast tetrad. The diploid +D4, which contains both Type I and Type II cistrons, was sporulated. The tetrads produced at meiosis were dissected. Spores were grown up individually and rDNA was isolated. The rDNA was treated with EcoR1 and the digestion products analyzed on agarose gels. The four gel patterns shown (duplicates of each sample are shown at different photographic exposures) are the result of the analysis of a single tetrad. Spores 1 and 2 show the Type I rDNA pattern; spores 3 and 4 show the Type II rDNA pattern.

The lack of heterogeneity in the spores also indicates that yeast ribosomal DNA cistrons do not recombine frequently. It can be calculated from genetic data (12) that each yeast cell recombines at least 52 times each meiosis. Since ribosomal DNA represents about 5% of the total DNA of the cell (1), the ribosomal cistrons would be expected to recombine 2 to 3 times in each cell in each meiosis. The pure segregation of Type I and Type II patterns in seven of the eight tetrads indicates that the frequency of recombination is much lower than predicted. The simplest interpretation of this result is that meiotic recombination of the yeast ribosomal DNA is suppressed. Alternative explanations, such as "master-slave" corrections of recombined chromosomes or amplification of a single germ line copy of ribosomal DNA following meiosis, have not yet been excluded.

ACKNOWLEDGEMENTS

T.D.P. was supported by a N.I.H. postdoctoral fellowship. L.M.H. was supported by a Helen Hay Whitney fellowship, a grant from the Medical Foundation and a N.I.H. grant (GM 23549) to Michael Rosbash. The work was also supported by N.I.H. grants to D.B. (GM 21253, GM 18973) and a Research Career Development Award to D.B.

REFERENCES

1. Schweizer, E., MacKechnie, C. and Halvorson, H.O. (1969) J. Mol. Biol. 40, 261.
2. Meyerink, J.H. and Retel, J. (1976) Nucleic Acids Research 3, 2697.
3. Nath, K. and Bollon, A.P. (1976) Mol. Gen. Genet. 147, 153.
4. Cramer, J.H., Farrelly, F.W. and Rownd, R.H. (1976) Mol. Gen. Genet. 148, 233.
5. Williamson, D.H. and Fennel, D.J. (1975) Methods in Cell Biology 12, 335.
6. Sharp, P.A., Sugden, B. and Sambrook, J. (1973). Biochem. 12, 3055.
7. Lobban, P.E. and Kaiser, A.D. (1973) J. Mol. Biol. 78, 453.
8. Stevens, B.J. and Moustachhi, E. (1972) Exptl. Cell Res. 64, 259.
9. Cramer, J.H., Bhargava, M.M. and Halvorson, H.O. (1972) J. Mol. Biol. 71, 11.
10. Finkelstein, D.B., Blamire, J. and Marmur, J. (1972) Nature New Biol. 240, 279.
11. Goldberg, S., Oyen, T., Idriss, J.M. and Halvorson, H.O. (1972) Mol. Gen. Genet. 116, 139.
12. Sherman, F. and Lawrence, C.W. (1974). Handbook of Genetics 1, 359.

TRICKS FOR RNA LABELING AND PARTIAL RESTRICTION DIGESTS

Nancy Maizels

The Biological Laboratories
Harvard University
Cambridge, Massachusetts 02138

ABSTRACT. Two techniques facilitated mapping the Dictyostelium ribosomal DNA. 1) RNAs were labeled in vitro at specific activities of about 3×10^7 cpm/μg, using T4 polynucleotide kinase to transfer the γ-phosphate from γ-^{32}P-ATP to a free 5' hydroxyl. 2) Digestion of DNA with eco Rl in the presence of the antibiotic distamycin A yielded partial digestion products useful for mapping and cloning.

INTRODUCTION

Analyses of eukaryotic ribosomal DNAs have shown that the transcribed region is embedded in a "non-transcribed spacer," and that the size of the entire region varies from organism to organism. The 5s DNA is linked to the rest of the coding region in bacteria and yeast, but not in higher eukaryotes. Restriction mapping and cloning techniques have generated a picture of the size and organization of the repeated region of the Dictyostelium genome which codes for rDNA. It is at least 38 kbp in length, the largest rDNA repeated unit yet reported, and the 5s DNA is linked to the 17s + 25s coding region. The mapping is described in detail in ref. 1, and has been confirmed by Cockburn et al. (2). The rDNA is repeated about 140 times per haploid nucleus (3), and the size of the Dictyostelium genome is 4.5×10^4 kbp, so about 12% of the genome is devoted to this repeat, and only 20% of each repeat is transcribed into rRNA.

Two techniques facilitated mapping the Dictyostelium rDNA. 1) Phage T4 polynucleotide kinase mediated in vitro labeling of RNAs with γ-^{32}P-ATP at specific activities 100-fold higher than those obtained by in vivo labeling. 2) The antibiotic distamycin A promoted incomplete digestion of Dictyostelium DNA by eco Rl, to yield partial digestion products useful for mapping and cloning. These techniques are described below.

RNA LABELING MEDIATED BY T4 POLYNUCLEOTIDE KINASE

Phage T4 polynucleotide kinase catalyzes phosphorylation of free 5' hydroxyls using the γ-phosphate of ATP as donor (4). Kinase will tag ends of nucleic acid molecules at a maximum theoretical specific activity of about 10,000 C/mmole of ends, the maximum specific activity of donor γ-^{32}P-ATP prepared as described by Glynn and Chappell (5); more typical specific activities of γ-^{32}P-ATP are about 3000 C/mmole, or about 2 x 10^{15} Cerenkov cpm/mmole. When, as in hybridizations, molecular integrity is not a concern, RNA molecules can be labeled at extremely high specific activity by preceding kinase labeling with mild alkaline hydrolysis. This degrades RNA non-specifically to produce fragments with free 5' hydroxyl termini which serve as sites for kinase-mediated labeling.

To label Dictyostelium rRNA, about 1 μg of a single purified species was hydrolyzed in 5 μl of 50 m*M* tris, pH 9.5, in a sealed capillary at 90°C. The appropriate timing for partial hydrolysis (5 to 30 min) was initially determined by varying the duration of hydrolysis of a small amount of RNA which had been lightly labeled following 1 min of hydrolysis. Degradation was assayed by electrophoresis on a 10% acrylamide gel in tris-borate-EDTA buffer (6), where molecules about 100 bases long migrate with a xylene cyanol dye marker. Each RNA preparation requires a different timing of hydrolysis, which very likely reflects both composition and purity of the RNA preparation. Every time RNA was hydrolyzed and labeled the reaction was monitored by gel electrophoresis.

Hydrolysis of 1 μg of RNA to an average length of 100 bases produces about 30 pm of 5' hydroxyl ends. This RNA is added to a tube containing 100 pm of γ-^{32}P-ATP which has been desiccated to dryness. Then the reaction is made up to 50 m*M* tris, pH 9.5, 10 m*M* $MgCl_2$, 5 m*M* dithiothreitol, 5% glycerol in a final volume of 10-15 μl and 3 μl of T4 polynucleotide kinase (purified as described by Panet et al. (7)) is added and the reaction incubated 30 min at 37°C. The labeled RNA is separated from unincorporated ^{32}P by chromatography on a 1 ml Sephadex G50 column in 0.1 m*M* EDTA, pH 8.0, or by repeated ethanol precipitations in the presence of 10 μg

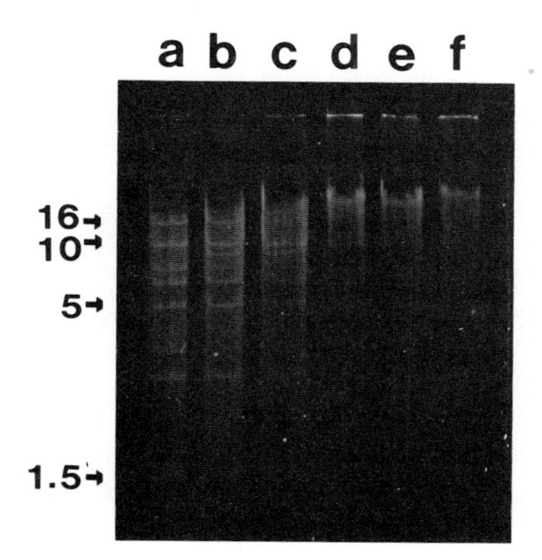

FIGURE. Agarose gel electrophoresis of Dictyostelium nuclear DNA digested with eco Rl in the presence of 0, 10, 20, 50, 100 and 200 μg/ml distamycin A, shown in lanes a through f, respectively. Arrows on the left indicate fragment sizes in kbp. Samples were extracted once with phenol and twice with ether prior to gel electrophoresis.

---

tRNA carrier. The specific activity of the labeled RNA is about 3×10^7 cpm/μg.

PARTIAL eco Rl DIGESTION MEDIATED BY DISTAMYCIN A

Braga et al. (8) first reported that the antibiotic distamycin A enhances partial digestion of phage λ DNA by eco Rl. Presumably the antibiotic binds to DNA, and thus protects certain regions from cleavage by the restriction enzyme. As shown in the figure, when Dictyostelium nuclear DNA at 100 μg/ml is digested with eco Rl in the presence of varying concentrations of distamycin A (Boehringer-Mannheim), the mean length of the digestion products increases with increasing antibiotic concentration. One can also encourage partial rather than complete digestion by using limiting amounts of enzyme, but the antibiotic has the advantages that its effect is stable while enzyme titers vary; and, since some enzyme preparations produce partial digests over only narrow ranges of concentration and incubation times, varying the concentration of antibiotic offers a convenient fine-tuning control on the size of digestion products. The dyes ethidium bromide and Hoechst 33258, and the antibiotic actinomycin D, are also effective in promoting partial eco R1 digestion.

To identify neighboring eco R1 fragments in the rDNA, products of distamycin A-induced partial digestion were fractionated by agarose gel electrophoresis, and then digested to completion and their component fragments identified. Partial eco R1 digestion products were also cloned, using T4 DNA ligase to insert them into the eco R1 site of the vector pMB9 (9). One plasmid thus generated, pDd64, carries a 9.5 kbp insertion made up of three adjacent R1 fragments (5, 3 and 1.5 kbp, respectively) which correspond to most of the transcribed region of the Dictyostelium rDNA. Long regions of DNA carrying multiple restriction sites are often desirable substrates for cloning, especially when one wishes to obtain an intact gene or operon. Clearly, when ligase inserts partial digestion products into a vector it may juxtapose fragments which are not neighbors in the genome, but this can easily be sorted out by analyzing the cloned DNA with other restriction enzymes.

ACKNOWLEDGEMENTS

I am a Junior Fellow of the Society of Fellows, Harvard University. This work was supported by NSF grant BMS75-21164.

REFERENCES

1. Maizels, N. (1976) Cell 9, 431.
2. Cockburn, A.F., Newkirk, M.J., and Firtel, R.A. (1976) Cell 9, 605.
3. Firtel, R.A., and Bonner, J. (1972) J. Mol. Biol. 66, 339.
4. Richardson, C.C. (1965) Proc. Nat. Acad. Sci. USA 54, 158.
5. Glynn, I.M., and Chappell, J.B. (1964) Biochem. J. 90, 147.
6. Peacock, A.C., and Dingman, C.W. (1968) Biochem. 7, 668.
7. Panet, A., van de Sande, J.H., Loewen, P.C., Khorana, G., Raae, A.J., Lillehaug, J., and Kleppe, K. (1973) Biochem. 12, 5045.
8. Braga, E.A., Nosikov, V.V., Tanyashin, V.I., Zhuze, A.L., and Polyanovski, O.L. Dokladi of Acad. of Sci., USSR, in press.
9. Rodriguez, R.L., Bolivar, F., Goodman, H.M., Boyer, H.W., and Betlach, M. (1976) ICN-UCLA Symp. on Molecular and Cellular Biology 5, 471.

GENETIC CONTROL OF PHOSPHORUS METABOLISM IN NEUROSPORA

Robert L. Metzenberg and Robert E. Nelson*
Department of Physiological Chemistry
University of Wisconsin, Madison, Wisconsin 53706

ABSTRACT. Several enzymes involved in the acquisition of phosphorus are not made in appreciable amounts by wild-type Neurospora when there is adequate phosphorus in the medium, but are made in large amounts when the cells are growing in low phosphate medium. The genes for alkaline and acid phosphatases have been mapped. They are unlinked to each other and to the regulatory genes. A model that satisfactorily fits the properties of the regulatory mutants may be summarized as follows.

(1) The normal product of the nuc-1 gene is required to turn on expression (transcription) of the structural genes of the phosphorus family.

(2) The normal product of the preg gene is required to nullify the activity of nuc-1 product.

(3) The normal product of the nuc-2 gene is required to nullify the activity of the preg product, or prevent its synthesis.

(4) Phosphate or something derived from it nullifies the activity of the nuc-2 product, or prevents its synthesis.

In this paper, we examine a constitutive mutant, $nuc\text{-}1^c$ (BC-152). In the presence of two doses of the normal preg product, the $nuc\text{-}1^c$ mutant becomes repressible by phosphate. $nuc\text{-}1^c$ (BC-152) may be an overproducer of normal nuc-1 product, and we suggest a way to isolate mutants that are extreme overproducers. We discuss the difficulties of identifying the nuc-1 product and note the value of the "slime" strain in the isolation of nuclei. Finally, we describe preliminary approaches that are being made toward a chemical detection of the nuc-1 product, and toward developing a bioassay that will allow this product to be measured in extracts.

INTRODUCTION

When wild-type Neurospora is starved of inorganic phosphate, it makes it least eight enzymes that are undetectable or barely detectable in phosphate-sufficient cultures. The

*Present address: School of Life Sciences
University of Nebraska-Lincoln
Lincoln, Nebraska 68583

known members of this family of enzymes all have obvious functions in helping the organism to acquire phosphate in times of dearth. These enzymes are alkaline phosphatase (10,11,22,26), acid phosphatase (10,21,26), three nucleases (7,9), a 5'-nucleotidase (8), a special high-affinity phosphate permease, (11,14), and a phosphoethanolamine permease (15). Each of these enzymes is located either in the cell membrane (the two permeases) or mostly between the cell membrane and wall (alkaline phosphatase, 2) or mostly in the medium (the other enzymes). None is in the cytosol. Since Neurospora is not generally permeable to phosphorus compounds other than phosphate itself and the two zwitterions, phosphoryl ethanolamine and phosphoryl choline, this fact makes teleologic sense; nucleic acids, nucleotides, and most phosphate esters must be hydrolyzed outside the cell membrane to serve as sources of phosphorus. Each of these phosphate-repressible enzymes (with the possible exception of phosphoryl ethanolamine permease) has at least one "housekeeping" counterpart that is not markedly affected by the level of phosphate in the medium, and is a different molecular species as judged by enzyme kinetics, fractionation properties, and in at least one case, by immunological criteria (see above references). These non-repressible enzymes are presumably coded by a different set of genes, and probably serve the economy of the cell in ways only distantly related to the adequacy of the phosphorus supply. In the case of the non-repressible phosphate permease, this enzyme is competent to transport phosphate into the cell when either the external phosphate concentration is high or the external pH is reasonably low. In medium of pH greater than 7.0 and external phosphate concentration of 50 μM or less, the repressible phosphate permease becomes essential for growth (11,14).

We know of five genes involved in the synthesis of these phosphate-repressible enzymes. Two of them, designated pho-2 and pho-3, are the structural genes for the repressible alkaline and acid phosphatases, and map in Linkage Groups IV and V, respectively (6,12,18). Mutants in these genes have normal activity of other enzymes in the family. Mutations in three genes, nuc-1, nuc-2, and preg ("pee-reg" for phosphorus regulation) affect several, and perhaps all members of the family in a parallel but not necessarily proportional fashion (8). These three regulatory genes map in Linkage Groups I, II, and II, respectively, with nuc-2 and preg being about 1-2 centimorgans apart. nuc-2 and preg are in functionally separate cistrons, and the genetic distance between them suggests that they are probably not adjacent genes (11). Thus the five genes map to four different linkage groups.

Two known kinds of mutations occur in nuc-1. Those first found by Ishikawa and his collaborators are simply called nuc-1, and are "null mutants" - they are unable to make significant amounts of any of the enzymes of the family even under conditions of phosphate starvation. They are thus unable to grow on nucleic acids as a phosphorus source, or on inorganic phosphate at low concentration and high pH. These mutants are recessive to their wild-type allele, *nuc-1*$^{+}$, both in heterocaryons (25) and in heterozygous partial diploids (16). The other kind of mutant is called *nuc-1*^{c}, and is physiologically the converse of *nuc-1*. It is constitutive, or non-repressible, for alkaline and acid phosphatase and high-affinity phosphate permease (3,16) and perhaps for the other enzymes of the family. *nuc-1*^{c} is dominant to *nuc-1*$^{+}$ in heterocaryons, but has not yet been tested in partial diploids.

Mutants in *nuc-2* are at first glance extremely similar to those in *nuc-1*. Null mutants in this gene do not make enzymes of the phosphorus family, and these *nuc-2* mutants are recessive to *nuc-2*$^{+}$ in heterocaryons and in partial diploids (17). Constitutive mutants in the same cistron, which we unfortunately named *pcon*^{c} for phosphorus-control constitutive before their allelism with *nuc-2* was appreciated, are, like *nuc-1*^{c} non-repressible for a number of enzymes of the family. *pcon*^{c} is dominant to *pcon*$^{+}$ (*nuc-2*$^{+}$) in both heterocaryons and in partial diploids.

Only one kind of *preg* mutant is known: constitutive mutants (*preg*^{c}) that are recessive to *preg*$^{+}$. Double mutants carrying both *nuc-1* and *preg*^{c} are null, and are physiologically indistinguishable from *nuc-1*, whereas those carrying *nuc-2* and *preg*^{c} are fully constitutive, and are indistinguishable from *preg*^{c}. Hence *nuc-1* and *nuc-2* which are superficially identical except for being on different chromosomes, have opposite epistasis/hypostasis properties when viewed in a *preg*^{c} background. A second asymmetry between *nuc-1* and *nuc-2* is seen in their epistatic interactions between a null allele of one and a constitutive in the other. The double mutant *nuc-1*; *pcon*^{c} is null (11,13) but the double mutant *nuc-1*^{c}; *nuc-2* is constitutive (3,16). All this can be summarized by saying that mutations in *nuc-1* are epistatic to those in *preg*, which in turn are epistatic to those in *nuc-2*. The dominance between different alleles and the epistasis between different gene mutations suggest a model in which a cascade of events results in the structural genes for the enzymes being turned on by a positive controller or "expressor", the *nuc-1*$^{+}$ product. Mutants that lack this are irretrievably null. Either the synthesis of the *nuc-1*$^{+}$ product or its activity is normally nullified by the *preg*$^{+}$ product. In turn, the synthesis or activity or effectiveness of the *preg*$^{+}$ product is opposed by the *nuc-2*$^{+}$ product. This, finally, is inhibited or repressed by phosphate or a corepressor derived from it. This rather complicated hierachy of regulatory genes is summarized in Fig. 1 (13).

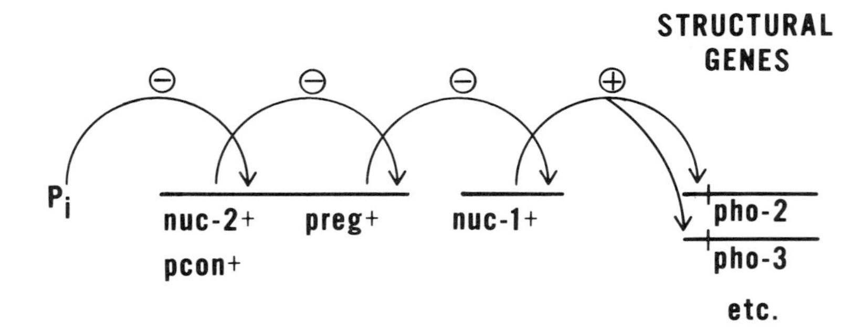

Fig. 1.-Model of the hierarchy of regulatory genes. A "+" over an arrow means "turns on", and a "-" means "turns off", or "represses" or "inhibits the action of". Enzyme production in a strain carrying a mutation in a given control gene can be "calculated" by multiplying the positive or negative signs of the regulatory products. Only those regulatory products that are connected to the structural genes by a sequence unbroken by mutation can be included in the calculation. For example, in nuc-2 strains, only the final two regulatory steps are operative ("-" times "+") and such strains are "-", i.e. null. Regardless of the phosphate concentrations, strains carrying $pcon^c$ mutations have three regulatory steps working ("-" times "-" times "+") and are "+", i.e. constitutive. Similarly, wild-type cells grown under conditions of phosphate starvation are "-" times "-" times "+", which multiplies to "+"; the strains produce alkaline phosphatase and its congeners. From (13), with permission.

In this paper, we first address a rather specific question about how the preg product opposes the nuc-1 product: does it prevent the synthesis of nuc-1 product, or does it somehow titrate out existing nuc-1 product, making it inactive? Second, we consider the longer term prospects for identifying the nuc-1 product as a chemical entity and demonstrating its activity in a bioassay.

METHODS

Assay of repressible alkaline phosphatase. This enzyme is usually assayed in the presence of EDTA to inhibit the constitutive, intracellular enzyme present in crude extracts (10). We have found that this inhibition is never complete, and some of the constitutive enzyme is always being measured at the same time (12). In derepressed cultures, this error is insignificant, but in cultures where the repressible enzyme is at very low levels, the great bulk of the apparent activity is due to the constitutive enzyme.

Strains and genetic manipulations. These are the standard strains that have been described previously (11,13,17) except for the $nuc\text{-}1^c$ mutant. The allele used in this work (BC-152) was isolated in a nuc-2 background by Bill Chia (3). When nuc-2 is mutagenized with UV light and plated to medium with 50 µM phosphate at pH 7.3, many of the colonies that are able to grow are constitutive for both the high affinity permease and for alkaline phosphatase. About 3/4 of these are new $preg^c$ mutants and about 1/4 are $nuc\text{-}1^c$. Chia outcrossed the new mutants to remove the unlinked nuc-2 mutant gene. Using a translocation, strains were constructed that were haploid for $nuc\text{-}1^c$ (BC-152) and diploid for preg and nuc-2, and in each case heterozygous for arg-12. The general method of preparation of these diploids is illustrated in Fig. 2.

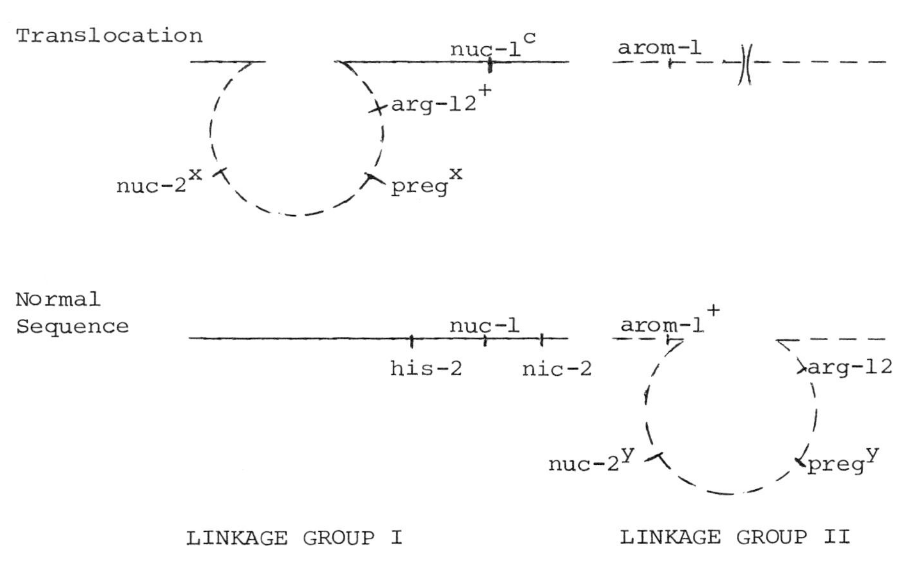

Fig. 2. A translocation strain (T(II→I)NM177), in which a piece of Linkage Group II is moved to Linkage Group I carries $nuc\text{-}1^c$ (BC-152) and the desired alleles of nuc-2 and preg, and in addition, arom-1, which causes a requirement for the aromatic amino acids. arom-1 maps very close to the breakpoint in Linkage Group II. The translocation strain is crossed to a Normal Sequence strain carrying the desired alleles of nuc-2 and preg and also genes for histidine requirement and for nicotinic acid requirement as flanking markers to $nuc\text{-}1^+$. This Normal Sequence strain also carries arg-12, a gene which causes arginine requirement and is recessive to $arg\text{-}12^+$. When ascospores from such a cross are plated to minimal medium, only the prototrophic recombinants are able to grow, and almost all of these are the partial diploids. The very rare

crossovers (which did not arise in this experiment) between *arom-1* and the breakpoint in Linkage Group II would give rise to euploid, prototrophic translocation strains, but these would be easily distinguished by their ascospore production from a partial diploid (24).

RESULTS AND DISCUSSION

Properties of nuc-1^{c}. The data in Table 1 shows an interesting gene dosage effect. In the euploid (haploid) condition, *nuc-1*^{c} (BC-152) is, of course, constitutive for the alkaline phosphatase when it carries one *preg*$^{+}$ and one *nuc-2*$^{+}$. It is likewise constitutive when carrying one *preg*$^{+}$ and no *nuc-2*$^{+}$, i.e., *nuc-1*^{c} is epistatic to *nuc-2*. However, *nuc-1*^{c} carrying two doses of *preg*$^{+}$ and two doses of *nuc-2*$^{+}$ exhibits lower activity of alkaline phosphatase when grown on the limiting phosphorus source (2 mM phosphoryl ethanolamine) and much lower activity when grown on phosphate-sufficient medium (7.35 mM inorganic phosphate. Such a strain is essentially repressible. When two doses of *preg*$^{+}$ and no doses of *nuc-2*$^{+}$ are present, the strain is repressed on high phosphate to a degree hardly different from wild-type. Even on phosphoryl ethanolamine the enzyme level achieved in this strain is only about 5% that of *nuc-1*^{c} carrying zero doses of *preg*$^{+}$. The specific activities achieved in the presence of one dose of *preg*$^{+}$ are roughly intermediate between these two sets of values. It is as though each incremental dose of *preg*$^{+}$ product subtracts in a rather orderly way from the specific activity of alkaline phosphatase.

We are all accustomed to thinking of a mutation that results in loss of function as being recessive to an allele that retains function, but this idea is not always true. It is useful to look more closely at the dominance/recessiveness of *preg*^{c}. In a *nuc-1*$^{+}$ background, *preg*^{c} is completely recessive to *preg*$^{+}$ (17). But let us imagine a world in which the "wild-type" repressible Neurospora is partially diploid and has the genetic constitution *nuc-1*^{c}*; nuc-2*$^{+}$ *preg*$^{+}$*/nuc-2*$^{+}$ *preg*$^{+}$. (See Table 1, Strain G). If we mutated one *preg* to *preg*^{c}, the resulting mutant, like Strain D, Table 1, would be constitutive, and we would have *preg*^{c} as a dominant constitutive mutant. Any model for the regulatory roles of *preg* and *nuc-1* products should explain this paradox.

Three interpretations for the molecular basis of *nuc-1*^{c} (BC-152) come to mind, all of them consistent with the hierarchy of genes shown in Fig. 1. These are:

(1) *preg*$^{+}$ might normally function as a repressor of the synthesis of *nuc-1*$^{+}$. *nuc-1*^{c} would then be interpreted as operator-constitutive for *nuc-1* product synthesis.

(2) *preg*$^{+}$ might normally function by titrating out the

TABLE 1. GENE DOSAGE EFFECTS ON SYNTHESIS OF ALKALINE PHOSPHATASE IN nuc-1^{c}(BC-152)

STRAIN	allele at	Normal Sequence allele at:		Translocated segment, allele at:		GENE DOSAGE OF		sp. act. alk. phosphatase	
	nuc-1	nuc-2	preg	nuc-2	preg	nuc-2^{+}	preg^{+}	phospho-ethanolamine 2 mM	inorganic phosphate 7.35 mM
A	BC-152	+	c-2	+	c-1	2	0	1482	873
B	BC-152	B1	c-2	+	c-1	1	0	1060	923
C	BC-152	B1	c-2	ts35	c-1	0	0	800	766
D	BC-152	+	+	+	c-1	2	1	292	293
E	BC-152	B1	+	+	c-1	1	1	307	268
F	BC-152	B1	+	ts35	c-1	0	1	252	297
G	BC-152	+	+	+	+	2	2	144	34.9
H	BC-152	B1	+	+	+	1	2	133	14.3
I	BC-152	B1	+	ts35	+	0	2	36.1	1.22
J	BC-152	+	+	not present		1	1	186	150
K	BC-152	not present		+	c-1	1	0	641	801
L	+	+	+	not present		1	1	100	1.08

All the partial diploid strains also carried arg-12 in heterozygous state (see legend of Fig. 2). The strains were grown in stationary culture for 40 hours at 35° (a temperature restrictive for ts35) on phosphoethanolamine or on inorganic phosphate, and assayed in the usual way. The specific activities reported are the mean values from duplicate cultures. nuc-2(B1) is the type strain and is the same as allele T28-M2(26). Strain L is our standard wild type, 74-OR8-1a.

nuc-1^{+} product, which is made constitutively. *nuc-1*^{c} could be thought of as making a *nuc-1* product of altered structure such that it is able to turn on the structural genes, but is blind or nearly blind to *preg*$^{+}$ product.

(3) Again, *preg*$^{+}$ might normally function by titrating out the *nuc-1*$^{+}$ product, which is made constitutively but *nuc-1*^{c} could be viewed as a mutant that makes an elevated amount of normal *nuc-1* product, such that it exceeds the stoichiometric amount of the *preg*$^{+}$ product.

The first of these hypotheses is not in mortal conflict with the data in Table 1, but nearly so. It requires that one dose of *preg*$^{+}$ product not be able to shut off the synthesis of *nuc-1* product very efficiently, but that two doses be able to do so. This is possible only if the association of "*nuc-1*^{c} operator" is sensitive to concentration of *preg*$^{+}$ product raised to a power greater than 1.

The second hypothesis is equally unsatisfactory. The altered *nuc-1*^{c} product is required to be nearly blind to one dose of *preg* product, but to see twice that amount very well.

The third hypothesis fits the data easily. Let us imagine that the normal, *nuc-1*$^{+}$ gene always makes 100 arbitrary units of product, and that *preg*$^{+}$ makes, say, 150 units. If these are unopposed by *nuc-2*$^{+}$ product, the *preg*$^{+}$ product will nullify the activity of the *nuc-1*$^{+}$ gene, with 50 units to spare. If *nuc-1*^{c} is a mutant that makes, say, 250 units of qualitatively normal *nuc-1* product, then *nuc-1*^{c} will be constitutive with 150 units of *preg*$^{+}$ product and repressible with 300 units of *preg*$^{+}$ product.

PROSPECTS FOR DETECTING AND ISOLATING THE *nuc-1* PRODUCT

The title of this symposium, Molecular *Approaches* to Eucaryotic Genetic Systems has a properly humble tone. At the same time, it invites us to talk about approaches that have not yet borne fruit, but that leave room for tempered optimism. We feel that the main problems in detecting and isolating a eucaryotic regulatory molecule, such as the *nuc-1* product are: (1) it is probably present in very small amounts, and (2) there is no chemical assay or bioassay for it. We will outline three projects that have been started and are in early stages of development.

An attempt to select for mutants with greatly increased production of nuc-1 product. If we are correctly interpreting our data on *nuc-1*^{c} (BC-152), this mutant makes an increased amount of *nuc-1* product. The increase is probably not extremely large, since two doses of *preg*$^{+}$ can render the strain repressible. It should be possible to improve on this by isolating *nuc-1*^{c} mutants that are constitutive even in a genetic background with two doses of <u>preg</u>$^{+}$. The selection involves using <u>nuc-1^{+}; nuc-2^{ts35}, preg^{+}/nuc-2, preg^{+}</u>. At nonpermissive temperatures, this strain, which has been constructed, cannot grow at high pH and low phosphate. Revertants are selected which can grow readily. Some of these will no doubt have eliminated the activity of one <u>preg</u>$^{+}$, either by point mutation or by deletion of part or all of the duplicated segment. These will be repressible. Other events (including perhaps mutation at hitherto undiscovered control genes) may give rise to strains that can grow and synthesize alkaline phosphatase constitutively. Among these constitutives, we hope to find some that are <u>nuc-1</u>^{c} and that overproduce the positive regulator to a much greater degree than does <u>nuc-1</u>^{c} (BC-152). A preliminary selection has been done, and genetic analysis has been started on the crop of about two dozen constitutives.

<u>An attempt at direct chemical detection of the nuc-1 product on two-dimensional gels</u>. Each of the seven chromosomes of Neurospora has about as much DNA as an <u>E. coli</u> chromosome. If we make the pessimistic assumption that most of this DNA codes for proteins and that half of these genes are expressed at any particular time, we can expect to find very roughly 10^{4} proteins, only one of which is the <u>nuc-1</u> product. This is an order of magnitude beyond the resolving power of even the gel method of O'Farrell (23). However, the results of inhibitor studies favor the notion that derepression requires transcription as well as translation (5), and it seems likely that the <u>nuc-1</u> product is concentrated in the nucleus and exerts its positive effect there. The nucleus of Neurospora contains roughly 5% of the cell protein. By isolating this organelle, one might hope to obtain material that is enriched up to 20-fold with respect to <u>nuc-1</u> product and containing perhaps 500 proteins.

Fungal nuclei are ordinarily extremely difficult to isolate because it is hard to break the cell wall without destroying the nucleus. In Neurospora, however, this can be avoided by using an extraordinary strain (really a triple mutant) discovered by Emerson in 1963 (4). This strain, called slime because of its odd growth behavior, multiplies as wall-less protoplasts. Slime can be lysed by detergents or by incautious agitation or dilution into hypotonic solutions, and nuclei can be prepared from it quickly and in excellent yield (20). Our best O'Farrell gel display of S^{35}-labelled nuclei prepared in this manner gave about 480 discrete spots on an autoradiogram. By some rather tedious techniques, one can cross mutant genes into the slime background, (19) and we have done so with <u>nuc-1</u>, with <u>nuc-1</u>^{c}, and with a number of other genes.

The goal is to compare the nuclear proteins of nuc-1 with those of nuc-1^c and to detect a protein present in nuc-1^c that is absent or of altered mobility in nuc-1. Clearly, no side-by-side comparison of two gel electrophoretograms will suffice for such a complex mixture. Therefore we have been developing a double-label method that employs a single electropherogram with subsequent resolution of the two isotopic signals. Slime nuc-1 cells are labelled by growing them in S^{35}-sulfate or in C^{14}-leucine. Slime nuc-1^c cells are labelled by growing them in H^3-leucine of high specific activity. Then cells of the two strains are mixed in proportions that correspond to about 200 times more d.p.m. of H^3 than there are d.p.m. of the harder beta emitter, S^{35} or C^{14}. Nuclei are prepared, nuclear proteins are displayed on an O'Farrell gel, and the gel is infiltrated with the scintillator, PPO (1), and is vacuum-dried on porous cellophane (Hoefer Scientific Co.). The side of the gel opposite the cellophane is covered with very thin aluminum leaf, about 0.7 μm thickness (Wehrung and Billmeier Co., 3206 Southport Ave., Chicago, Ill. 60657). The leaf stops all detectable photons, but allows the majority of the beta particles of the harder emitter to pass. Single-coated X-Ray film (Eastman SB-54) is put both on the cellophane side to give a fluorogram and on the aluminum leaf side to give a radiogram, and the sandwich is held at -70°C. The radiogram, which detects only the hard emitter, requires a much longer exposure than the fluorogram, which detects mostly the majority isotope, H^3. We have usually required about 16-24 hours for the H^3 fluorogram and 3 weeks for the S^{35} radiogram, but the ideal ratio of exposures will depend on gel thickness and other variables. The most important constraint is that there must be enough S^{35} or C^{14} to give a good radiogram within a reasonable length of time; yet there must be many times more H^3 or the photon trace on the fluorogram will have originated in substantial part from the harder emitter.

A comparison made in this fashion, (see Fig. 3) using S^{35} and H^3 as described above has proved intriguing, but not conclusive. First of all, the great majority of the spots are completely superimposable, as expected. However, there are substantial quantitative differences in intensity of positionally identical spots on the two films. These quantitative differences could be due either to variation in the amounts of certain proteins between the two strains, or to differences in the ratio of (methionine + cysteine)/leucine in different proteins. What is encouraging is that there seems to be a very few absolute differences, in which a spot that is present on the H^3 fluorogram of nuc-1^c is not visible at all on the S^{35} radiogram of nuc-1. One of these spots is in a lightly-populated region of the gel. The other three spots are far from the first in both the electrophoretic and the molecular

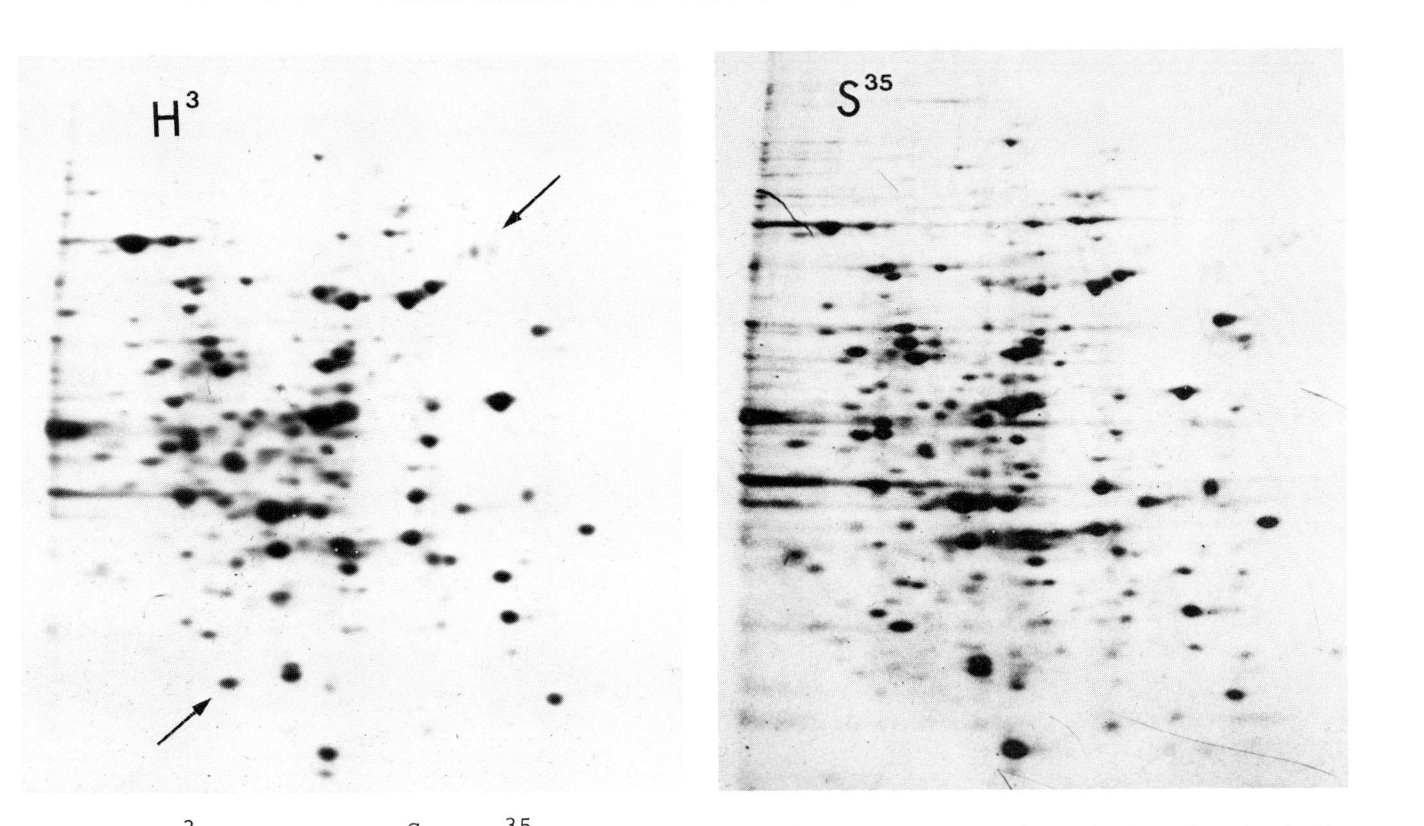

Fig.3 - H^3-labelled *nuc-1*^{c} and S^{35}-labelled *nuc-1* slime nuclear proteins displayed on a 2-dimensional gel and detected differentially as described in the text. Areas of possible interest are indicated by arrows. Isoelectric focusing run left to right, electrophoresis top to bottom.

weight dimension. These three are close together, and differ in isoelectric focusing properties but not in molecular weight. They might be three different proteins, but they might also be a single polypeptide that has been made heterogeneous by post-translational modification or by post-extraction artifact.

It would be extremely rash to suggest that any of these differences between the fluorogram and the radiogram constitute chemical detection of the *nuc-1* product. They might simply be proteins that do not contain any sulfur, for instance. We plan to repeat the experiment using C^{14} leucine in place of S^{35}. Alternatively, the differences could be due to the fact that the two slime strains cannot be made perfectly isogenic, and these differences at loci other than *nuc-1* may be detected. This difficulty can be circumvented by examining a considerable number of isolates of various *nuc-1* mutants, including putative nonsense and deletion mutants. Any difference that is consistent in all these strains is unlikely to be due to random genetic variation. Finally and more disturbingly. it is possible that the presence or absence of a particular spot may be consistently correlated with the status of the *nuc-1* gene without its spot being the *nuc-1* product. The spot might be a protein (such as alkaline phosphatase) that is coded by a structural gene under the regulatory control of *nuc-1*. The fact that all the known members of the family controlled by *nuc-1* are in the cell membrane or beyond its limit allows us to hope that there will not be some unknown member of the family in the nucleus; but it does not prove it. What is clearly needed is a biological assay of the *nuc-1* product that could be used in parallel with two-dimensional electrophoresis. If certain fractions from a purification scheme showed both enrichment for assayable material and for a chemical entity detected by gel electrophoresis, one could proceed with more confidence.

Preliminary steps toward a bioassay of the nuc-1 product. *nuc-1* mutants on low phosphate, like repressed wild-type on high phosphate (12) make only extremely small amounts of repressible alkaline phosphatase. The amount is at least 10^4 times lower than *nuc-1*$^+$ would make under the same conditions. If even a small amount of *nuc-1*$^+$ product from an extract of *nuc-1*$^+$ or *nuc-1*^{c} could be gotten into *nuc-1* cells, it might be possible to detect an increase in repressible alkaline phosphatase synthesis, since the basal level is so low. Furthermore, there is reason to believe that no special system is needed to deliver *nuc-1*$^+$ product from the cytoplasm into the nucleus, since heterocaryons between *nuc-1; pho-2*$^+$ and *nuc-1*$^+$*; pho-2* readily make alkaline phosphatase under permissive conditions (15). One trick that has been used to deliver other kinds of impermeant molecules into animal cells or other wall-

TABLE 2. STIMULATION OF THE INCORPORATION OF INORGANIC SULFATE INTO TCA-INSOLUBLE MATERIAL IN SLIME CELLS BY LIPOSOME-MEDIATED ATP-SULFURYLASE

Incubation Period (hr)	TCA-insoluble cpm in slime *cys-11*: Complete reaction mixture	Minus enzyme	Minus liposomes	Minus cells	TCA-insoluble cpm in slime *cys-11*$^{+}$: Complete reaction mixture	Minus enzyme	Minus liposomes	Minus cells
0	84	106	86	140	135	98	87	298
1	4334	396	264	137	21390	18814	18346	171
2	13136	1232	552	143	49284	45995	41326	151
3	23359	2445	1022	148	89321	91938	72922	156

Complete reaction mixtures (volume = 1 ml) contained $2.5x10^{7}$ slime *cys-11* or slime *cys-11*$^{+}$ cells grown and prepared as described below; liposomes prepared as described below, and containing 0.11 units of ATP sulfurylase, 1.2 μmoles of phosphatidyl choline, 0.3 μmoles of dicetyl phosphate, and 0.9 μmoles of ergosterol; and 0.1 μmoles of K_2SO_4 containing S^{35} ($3.2x10^{6}$ cpm). The reaction mixtures were incubated at 30°C with gentle rotary agitation (120 cycles/min.). Aliquots (0.1 ml) were discharged at the indicated times into 1 ml of cold 10% trichloracetic acid (TCA) containing 20 mM K_2SO_4. After 30 minutes, TCA-insoluble material was collected on Millipore filters, washed with TCA, dried, and counted in Bray's scintillator.

The details of growth and preparation of the cells were as follows. Slime cells of both strains were grown at 30°C into middle exponential phase (about 0.3 mg/ml of cell protein) in Fries' medium without sulfate ($MgSO_4$ being replaced by an equivalent of $MgCl_2$) and containing 1 mM taurine as the sulfur source, 1.5% sucrose as carbon source, and 7.5% sorbitol as a non-metabolizable osmotic stablizer. The cells were harvested by centrifugation at 30°C and

TABLE 2 legend, continued.

resuspended in fresh, pre-warmed medium to give 5×10^7 cells/ml. 0.5 ml. of this was combined with an equal volume of ATP-sulfurylase-containing liposomes prepared as follows. Phosphatidyl choline (Pabst Labs), 8 µmoles, dicetyl phosphate (ICN Pharmaceuticals, recryst.), 2 µmoles, and ergosterol (Sigma Chemical Co., recryst.), 6 µmoles were dried down from chloroform under nitrogen. Yeast ATP sulfurylase (Sigma), 0.5 mg. (3 units) in 2.0 ml. of 0.2M glucose-0.3M sorbitol was gassed with nitrogen and added to the dry lipid. The lipid was allowed to "swell" for 2 hours at 20°C and the mixture was then sonicated in a Heat Systems Ultrasonics bath for 10 min. at 18-20°C. The temperature was slowly raised to 30°C and the suspension was passed through a Sephadex G75 column at the same temperature (27). The material eluted in the void volume was collected and used without any attempt to separate the multivesicular from univesicular liposomes. The amount of ATP sulfurylase in this preparation was estimated by molybdolysis (25) in the presence of 0.2% Triton X-100 and in its absence. The activity in the presence of Triton was 0.22 units/ml. and was 12.5 times greater than in the absence of Triton - that is, the enzyme was 92.5% cryptic in the absence of detergent.

Complete reaction mixtures were made by mixing equal volumes of cells and liposomes. In the mixtures from which enzyme was omitted, liposomes prepared in the absence of enzyme were used. In the mixture from which liposomes were omitted, cells were mixed with enzyme that was in free solution in the same medium, and at the same concentration as the total amount apparently present in the liposome suspension.

less cells is to enclose them in the interior space of liposomes (see, for example, 17). If $nuc\text{-}1^+$ product could be delivered into nuc-1 slime via liposomes, it should be possible to detect and measure $nuc\text{-}1^+$ activity. We have done a feasibility test that suggests it is possible to deliver an enzyme, yeast ATP sulfurylase, into a slime auxotroph, slime cys-11, which lacks this enzyme. Since ATP sulfurylase is the principal if not the sole way of adenylating inorganic sulfate to adenosine phosphosulfate and subsequently making cysteine and methionine, cys-11 and its slime derivative grow only very slowly (probably by an alternate sulfur-reducing pathway) on minimal medium, and incorporate $S^{35}O_4^{=}$ only slowly into acid-insoluble materials, viz., proteins. The data in Table 2 show that yeast ATP sulfurylase sonicated along with a suitable mixture of lipids forms something, presumably liposomes containing the enzyme, that is capable of stimulating the incorporation of $S^{35}O_4^{=}$ into acid-insoluble materials in slime cys-11. There is no such stimulation if enzyme or lipids are omitted from the sonication mixture. No stimulation is seen in slime $cys\text{-}11^+$ cells, which already contain ATP sulfurylase. Since ATP, a substrate, is not found outside the cell, and adenosine phosphosulfate could not get into the cell if it were made outside, it seems very likely that ATP sulfurylase entered the cell. If $nuc\text{-}1^+$ product can be made to enter slime nuc-1 cells, the existing fluorometric and radiochemical assays for alkaline phosphatase should allow a bioassay of the $nuc\text{-}1^+$ product.

ACKNOWLEDGEMENT

This work was supported by an NIH grant, GM 08995. We thank Drs. James Dahlberg and Stephen Free for reading this manuscript.

REFERENCES

1. Bonner, W. M. and Laskey, R. A. (1974). Eur. J. Biochem. 46, 88.
2. Burton, E. G. and Metzenberg, R. L. (1974). J. Biol. Chem. 249, 4679.
3. Chia, W. (1976). Masters Thesis, University of Wisconsin.
4. Emerson, S. (1963). Genetica 34, 162.
5. Gleason, M. K. (1973) Masters Thesis, University of Wisconsin.
6. Gleason, M. K. and Metzenberg, R. L. (1974). Genetics 78, 645.
7. Hasunuma, K. (1973). Biochim. Biophys. Acta 319, 288.
8. Hasunuma, K. and Ishikawa, T. (1977). Biochim. Biophys.

Acta 432, 480, 178.

9. Hasunuma, K., Toh-E, A. and Ishikawa, T. (1976). Biochim. Biophys. Acta 432, 223.
10. Kadner, R. J., Nyc, J. F. and Brown, D. M. (1968). J. Biol. Chem. 243, 3076.
11. Lehman, J. F., Gleason, M. K. Ahlgren, S. K. and Metzenberg, R. L. (1973). Genetics 75, 61.
12. Lehman, J. F. and Metzenberg, R. L. (1976). Genetics 84, 175.
13. Littlewood, B. S., Chia, W. and Metzenberg, R. L. (1975). Genetics 79, 419.
14. Lowendorf, H. S. and Slayman, C. W. (1975). Biochim. Biophys. Acta 413, 95.
15. Metzenberg, R. L. (1977) unpublished.
16. Metzenberg, R. L. and Chia, W. (1977) in preparation.
17. Metzenberg, R. L., Gleason, M. K. and Littlewood, B. S. (1974). Genetics 77, 25.
18. Nelson, R. E., Lehman, J. F. and Metzenberg, R. L. (1976). Genetics 84, 183.
19. Nelson, R. E., Littlewood, B. S. and Metzenberg, R. L. (1975). Neurospora Newsl. 22, 15.
20. Nelson, R. E., Totten, R. E. and Metzenberg, R. L. (1977) in preparation.
21. Nyc, J. F. (1967). Biochem. Biophys. Res. Commun. 27, 183.
22. Nyc. J. F., Kadner, R. J. and Crocken, B. J. (1966). J. Biol. Chem. 241, 1468.
23. O'Farrell, P. H. (1975). J. Biol. Chem. 250, 4007.
24. Perkins, D. D. (1972). Genetics 71, 25.
25. Ragland, J. B. (1959). Arch. Biochem. Biophys 84, 541.
26. Toh-E, A. and Ishikawa, T. (1971). Genetics 69, 339.
27. Weissmann, G., Bloomgarden, D., Kaplan, R., Cohen, C., Hoffstein, S., Collins, T., Gotlieb, A. and Nagle, D. (1975). Proc. Nat. Acad. Sci. 72, 88.

REGULATION OF QUINATE CATABOLISM IN NEUROSPORA: THE QA GENE CLUSTER

J.W. Jacobson, J.A. Hautala, M.C. Lucas,*
W.R. Reinert, P. Strøman, J.L. Barea,
V.B. Patel, M.E. Case, and N.H. Giles

Genetics Program and Department of Zoology
University of Georgia, Athens, Ga. 30602

ABSTRACT. The expression of the three closely linked structural genes (*qa-2*, *qa-3*, and *qa-4*) encoding the enzymes functioning in quinic acid catabolism in *Neurospora* is controlled by the protein product of a tightly linked regulatory gene (*qa-1*) which appears to act with the inducer quinic acid as a positive regulatory effector. Mutants in the *qa-1* gene are non-inducible for all three enzyme activities. Two of the enzymes under the control of the *qa-1* gene have been purified. One has been fully characterized. Attempts are currently underway to determine the amino acid sequences of both enzymes. The biochemical and genetic characterization of this system has facilitated attempts to analyze the mechanisms functioning in the regulation of the expression of the *qa* gene cluster at the molecular level.

The identification and subsequent purification of the *qa-1* encoded regulatory protein are currently underway. Initial attempts to isolate this protein from whole cell extracts had little success since it is apparently present at very low intracellular concentrations. Since current complementation data suggest that the *qa-1* protein functions primarily in the nucleus in which it is encoded, experiments are underway to detect the regulatory protein in extracts of isolated nuclei and to see if it fits the criteria of a nonhistone chromatin protein.

Attempts are underway to isolate mRNA transcribed from the *qa* region. Messenger RNA from induced cultures has been isolated and characterized. Polyadenylated and non-polyadenylated mRNA's have been translated in the cell-free wheat germ system. Proteins synthesized *in vitro* range in M.W. from 10,000 to over 100,000 daltons. Experiments are underway to identify the fraction of RNA containing the specific *qa* mRNA by double immunoprecipitation of the *qa-2* gene product synthesized *in vitro*.

Recovery of a bacterial plasmid carrying the *qa* gene cluster is currently being attempted. *E. coli* which are

*Present address: Dept. of Biological Sciences, Florida State University, Tallahassee, Fla.32306

aro D^- and therefore lack biosynthetic dehydroquinase are transformed with the plasmid pBR322 which contains inserted Neurospora DNA. Relief of the aromatic amino acid requirement will occur if a Neurospora gene for dehydroquinase is present on the plasmid and is expressed.

INTRODUCTION

The inducible quinic acid catabolic pathway in *Neurospora crassa* is an excellent system for examining the molecular mechanisms functioning in the control of gene expression in eucaryotes (Fig. 1). The first three reactions in the pathway are catalysed by three inducible enzymes encoded in a tightly-linked gene cluster, the *qa* cluster (10). Gene *qa-2* encodes catabolic dehydroquinase (5-dehydroquinate hydrolase, EC 4.2.1.10); *qa-3* encodes quinate dehydrogenase (quinate:NAD^+ oxido-reductase, EC 1.1.1.1.24); and *qa-4* encodes dehydroshikimate dehydrase. The fourth gene, *qa-1*, encodes a regulatory protein which, in conjunction with the inducer, quinic acid acts in a positive fashion to initiate enzyme synthesis.

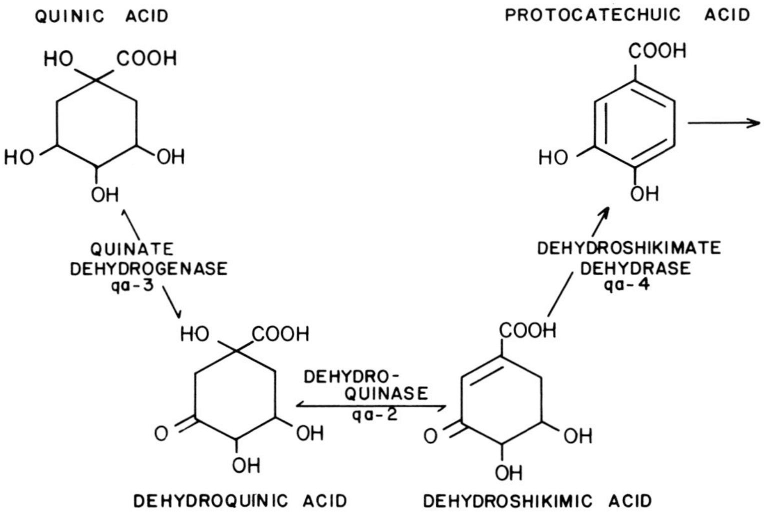

Fig. 1. Inducible quinic acid catabolic pathway in *Neurospora crassa*.

This system is well characterized both genetically and biochemically. Genetic analysis has resulted in the isolation of a series of mutants for each structural gene, as well as a number of regulatory mutants affecting the qa-1 gene. The biochemical characterization has involved the purification of two of the catabolic enzymes. One, dehydroquinase, has been extensively characterized; the second, quinate dehydrogenase, has been partially characterized. The third enzyme, dehydroshikimate dehydrase, is currently being purified. The combination of analytical tools provided by the extensive genetic and biochemical characterization of this pathway provide a very powerful system for analysis of the control of gene expression at the molecular level.

A multiple approach to the characterization of the molecular regulatory mechanisms functioning during induction of this pathway is currently under way. The final goal is to establish a coupled *in vitro* transcription-translation system where the role of the individual components in the system can be investigated. Attempts are currently underway to isolate and characterize the regulatory protein, to isolate and translate *in vitro* the mRNA specific for the pathway, and to isolate a bacterial plasmid carrying the gene cluster.

RESULTS AND DISCUSSION

Genetic Characterization of the Quinic Acid Catabolic Pathway

The qa gene cluster is positioned on the right arm of linkage groups VII. Recent genetic data have established the gene order in this tightly linked cluster as qa-1, qa-3, qa-4, qa-2 (4). Mutants in each of the three structural genes reduce or eliminate the enzyme activity encoded in the gene affected. Analysis of pairs of qa mutants in forced heterocaryons has demonstrated the occurrence of complementation among the three genes. In addition, allelic complementation has been demonstrated in the qa-2 gene (5).

Mutants in the qa-1 gene fail to synthesize all three catabolic enzymes. Complementation tests in heterocaryons with the individual structural gene mutants divide the qa-1 mutants into two complementation types which map in opposite regions of the gene (3). The qa-1^F mutants are recessive, complement rapidly with all structural gene mutants, and are interpreted as affecting that portion of the regulatory protein involved in binding the inducer quinic acid. The qa-1^S mutants are semi-dominant, complement very slowly, and are interpreted as affecting that portion of the regulatory protein involved in binding the site (or sites) of transcription initiation on the DNA. This interpretation is supported indirectly by the evidence that qa-1^S mutants (but not qa-1^F mutants) revert to a constitutive phenotype (qa-1^C) which facilitates enzyme synthesis in the absence of inducer. In heterocaryons the qa-1^C mutants are dominant. In addition, allelic complementation can be demonstrated between qa-1^F and qa-1^S mutants.

In heterocaryons of the double mutant qa-1^C, qa-3 and a qa-1^F, qa-3$^+$ strain, the synthesis of quinate dehydrogenase was detected. This synthesis demonstrates that the qa-1^C gene product functions in the other nucleus in the heterocaryon, i.e., in *trans*. However, the level of expression in the *trans* nucleus was very low compared to expression in a qa-1^C, qa-3$^+$ nucleus. Thus the function of the regulatory protein may be limited primarily to its own nucleus. This "nucleus limited" effect has been observed with the scon^c gene in Neurospora as well (2).

Biochemical Characterization of the Quinic Acid Catabolic Pathway

Biochemistry of induction. If at any time following conidial germination, Neurospora is placed in a medium containing quinic acid as sole carbon source, the quinate catabolic enzymes can be detected by assay after approximately 60 min of induction. Since the increase of all three enzyme activities is inhibited by cycloheximide, they appear to be synthesized *de novo*. However, this does not rule out the possibility that cycloheximide inhibits the synthesis of a protein which modifies or effects assembly of pre-existing enzyme subunits or precursors.

To clarify the question of *de novo* synthesis, mycelia were grown in sucrose minimal media and then induced by transferring cultures to minimal media containing quinic acid in either water or 95% deuterium oxide (7). Both catabolic dehydroquinase and quinic dehydrogenase were partially purified from each induced culture. The density of the enzymes was analysed by equilibrium centrifugation in cesium chloride. The results for dehydroquinase are shown in Fig. 2. In both cases the enzymes from the culture induced in deuterium banded at a higher density than the samples induced in water medium. This result was independent of whether the cultures were grown in water medium or in deuterium medium prior to induction. This is conclusive evidence that at least two of the enzymes are synthesized *de novo* during induction of the pathway.

It has previously been shown that the three enzymes are

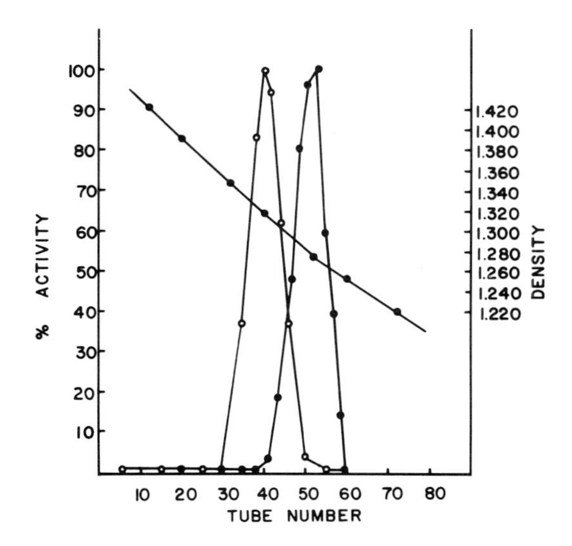

Fig. 2. Cesium chloride density gradient analysis of catabolic dehydroquinase extracted from cultures grown on minimal sucrose media and then shifted to quinic acid induction media in either H_2O (●-●) or 95% deuterium oxide (0-0).

co-ordinately induced (6). However, it has not been established whether they are translated from a polycistronic mRNA or from individual mRNA's. To date no genetic evidence is available on this point since no qa nonsense mutants have yet been obtained

Although the current working hypothesis states that the control in this system functions at the level of transcription, relatively little evidence is available on this point. The fact that the regulatory protein appears to function more effectively in the nucleus in which it is encoded can be interpreted to support transcriptional control. However, this effect can also be explained by a restriction on the rate of diffusion of the regulatory protein in the cytoplasm. Experiments designed to demonstrate the level at which control is exerted have so far been inconclusive. These studies are complicated by the long lag time between the introduction of inducer and the onset of enzyme synthesis and by the fact that induction of the pathway by quinic acid is subject to glucose repression and can only be examined under conditions of metabolic stress.

Characterization of the enzymes encoded in the qa gene cluster. Catabolic dehydroquinase, the qa-2 gene product, has been purified and thoroughly characterized (12). This enzyme is very unusual by a number of criteria. It is a 220,000 molecular weight aggregate composed of 20 to 22 identical subunits of approximately 10,000 daltons each. It is heat stable and resistant to denaturation by urea and guanidine hydrochloride at 25° C. It is dissociated by SDS, but can only be reduced to the monomeric form by boiling in 1% SDS for extended periods of time. The enzyme is fully dissociated and inactivated at pH 3.0, but reforms the aggregated structure and regains full activity when returned to pH 7.5. In addition, it is resistant to digestion with proteases and cleavage with cyanogen bromide. The enzyme has a very high specific activity and can be detected by assay at nanogram per ml concentrations. Finally, the amino acid composition of dehydroquinase is unusual in that it lacks cysteine and tryptophan. It has high concentrations of proline; the hydrophobic amino acids valine, isoleucine, and leucine; and the acidic amino acids glutamic acid and aspartic acid.

Since the subunit of dehydroquinase contains approximately ninety amino acids many attempts have been made to sequence this peptide. However, the stability of the enzyme has presented many problems. It is difficult to reduce the enzyme to its monomeric form and the amino terminal end of the peptide is not available in the aggregated state. In addition, the subunit itself has been very resistant to sequencing using a variety of conditions. The first two cycles of the Edman degradation have revealed an amino terminal proline followed by a leucine residue. For reasons not yet understood the third cycle is totally blocked and modifications of the sequencing procedure have not resolved this impasse.

The analysis of dehydroquinase from two complementing *qa-2* mutants and from the heterocaryon formed between them has provided information about the unusual aggregate structure of this enzyme (5). One of the two *qa-2* mutants has a very low level of enzyme activity and a very low concentration of multimeric dehydroquinase protein. The second has low activity but has a wild type concentration of multimeric protein as determined by immunological analysis. A heterocaryon between these two stains has an increased level of enzyme activity and the wild type level of native dehydroquinase protein. The interpretation of these data is that the first mutant is defective in that portion of the protein involved in aggregation but that it retains a fully functional catalytic site. The second mutant appears to be defective in catalytic function, but retains full capacity to aggregate. In the heterocaryon the subunits from the second mutant "rescue" those from the first to give a mixed aggregate containing many inactive subunits plus a few catalytically active subunits. It is clear from these data that two very distinct regions of the molecule are involved in enzyme activity and subunit aggregation.

The purification and characterization of the dehydroquinase has also provided information about the regulation of this pathway. One hypothesis of the mechanism by which the qa-1 regulatory protein functions was that it provided a common subunit to all three catabolic enzymes and was required for enzyme activity. Purification and extensive characterization of dehydroquinase from a temperature-sensitive $qa\text{-}1^{C}$ mutant could detect no differences between that enzyme and the enzyme purified from wild type (14). Thus the qa-1 gene product does not appear to be associated with the qa-2 gene product. Although similar data are not available for the other two enzymes in the system, it is unlikely that they are associated with the qa-1 gene product.

Quinate dehydrogenase has just recently been purified to homogeneity and is currently being thoroughly characterized. It appears to be composed of a single polypeptide with a molecular weight of 39,000 daltons as determined by SDS-polyacrylamide gel electrophoresis. In addition to the quinate dehydrogenase activity this bifunctional enzyme can also convert shikimate to dehydroshikimate, the second intermediate in the pathway. After polyacrylamide gel electrophoresis, three proteins bands having quinate dehydrogenase activity are present. This multiplicity is characteristic of the enzyme at all stages of purification, although the pure material is made up primarily of one of the electrophoretic species. An initial attempt to sequence this protein has been successful and a tentative sequence has been assigned to the first ten residues. Sequencing this molecule will be difficult, however, since it is a 39,000 dalton protein.

Dehydroshikimate dehydrase, the third enzyme encoded in the qa gene cluster appears to have a molecular weight of 35,000 daltons. The purification and characterization of this enzyme is currently in progress. No information is yet available on its subunit structure.

Molecular Approaches to Regulation of the qa Gene Cluster.

The regulatory protein. Strong genetic evidence supports the hypothesis that the qa-1 gene produce is a protein. The regulatory element is diffusible and acts in trans in a heterocaryon. Allelic complementation and temperature-sensitivity have been observed among mutants in the qa-1 gene. All of these data support the existence of a qa-1 regulatory protein.

In order to characterize the proposed in vitro transcription-translation system, the regulatory protein must be identified, isolated, and characterized. Initial attempts were made to isolate the qa regulatory protein from whole cell extracts. The location of the regulatory protein in fractionated extracts was determined by equilibrium dialysis against ^{3}H-quinic acid and by affinity chromatography on a column of quinic acid coupled to Sepharose 4B. In both systems significant binding was only observed when nucleic acids had not been removed from the protein preparations. A species was recovered from the quinate affinity column which had properties consistant with the qa-1 encoded regulatory protein. It could be isolated from induced cultures of wild type 74-A and from a constitutive mutant qa-1^C. It was recovered from induced cultures of a qa-1^S mutant which is postulated to lack DNA binding ability. It was not observed in induced cultures of several qa-1^F mutants which are thought to lack quinic acid binding capacity. Unfortunately, because it is present in low concentration, very little of the presumptive regulatory protein could be recovered. Insufficient material was available for further characterization.

Because of the difficulty in obtaining enough material, an attempt to isolate the regulatory protein by an alternate procedure is underway. The rationale for this procedure is based on the observation that the regulatory protein is largely nucleus limited in its function. In addition, current hypotheses predict that the regulatory protein may be a nonhistone-chromatin protein (22). Therefore, nuclei were isolated to be used as a source of chromatin. Nuclei are difficult to isolate from Neurospora because it is difficult to rupture the cells without rupturing the nuclei. The procedure developed uses either ungerminated or germinated conidia (11). Nuclei are released from the conidia by a French press of a frozen cell suspension and then freed from the cell debris by repeated homogenizations in an Omni-mixer. A buffer of sorbitol, Ficoll, glycerol, and Triton X-100 is used to protect the nuclei. Following removal of cell debris the nuclei were collected by centrifugation. This crude nuclear pellet is resuspended in Tris-sucrose buffer, briefly homogenized in a tissue homogenizer, and isolated from the contaminating mitochondria and cell debris by density banding in a Ludox gradient. The nuclei band is removed from the gradient, diluted with buffer, and collected by centrifugation. When observed by fluorescense microscopy using the DNA stain DAPI (23) no cytoplasmic attachments or mitochondrial contamination are observed. The recovery of purified nuclei was about 29% based on DNA content. They had a DNA/RNA/protein ratio of 1/3.5/7 and had endogenous RNA polymerase activity.

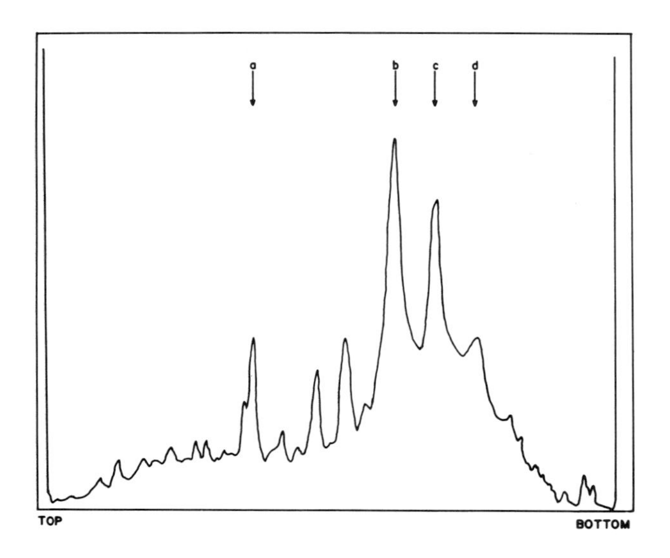

Fig. 3. SDS gel electrophoresis of basic nucleoproteins from Neurospora crassa. The histone bands are: (a) H1, (b) H3 and H2b, (c) H2a, and (d) H4. The histone bands were identified by comparison to rat liver histones.

The purified nuclei were fractionated into a soluble fraction, a residual membrane fraction, and chromatin. The chromatin was further fractionated on hydroxylapatite, following sonication, into basic proteins, acidic proteins, and nucleic acids. The two protein fractions were analyzed by SDS-gel electrophoresis (15). The basic protein fraction showed the normal complement of histones as compared to those extracted from rat liver (Fig. 3). The acidic proteins showed several distince peptides ranging in size from 15,000 to 70,000 daltons (Fig. 4B). Attempts to reconstitute dissociated chromatin were relatively unsuccessful with only 10% of the original DNA being reconstituted. The reconstituted material contained a full complement of histones and a less complex sample of acidic proteins than the dissociated material. (Fig. 4A).

Both the non-chromatin associated and the chromatin-associated proteins will be fractionated by DNA-cellulose and quinic acid affinity chromatography. The DNA-cellulose column is being prepared with homologous DNA. If a protein with

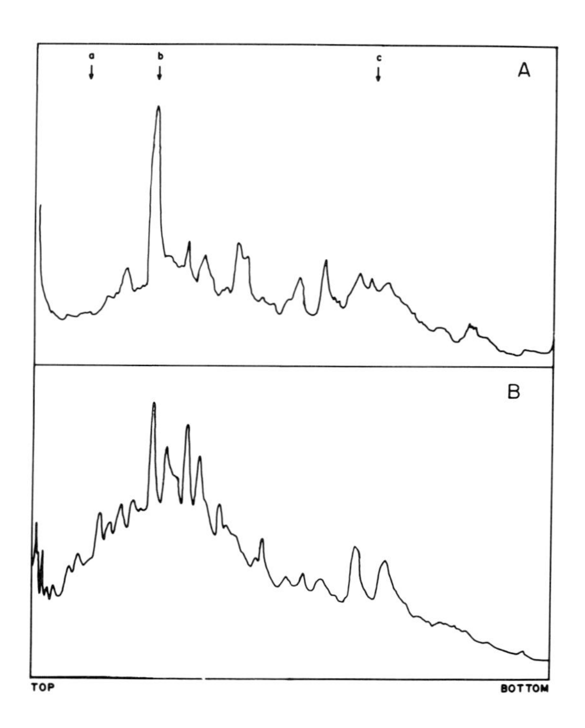

Fig. 4. SDS gel electrophoresis of acid nucleoproteins from Neurospora crassa. Samples were from reconstituted (A) and non-reconstituted chromatin (B). The arrows indicate the molecular weight markers (a) bovine serum albumin, (b) ovalbumin, and (c) lysozyme.

both high DNA and high quinic acid binding capacity can be identified, a series of qa-1 mutants will be analyzed, as before, to establish that the isolated protein is the qa-1 gene product.

Translation of Neurospora mRNA. A second component of the in vitro system which is being developed is the wheat germ cell-free protein-synthesizing system which was prepared as described by Roberts and Paterson (21). Neurospora mRNA is isolated from a lyophilized mycelial powder suspended in pH 5.0 sodium acetate buffer with 4% SDS. The suspension is extracted four times with an equal volume of phenol-chloroform/isoamyl alcohol and the RNA precipitated from the aqueous phase with 2.5 volumes of ethanol. The RNA was further fractionated by oligo(dT)-cellulose chromatography (1) or Sepharose 4B chromatography (24). The mRNA isolated by this procedure had an average size of approximately 17S. The poly (A) tacts isolated from this mRNA represented three size classes of approximately 30, 55, and 70 nucleotides. This poly(A) mRNA was used to optimize the wheat germ protein-synthesizing system for translation of Neurospora RNA. The optimums determined were: magnesium, 2.5 mM; potassium, 80 mM; and a pH of 7.4. The RNA which does not bind to oligo(dT)-cellulose also contains significant messenger activity. Although it has not yet been separated from contaminating ribosomal RNA, this RNA has a specific activity approximately 25% that of the poly(A) mRNA and therefore may represent a significant percent of total Neurospora mRNA. Presumably, it contains either no poly(A) or very short poly(A) tracts that are less than 15 nucleotides in length (17). Preliminary data indicate that translation of Neurospora mRNA is inhibited by 7-methylguanosine-5'-monophosphate which is evidence for the presence of a "cap" structure on the 5' end of the mRNA (13).

Analysis of the translation products by SDS-acrylamide gel electrophoresis (9) showed that they were between 10,000 and 100,000 daltons. Since many of the translation products are very small, which supports premature termination in the wheat germ system, the rabbit reticulocyte system is currently being developed as well. Once the endogenous messenger activity of this system has been removed, it translates exogenous mRNA's with good fidelity (19). Presently, endogenous messenger activity has been reduced to less than 10% of the original activity.

Attempts to identify _qa_ specific mRNA(s) are in progress. Antibody against catabolic dehydroquinase is being used to detect synthesis of this enzyme _in vitro_ by double immunoprecipitation followed by electrophoresis in the presence of the detergent cetyltrimethylammonium bromide (CTAB) (8). Since the concentration of _qa_ specific mRNA may be quite low, the percent of the total synthesis it directs may be low also. It is necessary, therefore, to reduce the non-specific precipitation to a level where very low levels of specific precipitation can be detected. In several assays apparent specific synthesis has occurred, but insufficient product has been formed to allow positive identification.

In an attempt to reduce the complexity of the mRNA being introduced into the _in vitro_ translation system, mRNA is now being isolated from polysomes extracted from germinating conidia growing under inducing conditions. The polysome fraction should be enriched for _qa_ specific mRNA. The polysomes are being isolated by the procedure of Mirkes (16).

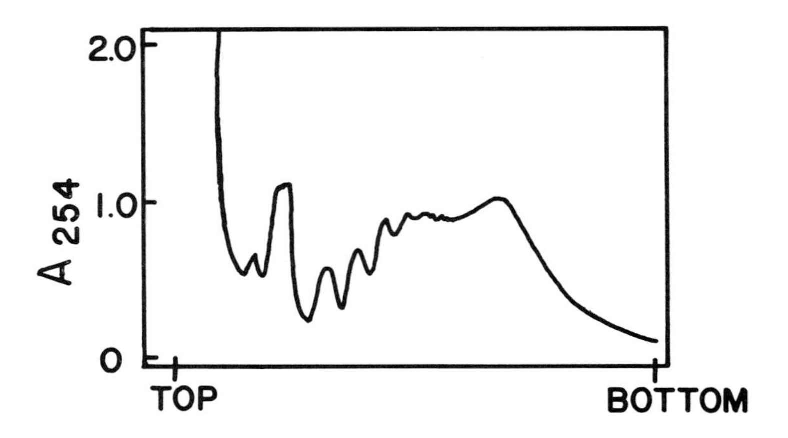

Fig. 5. Sucrose density gradient analysis of polysomes isolated from Neurospora crassa.

Fig. 5 shows the polysome profiles obtained when polysomes isolated in the presence of heparin (18) and cycloheximide to prevent degradation and runoff are analysed by sucrose gradient centrifugation. A large percentage of the polysomes is made up of five or more ribosomes. RNA is extracted from the polysomes as described by Rhodes et al (20). This mRNA will be used to direct the translation systems. Using this RNA should bypass the problem of determining whether qa specific mRNA is or is not polyadenylated.

Following identification of the qa specific mRNA, it will be isolated using a variety of RNA fractionation techniques. Once isolated, the qa mRNA and the complementary DNA (cDNA) made from it can be used as probes to quantitate the synthesis of message during induction and as probes for a DNA fragment carrying the qa gene cluster. Purification of the qa mRNA will also answer the question of whether the qa enzymes are translated from a polycistronic message.

Attempts to isolate the qa gene cluster on a bacterial plasmid. The difficulties inherent in understanding molecular mechanisms of regulation in eucaryotes can be greatly simplified if the DNA encoding the system of interest can be isolated from the remainder of the genome. This can be accomplished by cloning the genes of interest in a bacterium by using the techniques made available by research on recombinant DNA. The only requirement is that a selection technique be available so that the plasmid carrying the DNA of interest can be isolated.

The first requirement for this technique is that high molecular weight DNA can be isolated in good yield. For our purposes, the crude nuclear pellet is an excellent source of DNA (11). The crude pellet is resuspended in saline-EDTA with 1% SDS and treated with Protease K for 1 hr at 37° C. The sample is cleared by centrifugation, and nucleic acids are precipitated with 2 volumes of cold ethanol and collected by centrifugation after at least 1 hr at -20° C. The DNA is further purified by Sepharose 4B chromatography and banding in a cesium chloride gradient. From this procedure one recovers mg quantities of high molecular weight DNA.

This DNA is digested by the restriction endonuclease Hind III and ligated to the vector pBR322. *E. coli* cells are transformed with these vectors under conditions that allow only cells containing hybrid plasmids to survive. The *E. coli* strain transformed is an *aroD*$^-$ which lacks biosynthetic dehydroquinase. The cells should only grow on minimal media if a Neurospora gene for dehydroquinase is carried on the plasmid and expressed in the *E. coli* cytoplasm. Since the synthesis of the *qa* enzymes is regulated, the DNA used for the transformation experiments was isolated from a *qa-1*^{C} mutant which is constitutive for synthesis of the *qa* enzymes. Thus the *qa-2* gene would be expressed if either the regulatory protein is synthesized and functions correctly in the *E. coli* cell or if the *E. coli* RNA polymerase recognizes the transcriptional initiation site for the *qa-2* gene without the cooperation of the regulatory protein.

Recovery of the incorporated DNA from the plasmid would give a system with unlimited potential for analysis of regulation in this system. If the DNA fragment carries the entire *qa* gene cluster, the capacity for doing the *in vitro* transcription-translation experiments discussed earlier is immediately available. If only a portion of the *qa* cluster is recovered, it would serve as a probe to identify the *qa* gene cluster on fragments of DNA generated by other techniques. Also using the same cloning selection techniques, DNA fragments generated by a different restriction endonuclease could recover the entire qa gene cluster.

ACKNOWLEDGEMENTS

This work was supported by Research Contract E (38-1)-735 with the United States Energy Research and Development

Administration and Research Grants GM 22054 and GM 23051 from the Institute of General Medical Sciences of the U. S. Public Health Service. J.A.H. was supported by Public Health Service Postdoctoral Fellowship 5 F02 GM 55828 from the Institute of General Medical Sciences.

We would like to thank Barbara Conner, Fred Lewis, and Joann Lay for their excellent technical assistance.

REFERENCES

1. Aviv, H. and Leder, P. (1972) Proc. Nat. Acad. Sci. U.S.A. 69, 1408.
2. Burton, E. G. and Metzenberg, R. L. (1972) J. Bacteriol. 109, 669.
3. Case, M. E. and Giles, N. H. (1975) Proc. Nat. Acad. Sci. U.S.A. 72, 553.
4. Case, M. E. and Giles, N. H. (1976) Molec. gen. Genet. 147, 83.
5. Case, M. E., Hautala, J. A., and Giles, N. H. (1977) J. Bacteriol. 129, 166.
6. Chaleff, R. S. (1974) J. Gen. Microbiol. 81, 357.
7. Chrispeels, M. J. and Varner, J. E. (1973) In Molecular Techniques and Approaches in Developmental Biology p.79.
8. Fairbanks, G. and Avruch, J. (1972) J. Supramol. Structure 1, 66.
9. Fairbanks, G., Steck, T. L., and Wallach, D. F. H. (1971) Biochem. 10, 2606.
10. Giles, N. H., Case, M. E., and Jacobson, J. W. (1973) In Molecular Cytogenetics p.309.
11. Hautala, J. A., Conner, B. H., Jacobson, J. W., Patel, G. L., and Giles, N. H. (1977) Submitted to J. Bacteriol.
12. Hautala, J. A., Jacobson, J. W., Case, M. E., and Giles N. H. (1975) J. Biol. Chem. 250, 6008.
13. Hickey, E. D., Weber, L. A., and Baglioni, C. (1976) Proc. Nat. Acad. Sci. U.S.A. 73, 19.
14. Jacobson, J. W., Hautala, J. A., Case, M. E., and Giles, N. H. (1975) J. Bacteriol. 124, 491.
15. MacGillivray, A. J., Cameron, A., Krauze, R. J., Rickwood, D., and Paul, J. (1972) Biochim. Biophys. Acta 277, 384.
16. Mirkes, P. E. and McCalley, B. (1976) J. Bacteriol. 125, 174.
17. Nudel, U., Soreq, H., Littauer, U. Z., Marbaix, G., Huez, G., Leclercq, M., Hubert, E., and Chanterenne, H. (1976) Eur. J. Biochem. 64, 115.

18. Palacios, R. and Schimke, R. T. (1973) J. Biol. Chem. 248, 1424.
19. Pelham, H. R. B. and Jackson, R. J. (1976) Eur. J. Biochem. 67, 247.
20. Rhoads, R. E., McKnight, G. S., and Schimke, R. T. (1973) J. Biol. Chem. 248, 2031.
21. Roberts, B. E. and Paterson, B. M. (1973) Proc. Nat. Acad. Sci. U.S.A. 70, 2330.
22. Stein, G., Stein, J., Kleinsmith, L., Park, W., Jansing, R., and Thomson, J. (1976) Prog. Nucleic Acid Res. and Mol. Biol. 19, 421.
23. Williamson, D. H. and Fennell, D. J. (1975) Meth. Cell Biol. 12, 335.
24. Woo, S. L. C., Harris, S. E., Rosen, J. M., Chan, L., Sperry, P. J., Means, A. R., and O'Malley, B. W. (1974) Prep. Biochem. 4, 555.

Neurospora Workshop; presented at ICN-UCLA Symposium on the Eucaryotic Genetic Systems by Richard Weiss.

The current status of investigations into the nature of a number of regulatory systems in *Neurospora* were described. The desirability and feasibility of various molecular approaches were discussed.

Samson R. Gross, Duke University, described genetic and biochemical evidence for the existence of an internal promotor and a mitochondrial regulatory signal controlling the synthesis of the cytoplasmic and mitochondrial leucyl-tRNA synthetases. Mutants in the *leu-5* region have a structurally altered cytoplasmic enzyme, produce numerous translation errors, and have only a trace amount of unaltered mitochondrial enzyme. Partial revertants restore apparently normal cytoplasmic enzyme activity but continue to produce only a trace of the normal mitochondrial enzyme. Complete revertants have also been isolated. Inhibition of mitochondrial protein synthesis results in increased levels of the mitochondrial enzyme and a concomitant decrease in the cytoplasmic enzyme. A model was suggested involving a promotor for the mitochondrial enzyme located within the cytoplasmic enzyme structural gene, and a regulatory region between the structural genes which responds to a mitochondrial gene product to control transcription of the mitochondrial structural gene.

George A. Marzluf, Ohio State University, described evidence that the synthesis of many enzymes related to nitrogen metabolism in *Neurospora* are regulated in a complex manner by specific induction and/or nitrogen catabolite repression and by a major regulatory gene designated *amr* (allelic to *nit-2*). The nitrogen-regulated enzymes include allantoinase, uricase, allantoicase, nitrate reductase, and a general amino acid permease. Recent studies have shown that a permease for xanthine and uric acid and a distinct permease for hypoxanthine, adenine and guanine are both controlled by *amr*. Ammonia does not appear to be the corepressor for nitrogen repression since several amino acids cause such repression in *am* mutants lacking glutamate dehydrogenase. The *amr*$^+$ gene product appears to exert positive control, turning on the expression of numerous unlinked genes. Among the enzymes regulated by *amr* is an extracellular protease whose synthesis is also controlled by independent regulatory circuits for sulfur and carbon. It was suggested that the protease structural gene may be served by a complex control region sensitive to each of at least three distinct control signals. It was proposed that the protease gene and adjacent control region might be isolated by a series of selective hybridizations to mRNA molecules from *amr* and induced wild-type strains. A discussion of this proposal suggested several modifications and generated some skepticism as to its feasibility.

Richard L. Weiss, UCLA, described evidence that amino acid compartmentation plays a significant role in controlling arginine metabolism in fungi. Endogenous enzymes were shown to be capable of catabolism, but fail to operate in minimal medium because most of the substrate is sequestered in the vacuole. Accumulation of arginine catabolic enzymes was shown to occur 40 min. after the addition of inducer.

Experiments involving inducer removal and subsequent expression of enzyme forming capacity suggest that induction commences immediately upon inducer addition to the growth medium. It was suggested that the observed lag prior to enzyme accumulation might represent the time necessary for transcription, processing, transport and translation of mRNA molecules for the catabolic enzymes. It was proposed that gene isolation and cloning might provide a means of detecting mRNA synthesis and maturation by DNA-RNA hybridization. An alternative suggestion was made that detection of specific enzyme product from a wheat germ translation system might be easier. After some discussion, it was concluded that such a method was probably not feasible because of the high basal enzyme level and small (3-4x) magnitude of induction.

The workshop concluded with a concensus that further elucidation of regulatory mechanisms in *Neurospora* would benefit from isolation of specific regulatory gene products and isolation, amplification, and characterization of regulatory and structural genes (and associated regulatory regions). The relative simplicity of the genetic complement of *Neurospora* was thought to be an advantage for such studies.

The purpose of the slime mold workshop was to bring workers in and out of the field up to date on the approaches being used and problems encountered in analyzing differentiation in *Dictyostelium discoideum*. The workshop was divided into two sections: the first on the molecular approaches and the second on the genetic approaches.

Allan Jacobson (University of Massachusetts Medical School) described present knowledge on poly(A)$^+$ metabolism in *Dictyostelium* during growth and development and the relationship of this analysis to the problems involved in mRNA purification. Dr. Jacobson discussed several main points. He described evidence that poly(A) in mRNA turns over much more rapidly in developing than in vegetative cells. This rapid turnover includes poly(A) on both RNA synthesized during development and on mRNA synthesized during vegetative growth. Jacobson also stated that the mRNA containing long poly(A) in early developing cells is preferentially found on polysomes whereas mRNA containing small poly(A) is not. This could explain results previously reported by Alton and Lodish that a large fraction of mRNA during early development is not associated with polysomes. The relationship between the rate of poly(A) metabolism for specific mRNA or mRNA populations on the translational regulation of these messengers was discussed. In addition, Dr. Jacobson described work which supported early labeling studies that poly(A) decreased in size with time in vegetative cells and that steady-state mRNA contains mostly small poly(A). Work on the fractionation of mRNA using preferential elution from poly(U) Sepharose and poly(A) Sepharose (suggestive evidence for poly(U) stretches) was discussed. Moreover, other results suggested that histone mRNA appeared to be the only mRNA uniquely poly(A)$^-$.

Dr. Nancy Maizels (Harvard University) described the organization of the rDNA in *Dictyostelium* and the applications of the methods used to analyze these genes for the analysis of other gene sequences. Her results showed that the rRNA gene is contained within a large non-transcribed spacer region. Moreover, the 5S genes are also present in this rDNA repeat similar to the situation in yeast. Dr. Maizels also described methods which allow partial restriction enzyme digests reproducibly for the analysis of gene structure and for DNA sequencing. The general method involves digesting the DNA in the presence of drugs which specifically bind to the DNA and therefore inhibit the specific restriction enzyme. The applicability of distamycin A for *Eco*RI and actinomycin D for *Hind*II were described.

Dr. R. Firtel (UC, San Diego) described experiments in which nuclear and cDNA to mRNA were cloned in recombinant plasmids. Three nuclear DNA clones were described which are complementary to a large fraction of pulse-labeled mRNA. The first is complementary to a specific mRNA which has been tentatively found to code for actin protein. The second is complementary to approximately 10% of mRNA. The complementary mRNA shows a very heterodisperse size distribution. The hybrid regions formed between the plasmid DNA and mRNA are probably short and may contain poorly base-paired regions. A third plasmid has both of the hybridization patterns observed for the other two cloned DNA sequences. Evidence that there are approximately 15-20 non-tandemly repeated actin "genes" was discussed.

In addition, Firtel described the methodology and the applicability of using recombinant plasmids carrying cDNA to mRNA as a bootstrapping method for analyzing developmentally regulated genes for abundant mRNAs. The advantages of a method using the poly(dG) - poly(dC) tailing proceeding in which the heterologous DNA is inserted into a PstI site in a plasmid was discussed. In this method, the cloned DNA can be re-excised from the vehicle using PstI with high efficiency.

Dr. William Loomis (UC, San Diego) discussed the evidence that less than 400 genes play essential roles in morphogenesis and are dispensable for growth in *Dictyostelium*. Three independent approaches arrive at similar values.

1) Although early hybridization experiments indicated that about 5,000 RNA sequences of 1,000 bases (sufficient to be a gene) were present uniquely during development, these studies followed total cellular RNA, both nuclear and cytoplasmic, and measured RNAs present only once in seven cells. Firtel (unpublished) has extended this study to focus only on polysomal poly $(A)^+$ RNAs and finds that only 200-400 such sequences of 1,000 bases are unique to development.

2) Complementation analysis of independently isolated mutations blocking aggregation by Newell's and Coukell's groups have indicated that less than 150 loci are essential for aggregation.

3) The mutational target size for aggregation and multicellular morphogenesis genes was compared to the single gene target size. The frequency of recovery of structural gene mutations in each of five genes coding for specific enzymes was found to be 5.7×10^{-4}. Since aggregation mutants were found in these same populations at a frequency of 0.11, the number of aggregation genes appears to be about 154. By a similar analysis, the number of genes involved in multicellular morphogenesis was found to be about 140. Thus, these three analyses all indicate that only 300±100 genes are needed for development of this organism.

Dr. Randall Dimond (MIT) reported studies by Drs. Tom Alton and Harvey Lodish on changes in the major proteins during development. By using two-dimensional gel separation of ^{35}S methionine labeled proteins, 400 proteins could be followed. The relative rate of synthesis of 40 decreased during development while that of 50 increased. Twelve proteins appeared to be synthesized uniquely during development. The advantages and disadvantages of the technique were discussed.

Dr. Barrie Coukell (York University) described in detail his analysis of complementation groups involved in aggregation. From about 400 diploids formed from pairs of independently isolated aggregateless mutants of a specific subclass, 10 were found in which the mutations did not complement. They arose from 14 mutations which fell in five complementation groups. Two groups had four members and three had two. Statistical analysis of these data suggest that 38 genes are needed at this stage of development. Dr. Coukell went on to describe the mapping of these loci to five linkage groups in *Dictyostelium*.

The discussion of these presentations brought out many of the particular attributes of *Dictyostelium* as the subject for intense genetic analysis both by molecular techniques and by isolation, characterization and genetic manipulation of specific mutations. The fusion of these approaches holds the promise of detailed understanding of various developmental events.

Richard A. Firtel
William F. Loomis

POLY(A) METABOLISM IN *DICTYOSTELIUM DISCOIDEUM*

Allan Jacobson, Carl Mathew Palatnik, Cheryl T. Mabie
and Carol Wilkins

Department of Microbiology, University of Massachusetts
Medical School, Worcester, Massachusetts 01605

ABSTRACT. We have developed a procedure for fractionating mRNA according to its poly(A) content. Using this procedure, we have studied poly(A) metabolism during vegetative growth and development in *Dictyostelium discoideum*. We have found that most newly synthesized mRNA in vegetative cells contains long poly(A) tracts which shorten with age. We have also detected a second, less stable class of mRNA which contains short poly(A) tracts. Finally, we have found that there are dramatic changes in poly(A) metabolism associated with the transition from growth to development. These changes may be involved in the release of mRNA from polysomes, which has been shown to occur during this transition.

INTRODUCTION

Although it has been several years since the discovery of poly(A) sequences in eukaryotic mRNA (1-4), the physiological role of these sequences has yet to be determined. It has recently been shown that mRNA stability can be radically affected by the presence or absence of long (greater than 25-30 nucleotides) poly(A) stretches (5). It is not known, however, whether a cell uses poly(A) to regulate mRNA stability or whether removal of poly(A) from mRNA affects some other physiological property which renders the mRNA unstable. Furthermore, its effect on mRNA stability does not preclude other postulated functions for poly(A).

We have been studying poly(A) metabolism in the cellular slime mold *Dictyostelium discoideum*. In order for us to determine poly(A) size without enzymatically degrading the RNA, we have devised a method for fractionating mRNA according to its poly(A) content. This has enabled us to do subsequent analyses of mRNAs of particular poly(A) size classes. Using this technique, we have analyzed poly(A) metabolism during different developmental stages. We have found that there are at least two classes of poly(A)-containing RNA in *Dictyostelium*. In addition, we have found that there are dramatic changes in poly(A) metabolism associated with the transition from vegetative growth to development. It has previously been shown that a large fraction of vegetative mRNA is released from polysomes during this transition (6). We suggest that poly(A) tracts may be intimately involved in this process.

RESULTS

Fractionation of mRNA by Thermal Elution from Poly(U) Sepharose. In order to fractionate mRNA according to its poly(A) content, we developed the following method. Using water jacketed columns, we bound mRNA to poly(U) Sepharose at 25°C in the presence of 0.7M NaCl and 25% formamide. To elute the bound mRNA, the salt concentration was reduced to 0.1M NaCl. We then washed the column with the same buffer at 25°C, 35°C, 45°C and 55°C. In a final wash at 55°C, we raised the formamide concentration to 90%. Figure 1 shows an elution profile from such a column which was run with

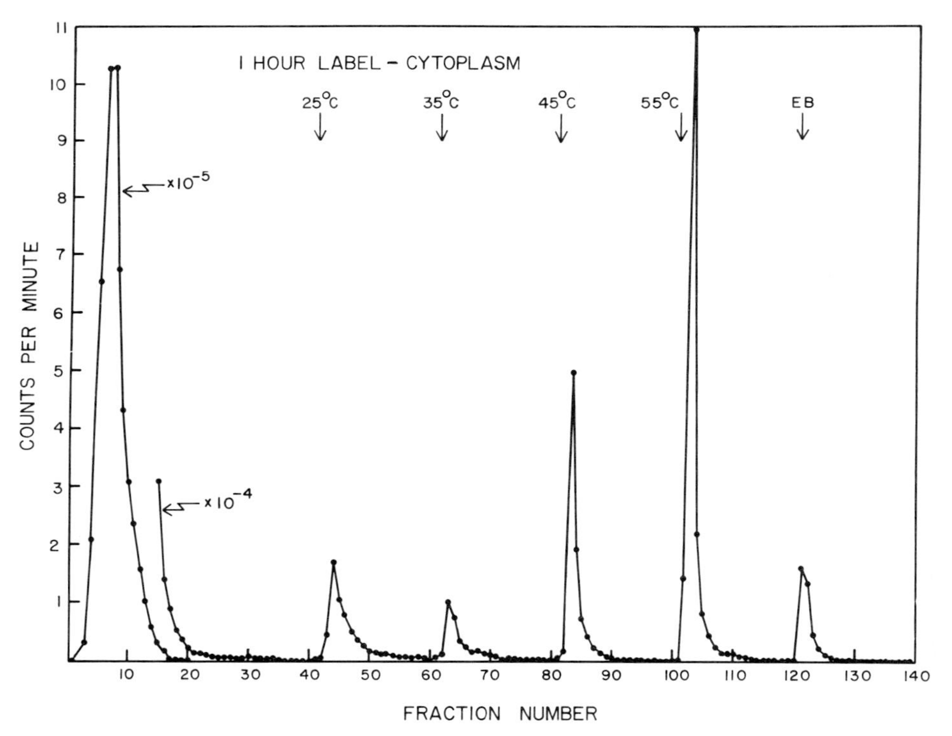

Fig. 1. Poly(U) Sepharose thermal elution profile of vegetative cytoplasmic RNA labeled for 1 hour with $^{32}PO_4$.

$^{32}PO_4$-labeled vegetative cytoplasmic RNA. Analysis of poly (A) tracts on polyacrylamide gels has shown that the mean poly(A) size of the different thermal eluates, in order of elution, is 25, 55, 80, 110 and 120 nucleotides, respectively.

Shortening of Poly(A) With Age. Figure 2 compares the elution profiles of three different RNA samples isolated from vegetative cells. Figures 2A and 2B are profiles of cytoplasmic RNA extracted from cells labeled with $^{32}PO_4$ for 1 and 4 hours, respectively. Figure 2C is the "steady state" elution profile of whole cell RNA, obtained by monitoring the optical densities of the various fractions. As shown, newly synthesized poly(A)-containing RNA contains predominantly long poly(A) tracts. By 4 hours of labeling, however, the size of the poly(A) tracts has shortened considerably and has begun to approach a steady state size distribution. Figure 4E illustrates the poly(A) size distribution of RNA isolated from polysomes labeled for 1 hour and "chased" for four hours during vegetative growth. This profile is extremely similar to that obtained from cytoplasmic RNA labeled continuously for 4 hours (Fig. 2B) and lends additional support to a shortening of poly(A) with age.

Translation Products of Vegetative mRNAs Containing Different Size Poly(A) Tracts. Figure 3 shows in vitro protein synthetic patterns of RNAs isolated from various poly(U) Sepharose thermal eluates. Although this analysis has been initially made on a one-dimensional gel system, it can readily be seen that most translatable mRNAs (approximately 90%) are distributed throughout the various eluates, presumably because they originated with long poly(A) tracts which metabolized with age. We have, however, been able to detect at least 10 reproducible differences between the various eluates. For example, although it has been previously reported that poly(A) deficient mRNA is representative of all cellular mRNAs (7), examination of the translation products of the poly(U)-Sepharose flow through, indicates that translation activities for two proteins, actin and histone, predominate. In addition, translation activities for two proteins also predominate in the 25°C eluate, actin and an unidentified protein of about 20,000 molecular weight. Furthermore, although it cannot be clearly seen from Figure 3, we have been able to detect translation activities in the 25°C and 35°C eluates which are absent from the 45°C and 55°C eluates. Preliminary experiments with inhibitors of RNA synthesis suggest that these RNAs are

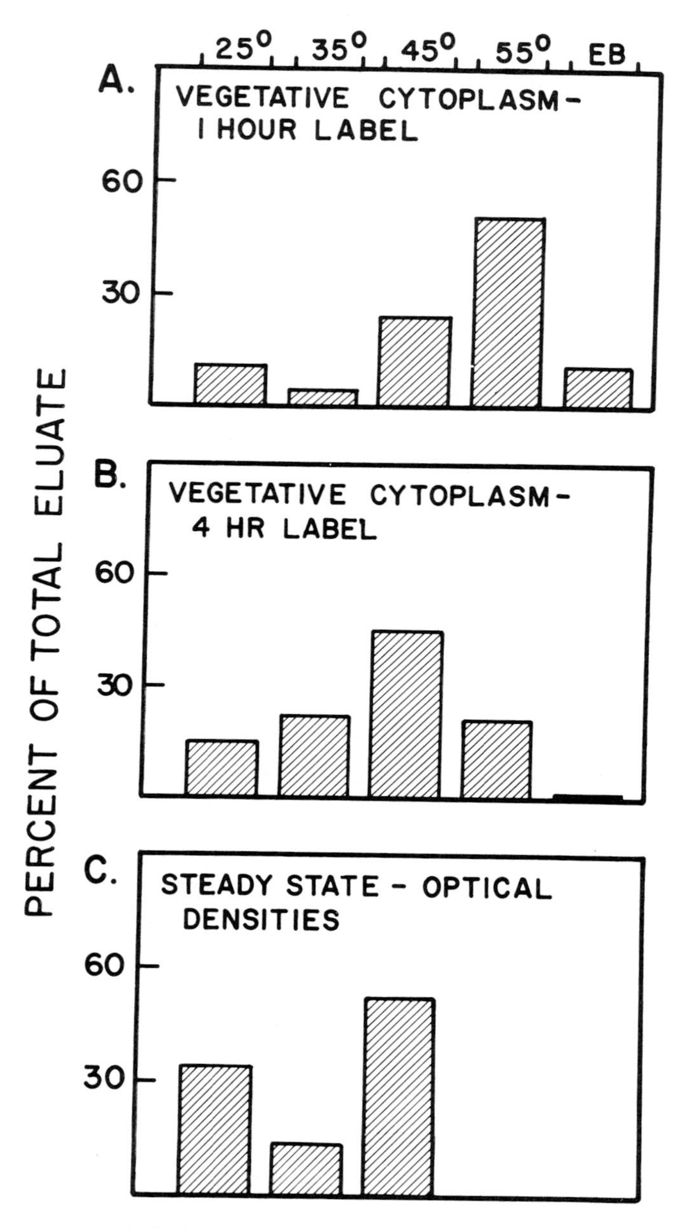

Fig. 2. Poly(U) Sepharose thermal elution profiles of vegetative RNA: (a) cytoplasmic RNA labeled with $^{32}PO_4$ for 1 hour, (b) cytoplasmic RNA labeled with $^{32}PO_4$ for 4 hours and (c) whole cell RNA monitored at A_{254}.

much less stable than the majority of cellular mRNAs. Since translation activities for these less stable mRNAs are not apparent in higher thermal eluates, they probably represent a second class of poly(A)-containing RNA which is synthesized and metabolized in a unique fashion.

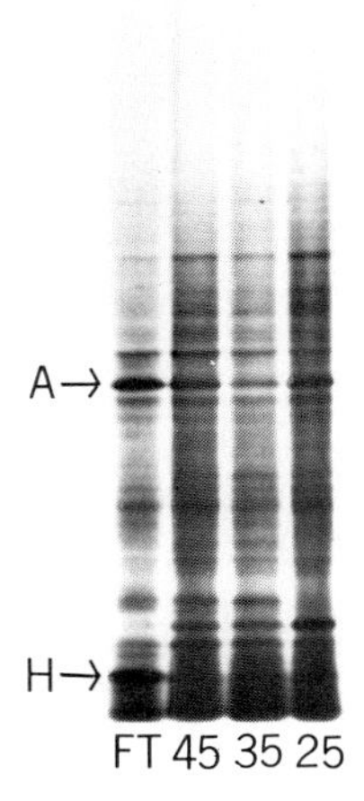

Fig. 3. Translation products of vegetative whole cell RNA fractionated by thermal elution from poly(U) Sepharose. Individual eluates were translated in wheat germ extracts, displayed on 6-15% linear gradient polyacrylamide gels and analyzed by fluorography. The actin and histone bands are labeled "A" and "H" respectively.

Changes in Poly(A) Metabolism Associated with Early Development. In early experiments in which we isolated RNA from cells in early development, we repeatedly noticed that a significantly smaller fraction bound to either oligo (dT) cellulose or poly(U) Sepharose than did vegetative RNA isolated under similar conditions. Since it had been previously shown that rRNA content decreases continuously throughout development (8), these experiments suggested that either poly(A)-containing RNA or the poly(A) itself was disappearing at a more rapid rate than rRNA. We therefore pulse-labeled cells with $^{32}PO_4$ for the first hour of development and analyzed the labeled RNA on poly(U) Sepharose. As shown in Figure 4B, the distribution of cytoplasmic poly(A)-containing RNA synthesized during this time is very different from the distribution obtained from vegetative cytoplasmic RNA (Fig. 4A). Most of the newly synthesized mRNA has very short poly(A).

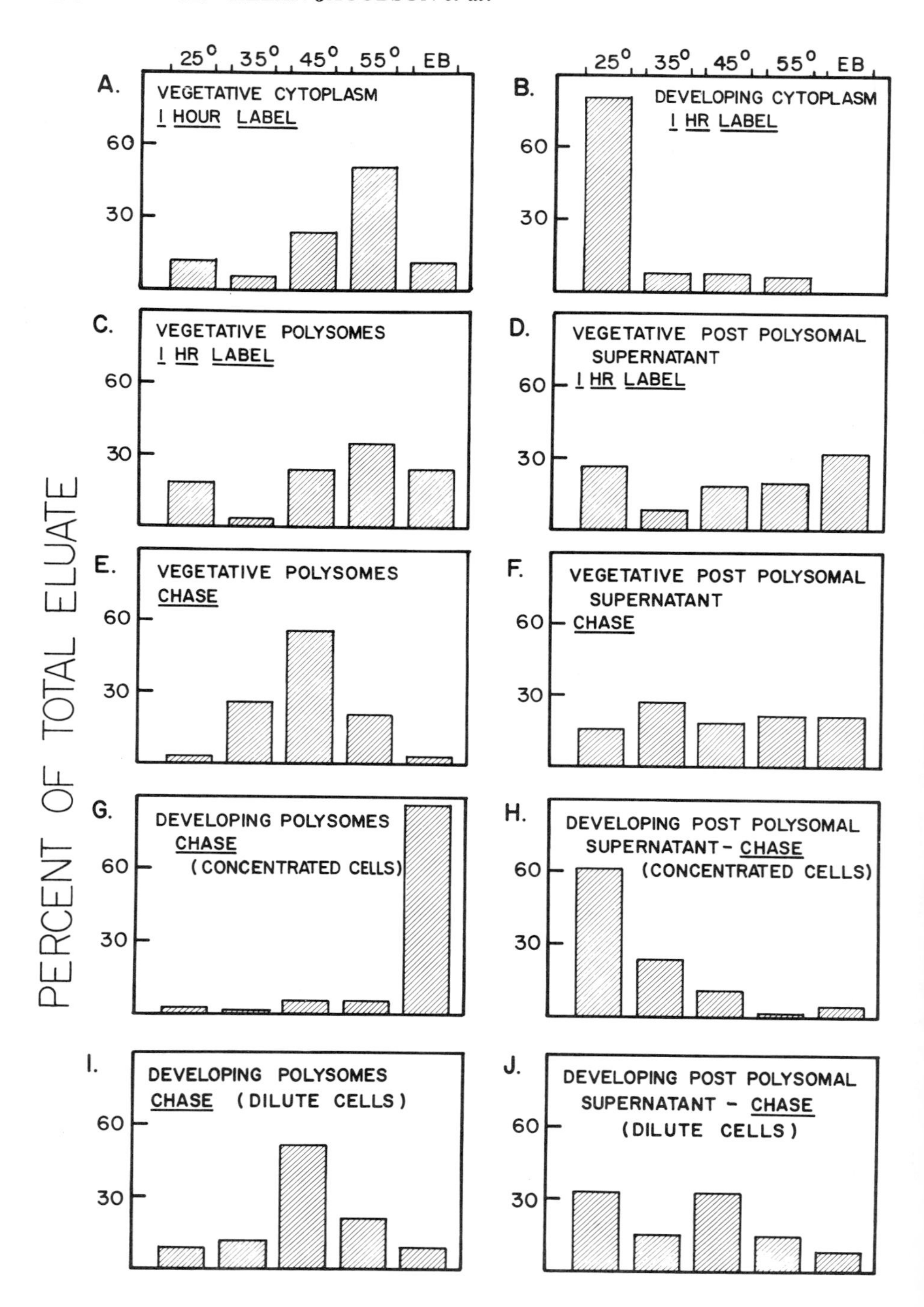
PERCENT OF TOTAL ELUATE
25° 35° 45° 55° EB
A.
VEGETATIVE CYTOPLASM
I HOUR LABEL
60
30
B.
DEVELOPING CYTOPLASM
I HR LABEL
60
30
C.
VEGETATIVE POLYSOMES
I HR LABEL
60
30
D.
VEGETATIVE POST POLYSOMAL SUPERNATANT
I HR LABEL
60
30
E.
VEGETATIVE POLYSOMES
CHASE
60
30
F.
VEGETATIVE POST POLYSOMAL SUPERNATANT
CHASE
60
30
G.
DEVELOPING POLYSOMES
CHASE
(CONCENTRATED CELLS)
60
30
H.
DEVELOPING POST POLYSOMAL SUPERNATANT - CHASE
(CONCENTRATED CELLS)
60
30
I.
DEVELOPING POLYSOMES
CHASE (DILUTE CELLS)
60
30
J.
DEVELOPING POST POLYSOMAL SUPERNATANT - CHASE
(DILUTE CELLS)
60
30

We then labeled cells with $^{32}PO_4$ for 1 hour during vegetative growth, "chased" the label for 4 hours under starvation conditions (one of the requirements for initiating the developmental program in *Dictyostelium*), isolated RNA from polysomes, and again analyzed the distribution of poly(A)-containing RNA. We did the "chase" under two sets of conditions: (1) with cells concentrated to 1-2 X 10^8 cells/ml, which Alton has shown (6) faithfully reproduces mRNA metabolism in early development and (2) under conditions in which the cells were too dilute to initiate the early developmental program (10^7 cells/ml). As can be seen, the distribution of poly(A)-containing RNA in the "chased" dilute cells (Fig. 4I) was similar to the distribution obtained from similarly "chased" vegetative cells (Fig. 4E), implying a simple aging type phenomenon as observed in vegetative cells. The concentrated cells showed a dramatically different profile (Fig. 4G). Most of the "chased" RNA which remained associated with polysomes had very long poly(A). Furthermore, examination of the post-polysomal supernatant (Fig. 4H) showed a large amount of RNA with short poly(A). These results indicate that during early development the cell selectively translates those mRNAs containing long poly(A). In addition, the results shown in Figure 4H raise the possibility that "chunks" of poly(A) are cleaved from previously existing poly(A)-containing RNA at this time.

DISCUSSION

The ability to fractionate mRNAs according to their poly(A) content has enabled us to gain insight into several

Fig. 4. Poly(U) Sepharose thermal elution profiles of various $^{32}PO_4$ labeled RNA samples: (a) vegetative cytoplasmic RNA labeled for 1 hour; (b) cytoplasmic RNA from cells labeled during the first hour of development; (c) vegetative polysomal RNA labeled for 1 hour; (d) the post-polysomal supernatant from (c); (e) vegetative polysomal RNA labeled for 1 hour and "chased" for 4 hours; (f) the post-polysomal supernatant from (e); (g) polysomal RNA from cells labeled for 1 hour during growth and "chased" for 4 hours into development; (h) the post-polysomal supernatant from (g); (i) polysomal RNA from cells labeled and "chased" as in (g) but with cells "chased" in dilute suspension; and (j) the post-polysomal supernatant from (i).

important aspects of poly(A) metabolism in Dictyostelium discoideum. Using this technique, we have shown that most of the newly synthesized mRNA is vegetative cells contains long poly(A) tracts which shorten with age. We have also been able to detect a less stable class of mRNA, which contains much shorter poly(A) tracts and which cannot originate from the first class. We have further shown (1) that a cataclysmic shortening of poly(A) may occur during early development; (2) that those mRNAs with long poly(A) tracts are preferentially translated during this time; and (3) that, in early development, most newly synthesized mRNA contains short poly(A) and is, consequently, excluded from polysomes. Alton has shown that within the first 5 minutes of development, a large fraction of mRNA is released from polysomes and stored in the cytoplasm (6). The changes in poly(A) metabolism which we observe could be responsible for the differential release of mRNA from polysomes. We have preliminary data which suggests that the distribution of mRNAs in the various eluates changes during early development.

Finally, although poly(A) metabolism may be related to the distribution of mRNAs on polysomes in early development, we have additional evidence (Palatnik, Wilkins and Jacobson, in preparation) which indicates that there are significant changes in transcription associated with the first 5 minutes of development. It appears, therefore, that developmental controls operating at both the transcriptional and translational levels are initiated within this short period of time.

ACKNOWLEDGEMENTS

This work was supported by a grant to A.J. from the American Cancer Society. A.J. is a recipient of a Faculty Research Award from the American Cancer Society. C.M.P. is a postdoctoral fellow of the National Institutes of Health.

REFERENCES

1. Kates, J. (1970) Cold Spring Harbor Symp. Quant. Biol. 35, 743.
2. Darnell, J.E., Wall, R., and Tushinski, R.J. (1971) Proc. Nat. Acad. Sci. U.S.A. 68, 1321.
3. Lee, S.Y., Mendecki, J., and Brawerman, G. (1971) Proc. Nat. Acad. Sci. U.S.A. 68, 1331.
4. Edmonds, M., Vaughan, M.H., Jr., and Nakazato, H. (1971) Proc. Nat. Acad. Sci. U.S.A. 68, 1336.
5. Nudel, V., Soreq, H., Littauer, U.Z., Marbaix, G., Huez, G., Lecleuq, M., Hubert, E., and Chantrenne, H. (1976)

Eur. J. Biochem. 64, 115.
6. Alton, T. (1977) Ph.D. Thesis, Department of Biology, Massachusetts Institute of Technology.
7. Lodish, H.F., Jacobson, A., Firtel, R., Alton, T., and Tuchman, J. (1974) Proc. Nat. Acad. Sci. U.S.A. 71, 5103.
8. Sussman, R.R. (1967) Biochim. Biophys. Acta 149, 407.

ANALYSIS OF RECOMBINANT PLASMIDS CARRYING GENE SEQUENCES FROM *DICTYOSTELIUM*

Karen L. Kindle and Richard A. Firtel

Department of Biology
University of California, San Diego
La Jolla, California 92093

ABSTRACT. *Dictostelium* nuclear DNA containing plasmids coding for high frequency message have been analyzed. One of these is complementary to a single messenger which appears to code for *Dictyostelium* actin. The actin clone has been mapped and fine structure analysis indicates that the coding sequence is repeated some 15-20 times in the *Dictyostelium* genome. Moreover, there are two forms of the message which differ in molecular weight by about 100 nucleotides. The other two clones which have been analyzed hybridize a heterogeneous population of messengers apparently by a short sequence common to a class of messages. One of these also carries the gene for a specific mRNA.

INTRODUCTION

The cellular slime mold, *Dictyostelium discoideum*, has been widely used as a model for developmental systems, since vegetative cells can be induced to undergo synchronous differentiation in approximately 24 hr. The organism is simple, relative to mammalian cells, containing approximately 1/100th the amount of DNA or 12 times that found in *E. coli* (1). In addition, it has a genome organization which is similar to that of higher eukaryotes (2) and some, but not all, features of its RNA transcription and processing are similar (3,4).

PROPERTIES OF mRNA COMPLEMENTARY TO CLONED SEQUENCES

We have used molecular cloning techniques to examine the structural organization of purified genes and to examine the differential transcription of these genes during *Dictyostelium* development. We have inserted randomly sheared *Dictyostelium* nuclear DNA into plasmid pMB9 using poly(dA)-poly(dT) tailing method (5). Using the colony filter hybridization technique (6) we selected a series of recombinant plasmids which are complementary to a relative large fraction of [^{32}P] *in vivo* labeled poly(A)$^+$ RNA and which are not complementary to either ribosomal RNA or to mitochondrial DNA. Our main

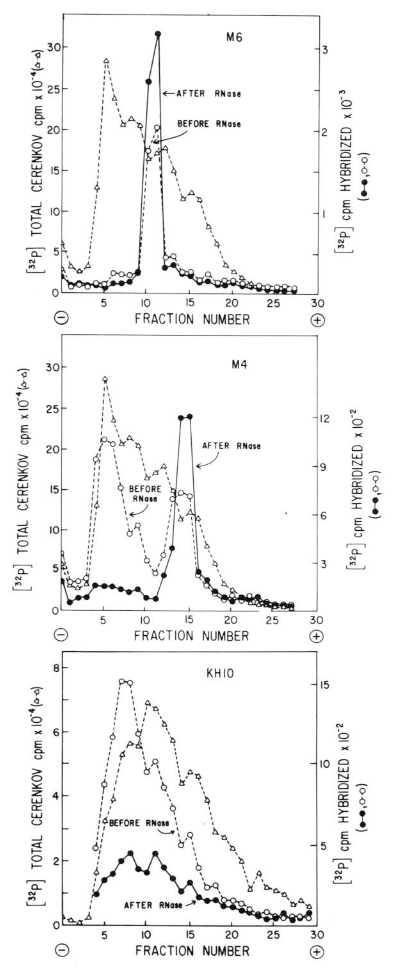

Fig. 1. Hybridization of size fractionate RNA to M4, M6, KH10 DNA filters. [^{32}P] *in vivo* labeled poly(A)$^+$ RNA was size-fractionated on 99% formamide 3.6% acrylamide gels. The gel was sliced and the RNA eluted by sonication as suggested by G. Rubin. Each fraction was hybridized to DNA filters, which were extensively washed, counted by Cerenkov radiation, treated with 20 µg/ml RNase A in 2 x 55S for 30' at room temperature, and then counted again. Δ--Δ Total Cerenkov cpm; O--O Cerenkov hybridized cpm before ribonuclease; ●--● [^{32}P] Hybridized cpm after ribonuclease.

analysis has been of 3 clones: M4, M6, and KH10, which hybridize 1.5, 0.5, and 5% of total poly(A)$^+$ respectively.

To determine the size and nature of the RNA complementary to the various clones, *in vivo* labeled poly(A)$^+$ RNA was size fractionated on polyacrylamide gels in 99% formamide. The relative mobility of the RNA complementary to various clones was determined as described in the legend to Fig. 1. The results showed that plasmid M6 hybridizes to a discrete sized peak of RNA which is largely resistant to ribonuclease treatment. In contrast, plasmid KH10 hybridizes to a broad size range of mRNAs, and a major fraction of the RNA hybrids in ribonuclease sensitive. Plasmid M4 shows hybridization similar to both M6 and KH10. It hybridizes to a large heterogeneous size class of RNA, the hybrids of which are ribonuclease sensitive, and to a specific peak of low molecular weight mRNA. This peak of hybridization is relatively RNase resistant and probably corresponds to the hybridization of a specific mRNA to its complementary gene sequence.

The quality of the hybrids observed in the above experiments were analyzed by melting profile. The RNA-DNA hybrid formed with M6 DNA shows a relative high melting profile while that of KH10 shows a broad melting profile at a relatively low Tm. The hybrids formed by the large population of RNA molecules which are complementary to plasmid M4 shows a melting profile similar to that observed in plasmid KH10 while the specific peak of hybridization to plasmid M4 showed a melting profile similar to that observed for plasmid M6.

We interpret the hybridization results with the heterogeneous population of RNA to plasmid M4 and KH10 as resulting from short or poorly base-paired region (5) present on these plasmids which are complementary to a large population of different mRNAs. This interpretation is supported by experiments that show the size of RNase resistant hybrids for plasmid KH10 and the heterogeneous RNA hybrids to plasmid M4 are only approximately 80-100 nucleotides in length.

MAPPING OF CODING SEQUENCES AND REITERATED SEGMENTS

Restriction enzyme maps of the *Dictyostelium* inserts have been determined for the 3 plasmids and the gene coding sequences for plasmid M6 and M4 have been localized. In both M4 and M6, the entire gene coding region is contained within the *Dictyostelium* insert and thus any potential regulatory elements which may be closely linked to either the 3' or 5' end of the gene will be present.

To examine whether the *Dictyostelium* insert contains repeat or single copy sequence elements, two sets of experiments were performed: (1) hybridization kinetics of nick translated DNA (7) to an excess of *Dictyostelium* DNA; (2) hybridization of plasmid probe to restriction digest of *Dictyostelium* DNA according to the method of Southern (8). Hybridization kinetic results show that plasmid M4 contains mostly single-copy sequences with a small fraction of the DNA complementary to repeat sequences. In plasmid M6, approximately 30% of the *Dictyostelium* insert hybridizes to sequences that are repeated approximately 15-20 times per genome and the remainder renatures to single-copy sequences. Reiterated sequences in plasmid M6 have been localized and several experiments show that the Hae III-Hap II 1.7 kb restriction fragment, which contains the encoded gene, is reiterated approximately 15-20 times in the genome while the outside sequences are single-copy.

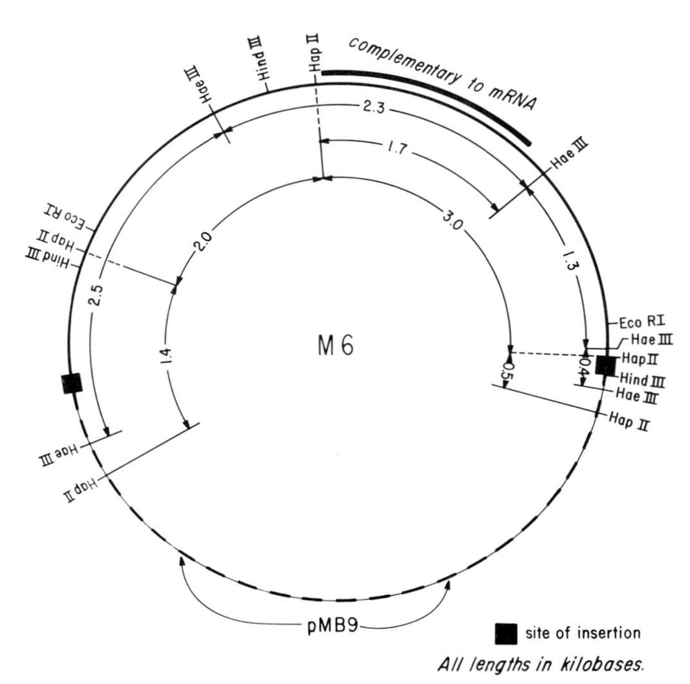

Fig. 2. Restriction map of M6. Restriction map was constructed using standard techniques for restriction digestion of plasmid DNA, running agarose gels, and locating coding by hybridization of [^{32}P] RNA to Southern filter blots. [▬] coding region.

Results of the DNA excess hybridization were confirmed using the Southern DNA blot technique. cRNA (9) or nick translated DNA (7) probes made to the various Hae III or Hap II restriction fragments of plasmid M6 were hybridized to Southern DNA blots of restricted *Dictyostelium* nuclear DNA. The results showed that restriction fragments from noncoding regions hybridized only to 1-2 restriction fragments on the Southern DNA blots, while those fragments carrying the gene sequence hybridize to approximately 15-20 different size fragments. These and additional experiments not reported here indicate the gene sequence itself is reiterated and that the outside noncoding sequences are single copy. In similar experiments using plasmid M4, it was shown that only one or two restriction fragments show hybridization as expected since the majority of the plasmid is a single copy when analyzed kinetically.

EVIDENCE THAT PLASMID M6 CODES FOR ACTIN

There are a number of results which suggest that M6 may code for *Dictyostelium* actin. mRNA complementary to M6 was purified by specific hybridization and elution from M6 DNA cellulose (10). It directs the synthesis of a protein in a wheat germ cell-free protein synthesizing system which co-migrates with *Dictyostelium* actin on 12.5% polyacrylamide gels. In addition, in *E. coli* mini-cells, the plasmid appears to direct the synthesis of a protein which co-migrates with *Dictyostelium* actin. Control experiments using the parental plasmid, pMB9, do not direct the synthesis of this protein. We have already shown that the *in vitro* synthesized product made in the wheat germ system specifically binds to DNase I columns (11), suggestive evidence that the protein is actin. Peptide maps are presently being done to confirm the identity of these proteins.

Initial studies on the differential transcription of this particular gene during *Dictyostelium* development have been done. The results show a pattern of transcription that would be expected from a known increase in synthesis of actin protein in the first few hours of the developmental cycle (12). These experiments along with the fact that this message represents almost 1% of the vegetative mRNA as determined by RNA excess hybridization kinetics suggest that the clone probably codes for actin.

SIZE ANALYSIS OF ACTIN MESSAGE

[^{32}P] *in vivo* labeled RNA was hybridized and eluted from M6 DNA filters and analyzed on urea polyacrylamide (13) and methyl mercury (14) agarose gels. In both gel systems, two discrete mRNA bands of approximately 1300 and 1400 nucleotides in length were observed. Because methyl mercury gels are thought to denature nuclei acids completely, we assume that this is not an artifact due to conformational changes of the RNA. In additional experiments, the specific RNA which which hybridizes to plasmid M4 shows a single band migrating with a molecular weight of approximately 800-900 nucleotides.

We are presently continuing the analysis of these 3 particular plasmids in order to further our understanding of gene structure in *Dictyostelium* and to start to understand the mechanisms of differential gene activity and its relationship to *Dictyostelium* differentiation.

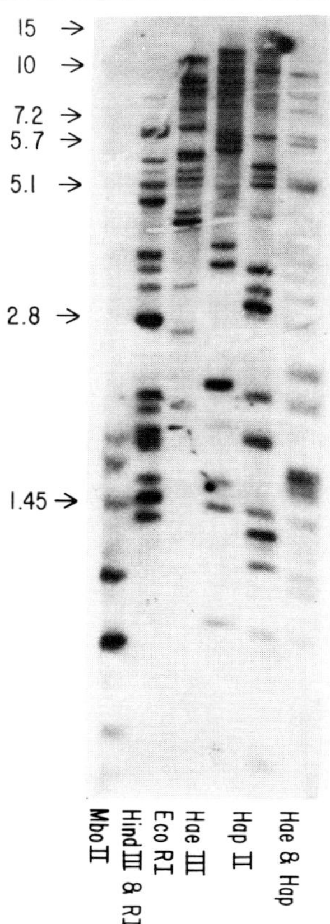

Fig. 3. Hybridization of M6 coding DNA to total *Dictyostelium* DNA. *Dictyostelium* DNA was digested with various restriction enzymes and run on a 0.8% agarose gel. Nitrocellulose filter blots were made by the method of Southern (8) and hybridized to a nick translated probe from Hae III-Hap II M6 coding (6) and DNA.

ACKNOWLEDGEMENT

Karen Kindle is a predoctoral fellow of the National Science Foundation.

This work was supported by grant from the National Science Foundation, The American Cancer Society and the Cancer Coordinating Committee of the University of California.

REFERENCES

1. Loomis, W.F., *Dictyostelium discoideum*: A Developmental System, *Academic Press* (1975).
2. Firtel, R.A. and Kindle, K. *Cell* 5, 401 (1975).
3. Firtel, R.A. and Lodish, H.F. *J. Mol. Biol.*79, 295 (1973).
4. Dottin, R.P., Weiner, A.M. and Lodish, H.F. *Cell* 8,233 (1976).
5. Lobban, P. and Kaiser, A.O. *J. Mol. Biol.* 78, 453 (1973).
6. Grunstein, M. and Hogness, D. *Proc. Nat. Acad. Sci. USA* 72,3961 (1975).
7. Maniatis, T., Jeffrey, A., and Kleid, D.G. *Proc. Nat. Acad. Sci. USA* 72,1184 (1975).
8. Southern, E.M. *J. Mol. Biol.* 98,503 (1975).
9. Barnes, W.M., Reznikoff, W.S., Blattner, F.R., Dickson, R.C. and Abelson, J. *J. Biol. Chem.* 250,8184 (1975).
10. Noyes, B.E. and Stark, G.R. *Cell* 5, 301 (1975).
11. Lazarides, E. and Lindberg, U. *Proc. Nat. Acad. Sci. USA* 71,4742 (1974).
12. Tuchman, J., Alton, T. and Lodish, H.F. *Developmental Bio.* 40,116 (1974).
13. A. Spradling (personal communication).
14. Bailey, J. and Davidson, N. *Anal. Biochem.* 70, 75 (1976).

ORGANIZATION OF RIBOSOMAL AND 5S RNA CODING REGIONS IN *DICTYOSTELIUM DISCOIDEUM*

William C. Taylor, Andrew F. Cockburn, Gary A. Frankel,
Mary Jane Newkirk and Richard A. Firtel

Department of Biology
University of California, San Diego
La Jolla, California 92093

ABSTRACT. The organization of the ribosomal repeats in *Dictyostelium* have been determined. The repeats which are 44 kb long, are linked head-to-head as palindromic dimers. Each repeat contains one coding region for the 36S rRNA precursor and the 5S rRNA. The dimers are isolated as free molecules unlinked to chromosomal DNA.

INTRODUCTION

Previous work from this lab (1-5) has shown that the rDNA (ribosomal DNA) comprises 18% of the nuclear DNA of *Dictyostelium discoideum* or approximately 180 identical copies per haploid genome. Each repeat is 44 kb (1 kb = 10^3 base pairs) long and consists of a transcribed 8 kb region coding for the 36S rRNA precursor bounded by a 28 kb non-transcribed spacer to the 5' and a 9 kb nontranscribed spacer to the 3' end (Fig. 1). A single coding sequence for 5S rRNA is contained within the latter region. The 17S coding sequences are located to the 5' end of the 8 kb transcribed strand and the 26S to the 3' end. The 5.8S rRNA is located within the internal transcribed spacer, probably close to the 26S rRNA. In addition, there is a transcribed spacer at the 5' end of the 36S transcript. This pattern is similar to that found in other eukaryotes (6-8). Our mapping data has been generally corroborated by work in another laboratory (9).

LINKAGE OF rDNA REPEATS

We have demonstrated that the 44 kb units occur as palindromic dimers. Palindromic DNA should snap-back to form hairpin molecules when denatured and quick-cooled. *Dictyostelium* nuclear DNA was denatured and quick-cooled to allow snap-backs to form. Single stranded DNA was removed with S1 nuclease and the remaining DNA was digested with restriction endonuclease Eco R1 and analyzed on agarose gels. Two bands were obtained: one, 7.5 kb, which was 1/2 the lenth of Eco R1-band I and a second which co-migrated with band-IV (5.7 kb).

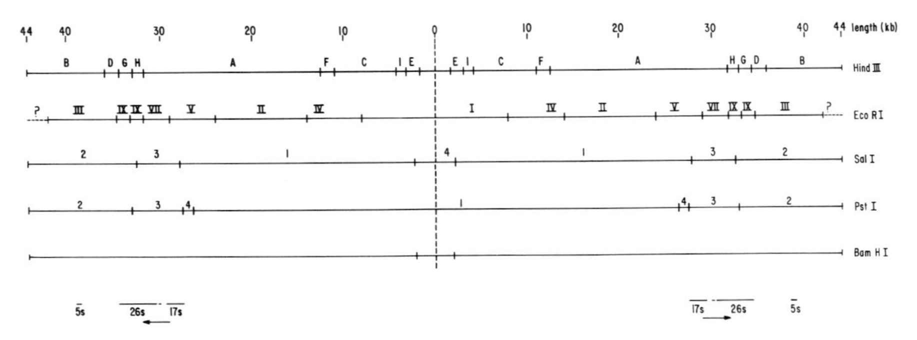

Fig. 1. Restriction map of the palindromic rDNA repeat dimer in *Dictyostelium discoideum.*

If DNA was digested with Eco R1 before the other procedures then only the 7.5 kb band was observed. cRNA (complementary RNA) made from the 7.5 kb snap-back DNA hybridized specifically to Eco R1-band I. These results demonstrate that band I is itself a palindrome and that it is the central region of a long palindrome. This conclusion is supported by similar experiments on snap-back DNA digested with Sal-I and Hind III.

The rDNA gives a linear restriction map. When the outside portion of the rDNA is analyzed by double digests with several restriction enzymes, the maps produced by the individual endonucleases (except for Eco R1) end coincidentally within the resolution of our agarose gels. The simplest explanation of these results is that the rDNA dimers, as isolated, are homogeneous lengthed, linear molecules of 88 kb which are not covalently associated with other DNA. Redigestion of isolated Pst-2, Sal-3 and Hind III-B bands with Eco R1 suggests that there may be several Eco R1 sites to the outside of the Eco R1 band III which are too small to be detected in our gel system.

FREE rDNA

Two lines of evidence indicate that most of the rDNA in *Dictyostelium* is isolated as free palindromic dimers. Log phase vegetative cells labeled with [^{3}H] thymidine were gently lysed in 25 mM EDTA, 5 mM EGTA, 12 mg/ml predigested pronase, and 0.5% SDS. The lysate was analyzed on a 0.15% horizontal agarose gel. The gel was sliced, the DNA eluted from each fraction and immobilized on nitrocellulose filters (10). 70% of the DNA remained at or very near the origin (Fig. 2). About 30% of the [^{3}H] label was found in a broad peak in the region of phage DNA markers, at a molecular weight range of

45-200 kb. Hybridization of [^{32}P] 26S rRNA and [^{32}P] cRNA to mt DNA (mitochondrial DNA) occurred primarily to the latter peak of DNA. The rRNA hybridized to DNA somewhat larger than the mt cRNA. [mt DNA is 45-50 kb long (1,11).] Both RNAs exhibited some hybridization to the DNA at the origin of the gel, however the [^{32}P] to [^{3}H] ratio shows that the concentration of mt DNA and rDNA was appreciably higher in the peak within the gel compared with that at the origin. It is probable that the mt DNA remaining at the origin was trapped by the long fibers of chromosomal DNA. We conclude that a large fraction of the rDNA consists of free molecules and that it is likely that at least some of the hybridization of rRNA to DNA at the origin is due to trapped episomal dimers. It is also possible that some rDNA repeats are contained in the chromosomal DNA.

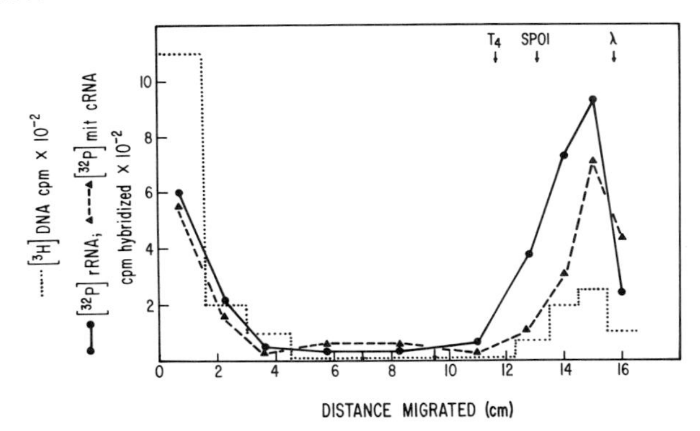

Fig. 2. Hybridization of [^{32}P] rRNA and mt cRNA to DNA from gently lysed *Dictyostelium* cells analyzed on a 0.15% agarose gel.

A second piece of evidence comes from the electron microscopic examination of DNA from cells gently lysed by the above method. The lysate was diluted 100-fold into 50% formamide, 0.1 M Tris and then spread for electron microscopy by a method similar to standard techniques (12). Most of the DNA was present in extremely long and tangled strands that were not possible to measure. There were, in addition, three classes of molecules which were found separate from the large tangles at fairly high frequencies (Fig. 3). The first class consists of molecules averaging 40-50 kb in length and are presumably mt DNA. 60% of the molecules in this class were open circles. Molecules in the second class were all linear with an average length of 80-90 kb, the expected length of the episomal rDNA dimers. The third class was made up of heterogeneous lengthed molecules not belonging to either of the first two classes and are presumably fragments produced during cell lysis and subsequent handling.

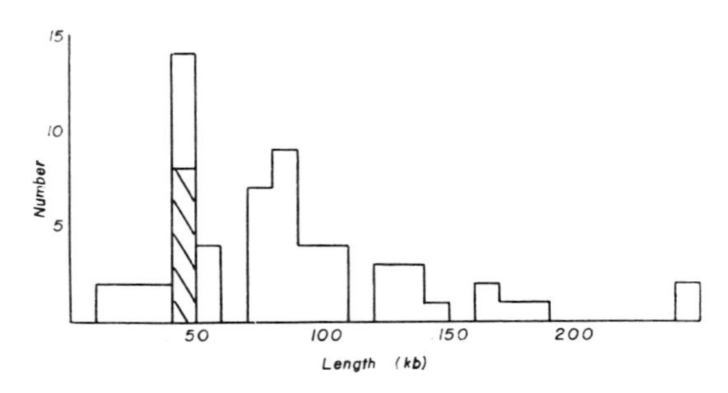

Fig. 3. Lengths of DNA molecules from gently lysed *Dictyostelium* cells. The shaded area represents those molecules which were open circles.

MAPPING OF THE 5S GENE

The 5S gene is found in Eco R1-band III. Eight enzymes (Hga I, Hap II, Sma I, Mbo I, Alu I, Pst I, Sal I, and Hha I) did not cut within Eco R1-band III. Two other enzymes (Hae III and Hinf I) produced small fragments, about 0.5-0.7 kb in length, which contain the 5S coding region. Hind III makes a single cut 1.2 kb from the left end of Eco R1-band III. 5S rRNA hybridizes to the 6 kb fragment produced by double digestion with both Hind III and Eco R1.

The Eco R1-III fragment was inserted into the plasmid pSC 101 by the ligation method after opening the plasmid with Eco R1. The original clone was found to be extremely unstable. Large regions of the inserted *Dictyostelium* DNA were lost during subsequent replication. Five new clones were isolated, containing from 1.5 to 3.3 kb out of the original 7.2 kb insert. Hybridization of 5S RNA to nitrocellulose bound restricted plasmid DNA eluted from agarose gels showed that the 5S coding region and the two Eco R1 sites at which the inserted DNA is linked to the vehicles are retained in all five derivative clones. Heteroduplexes were made between these five clones and analyzed in the electron microscope. Analysis showed that various portions of Eco R1-band III were lost and that the deletions were not localized at either end of the original fragment. These results suggest that the 5S region is not within at least 3 kb of the right end of fragment Eco R1-III. We can therefore tentatively say that the 5S coding region is located somewhere near the middle of Eco R1-III, at a distance of 1-4 kb from the end of the 36S coding regions.

CONCLUSIONS

We have demonstrated that the rDNA in *Dictyostelium* consists of episomal palindromic dimers, as does the rDNA in another protist studied, *Tetrahymena* (13). Each repeating unit is 44 kb long and contains a 36S ribosomal precursor region located 28 kb from the point of symmetry of the palindrome plus a 5S rRNA coding region located 1 to 4 kb distal to the 36S region. We do not know whether there are one or more chromosomal copies of the rDNA repeats.

ACKNOWLEDGEMENTS

William Taylor is a Fellow of the American Cancer Society; Andrew Cockburn is a National Science Foundation Predoctoral Fellow; Gary Frankel is a National Institutes of Health Predoctoral Trainee.

This work was supported by a National Science Foundation grant to Richard A. Firtel.

REFERENCES

1. Firtel, R.A., Cockburn, A., Frankel, G. and Hershfield, V. (1976). *J. Mol. Biol.* 102,831.
2. Cockburn, A., Frankel, G., Firtel, R.A., Kindle, K. and Newkirk, M.J. (1976). *ICN-UCLA Symposia on Molecular and Cellular Biology*. Vol. V. Nierlich, D.P., Rutter, W.J., and Fox, C. Fred (eds.) Academic Press, p. 599-603.
3. Cockburn, A.F., Newkirk, M.J. and Firtel, R.A. (1976). *Cell* 9,605.
4. Frankel, G., Cockburn, A.F., Kindle, K.L. and Firtel, R. A. (1977). *J. Mol. Biol.* in press.
5. Cockburn, A.F., Taylor, W.C. and Firtel, R.A. (1977). Manuscript in preparation.
6. Trapman, J. and Planta, R.J. (1975). *Biochim. Biophys. Acta.* 414,115.
7. Hackett, P.B. and Sauerbrier, W. (1975). *J. Mol. Biol.* 91,235.
8. Dawid, I.B. and Wellauer, P.K. (1976). *Cell* 8,443.
9. Maizels, N. (1976). *Cell* 9, 431.
10. Southern, E.M. (1975). *J. Mol. Biol.* 98,503.
11. Firtel, R.A. and Bonner, J. (1972). *J. Mol. Biol.* 66,339.
12. Davis, R.W., Simon, M. and Davidson, N. (1971). *Methods in Enzymology* XXI, part D, 413.
13. Gall, J.G. (1974). *Proc. Nat. Acad. Sci. USA* 71,3078.

A NEW APPROACH FOR IDENTIFYING AND MAPPING STRUCTURAL GENES IN *DROSOPHILA MELANOGASTER*

Michael W. Young* and David S. Hogness†

Department of Biochemistry, Stanford University
School of Medicine, Stanford, CA 94305

ABSTRACT. We describe a procedure whereby one can define, isolate and map a variety of structural genes without recourse to mutations, polypeptides or the prior isolation of specific mRNAs. The structural genes are those that code for the abundant mRNAs found in cultured cells of *Drosophila melanogaster* although the procedure is generally applicable to genes that code for the abundant mRNAs in any cell type. The procedure consists of cloning random segments of *D. melanogaster* DNA (Dm segments), and then screening the cloned segments for those that contain DNA sequences homologous to the RNA sequences present in a heterogeneous source of abundant mRNAs. A collection of 103 independently cloned Dm segments containing the structural genes that code for these mRNAs has been obtained. A sampling of this collection has revealed two classes of genes. One consists of genes that are repeated and located at many chromosomal sites which are widely dispersed throughout the genome, while a gene in the second class is confined to a single site without apparent repetition. We use a gene from the second of these two classes to show how the procedure allows one to map a structural gene, both within the cloned Dm segment and within the total genome. Extensions of the basic procedure are considered in the Discussion, as, for example, the purification of the specific mRNAs that correspond to the isolated genes.

INTRODUCTION

Units of gene function in *Drosophila melanogaster* are commonly identified by complementation tests and genetic mapping of mutations. However, additional information is required to determine how the mutations interact to form complementation groups. In particular, the identification

*Present address: The Rockefeller University, New York, N.Y. 10021

†To whom reprint requests should be sent.

of complementation groups as structural genes has usually depended upon the demonstration that the mutations alter the primary structure of polypeptides. While this procedure for identifying and mapping structural genes in _D. melanogaster_ has been employed in a limited number of cases with varying degrees of rigor (see ref. 26 for a review), it is generally too cumbersome for the analysis of genes at high resolution or in large numbers. Furthermore, it is applicable only to that fraction of the genome for which mutations are easily detected.

In this paper, we examine a new procedure for defining and mapping structural genes that does not depend upon the analysis of their mutations or their polypeptides. These dependancies are avoided by defining a structural gene as that part of the chromosomal DNA which consists of a sequence of base pairs homologous to the sequence of bases in a given mRNA*, and by isolating segments of _D. melanogaster_ chromosomal DNA (Dm segments) that contain such sequences. Isolation is accomplished by first cloning random Dm segments in the form of hybrid plasmids that replicate in _E. coli_ K12 (34), and then screening these segments for those that can hybridize with a given mRNA, using the colony hybridization technique of Grunstein and Hogness (12).

Structural genes can be mapped within Dm segments by further RNA-DNA hybridization - either by electron microscopic techniques such as R-loop mapping (35), or, as in this paper, by hybridization of the mRNA to an ordered set of DNA fragments formed by cleavage of the Dm segment with restriction endonucleases. A resolution somewhat better than 100 base pairs can be obtained by these techniques, and this can easily be extended to individual base pairs by recent advances in nucleotide sequence analysis (20). Genes that have been mapped within segments can also be mapped in relation to the remainder of the genome by _in situ_ hybridization of polytene chromosomes with ^{3}H-labeled RNA or DNA copies of restriction fragments contained within a cloned gene (10). Here the resolution corresponds to one or a few polytene bands, where the average band represents 30 kb out of the 170,000 kb in the haploid genome (25, 30; kb = 1000 bases or base pairs in single- or double-stranded nucleic acids, respectively). Structural genes can thus be mapped

*This definition closely parallels that originally given by Jacob and Monod (14, 15, 16), that a structural gene contains the genetic information determining the structure of a protein and that its primary product is an mRNA.

in two complementary ways: one of very high resolution but of limited range, and the other of very long range but of limited resolution.

At first sight, it would appear that this procedure depends upon the prior purification of a particular mRNA. This, however, is not the case; indeed, a major purpose of this paper is to demonstrate that the procedure is not constrained in this manner. For this purpose, we describe the isolation and mapping of a gene that codes for one of the abundant mRNAs found in cultured *D. melanogaster* cells. This gene was isolated, along with many other members of this class, from randomly cloned Dm segments that were screened by colony hybridization with a heterogeneous probe consisting of ^{32}P-labeled polyA-containing cytoplasmic RNA, which we presume are mRNAs. Each colony that exhibits a positive response carries a cloned Dm segment containing a sequence homologous to that in some of the RNA molecules in the probe; and, since the hybridization takes place in DNA excess (12), the level of the response reflects the fraction of such RNA molecules in the probe. Consequently, Dm segments containing genes that code for the abundant mRNAs are easily detected by the screen, whereas those that code for the rare mRNAs are not scored.

The important point, however, is that a heterogeneous mRNA probe can be used to isolate homogeneous Dm segments that contain specific structural genes. Furthermore, the same heterogeneous probe can be used to map the structural gene within the Dm segment, thereby defining a homogeneous source of the nucleotide sequence within the gene - sequences that can be used to map the gene within the polytene chromosomes. Quite the opposite from being required, purification of a specific mRNA is one of the by-products of the procedure, since this can be accomplished with the aid of the cloned gene by standard RNA-DNA hybridization procedures.

RESULTS

cDm plasmids containing abundant mRNA genes are recognized by colony hybridization with a polyA$^+$ cytoplasmic RNA probe. - The cDm plasmids consist of Dm segments inserted into the colicinogenic plasmid, ColE1. The cDm plasmids used here are part of the collection whose construction and cloning is described in detail elsewhere (Finnegan, Rubin, Bower and Hogness - in preparation). Briefly, Dm segments formed by shear breakage of Oregon R embryonic nuclear DNA were inserted at the EcoR1 site of ColE1 (13) by the polydA/polydT method of Wensink *et al.*(34),

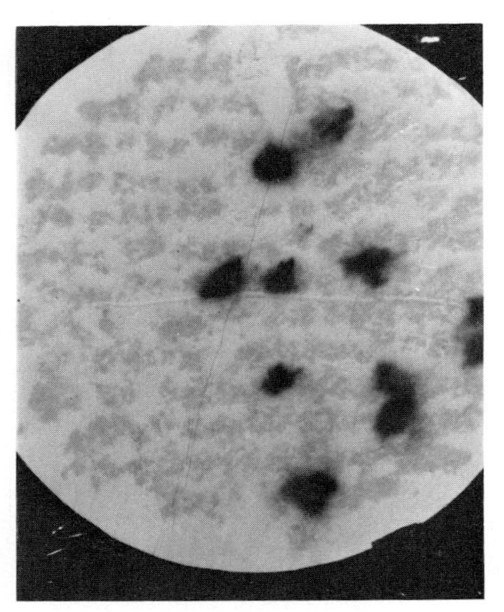

Fig. 1. Colony hybridization of [^{32}P]polyA$^+$ cytoplasmic RNA to cDm clones. In this figure, the x-ray film showing the autoradiographic response of the hybridization has been placed over the filter that contains the lysed bacterial colonies to which the [^{32}P] RNA was hybridized (Materials and Methods). Colonies exhibiting the intense response that is obvious in the figure result from cDm clones that contain either the copia or rRNA genes, which were subsequently distinguished by a second colony hybridization that is defined in the text. This form of presentation, with the filter behind the film, makes it difficult to see the weaker autoradiographic responses of the second group (see text).

and the resulting hybrid DNAs were then cloned by transformation of the HB101 strain of E. coli K12 (3) to colicin E1 immunity (9).

Approximately three thousand independently cloned cDm plasmids were screened by colony hybridization with a probe consisting of ^{32}P-labeled polyA$^+$ cytoplasmic RNA prepared from Eschalier's Kc line of D. melanogaster cells (Materials and Methods). Figure 1 shows the autoradiographic response obtained from the hybridization of a fraction of these clones. Some of the clones contain the genes for the 18S and 28S rRNAs, and exhibit a positive autoradiographic response because the probe is contaminated with these rRNAs. These clones were identified by a second colony hybridization equivalent to that used by Grunstein and Hogness (12) to screen for cDm plasmids containing rDNA. The remainder of the positively responding clones accounts for approximately three percent of the population (i.e., 103/3052), and these are presumed to contain genes that code for the abundant mRNAs in the probe.

These 103 clones divide into two groups according to the intensity of their autoradiographic response. One group consists of eight clones that exhibit a relatively uniform, strong response. This group has been examined in considerable detail and will be described elsewhere (Young and Hogness, in preparation). The uniformity and strength of the colony hybridization results from the fact that each of the eight independently cloned Dm segments contains the same structural gene, which is called copia (abbreviated cpa) since it codes for a particularly abundant mRNA. The eight Dm segments differ in that each is derived from a different one of the many widely dispersed chromosomal sites occupied by the repeated copia genes. This dispersed gene repetition is demonstrated by the in situ hybridization pattern shown in Figure 2A, which was obtained by hybridizing polytene chromosomes with [^{3}H]cRNA transcribed in vitro from a copia gene.

The weaker and less uniform colony hybridization exhibited by the cDm clones in the second group suggests that they carry genes coding for less abundant mRNAs. This was confirmed by quantitative filter hybridization of the [^{32}P]polyA$^+$ cytoplasmic RNA probe to saturating amounts of cDm DNAs purified from clones in both groups [see Materials and Methods for the hybridization conditions]. cDm DNAs from seven representative clones in the second group hybridized an average of 0.3% of the probe, with values ranging from $<$0.1% to 0.6%, in rough correspondence with the level of colony hybridization. By contrast, the values obtained for cDm DNAs containing a complete copia gene were at least 10-fold greater than the average value for the second group, the ratio depending somewhat upon the preparation of polyA$^+$ cytoplasmic RNA used.

These results exhibit an interesting correlation with the kinetic complexity of the polyA$^+$ cytoplasmic RNA, as determined from the rate of its hybridization with homologous cDNA. For example, Levy and McCarthy (19) found a complex set of rates that fit three frequency classes having kinetic complexities of 5 kb, 128 kb and 8.0 x 10^3 kb, or, if one assumes an average length of 2 kb per RNA molecule, consisting of 2.5, 109 and 4 x 10^3 different kinds of mRNA molecules, respectively. From these numbers and the fraction of the mRNA mass represented by each class, one can compute each kind of RNA molecule in the first class should represent an average of 2-to-3% of the polyA$^+$ cytoplasmic RNA, that each kind in the second should represent about 0.5%, and that each kind in the third, about 0.01%. These results suggest that the genes in the two groups of cDm clones detected by colony hybridization code for the mRNAs present in the first two frequency classes defined by the hybdridization kinetics.

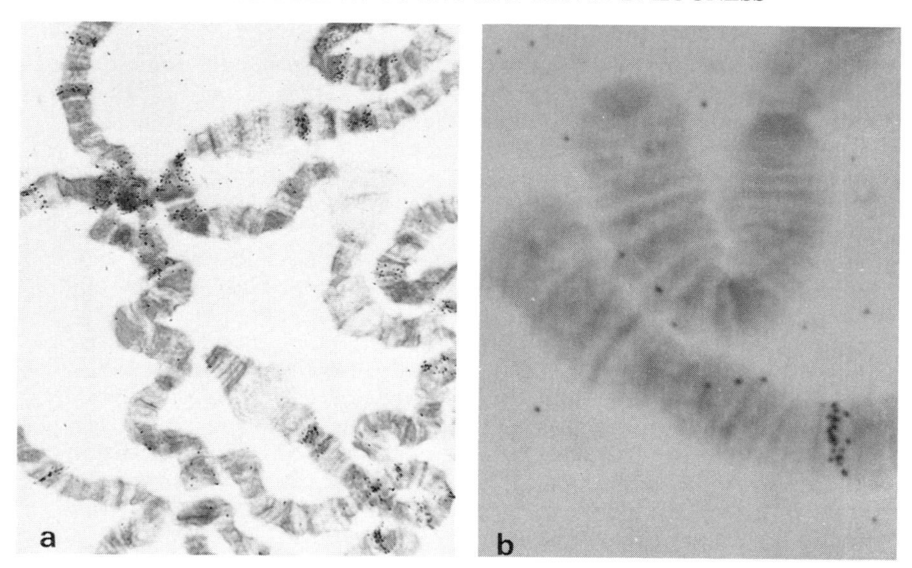

Fig. 2. (A) Mapping the copia sites in polytene chromosomes by in situ hybridization. This autoradiograph was obtained by hybridizing polytene chromosomes with [^{3}H] cRNA transcribed from pkDm1215 (Materials and Methods). The pkDm1215 plasmid is a hybrid between pSC105 and a HhaI restriction fragment that is entirely contained within the copia gene carried by cDm1142, one of the cDm clones that exhibited an intense response to colony hybridization with the polyA$^+$ cytoplasmic RNA probe (Young and Hogness, in preparation). pkDm1215 was constructed by inserting this HhaI fragment at the BamI site of pSC105, using the polydA/polydT procedures described in Materials and Methods. Since pSC105 sequences exhibit no in situ hybridization with these polytene chromosomes, the many sites that are marked by silver grains must contain sequences from the copia gene. Exposure time = 28 days.

(B) The location of the sam gene. [^{3}H]cRNA to cDm1020 was hybridized to polytene chromosomes in situ. After 28 days of exposure, the resulting autoradiograph revealed only one site of hybridization. This is shown in the figure and is located in the left arm of the 3rd chromosome at 62E.

If this is true, then approximately one hundred different kinds of structural genes should be detectable by colony hybridization with the probe used here. Only about one-fifth of these should be present in the 103 independently cloned Dm segments obtained thus far, since they derived from a screen that represents only one-fifth of the genome.

The implication of this comparison is that at least some genes in the second group of clones are repeated, much as is copia, the only gene we have found in the first group. This is certainly the case, since several members of the first dispersed multigene family to be detected, the 412 family, are present in the second group of clones (Rubin, Finnegan and Hogness, 1976 and in preparation). However, repetition is not a necessary characteristic of the genes that produce the abundant mRNAs, and in the remainder of this article we describe a gene from the second group that does not appear to be repeated. It is called sam and is carried by the cDm1020 plasmid. In the quantitative filter hybridization mentioned above, cDm1020 DNA hybridized 0.3% of the polyA$^+$ cytoplasmic RNA. This value can be compared to the 0.6% hybridized by the 412 gene, which is repeated at least 30 times per haploid genome (Rubin, Finnegan and Hogness, 1976 and in preparation).

The sequences in the Dm1020 segment are confined to a single chromosomal site and appear to be nonrepetitive. - Figure 2B shows the single chromosomal site at which the [^{3}H]cRNA to cDm1020 hybridizes. This site is in the 62E region of the left arm of chromosome 3. No other site, either in the chromosome arms or in the chromocenter, is labeled with this probe. Given the exposure time of the autoradiograph and the length of the Dm1020 segment (15 kb), the concentration of silver grains suggests that the Dm1020 sequences are not repeated at the site. This suggestion is supported by the following additional lines of evidence.

A map of the sites at which the BamI restriction endonuclease cleaves the cDm1020 DNA is given at the bottom of Figure 3. The five BamI sites provide four fragments, A-D, that are entirely contained within the Dm1020 segment, leaving the sequences in the Y and Z regions still attached to the ColE1 DNA. If Dm1020 were derived by shear breakage of a block of nonrepetitive sequences located at 62E, then cleavage of total D. melanogaster DNA with BamI should provide only six restriction fragments carrying Dm1020 sequences - four equivalent to fragments A-D in Dm1020, and two that contain the sequences in the Y and Z regions but exhibit lengths determined by the position of the two chromosomal BamI sites that are closest to, but not included in, the group of sites within Dm1020.

Total D. melanogaster (Oregon R) embryonic DNA was therefore digested to completion with BamI, and the resulting fragments separated according to length by agarose gel electrophoresis. After denaturation and transfer of the fragment strands to a nitrocellulose membrane by Southern's

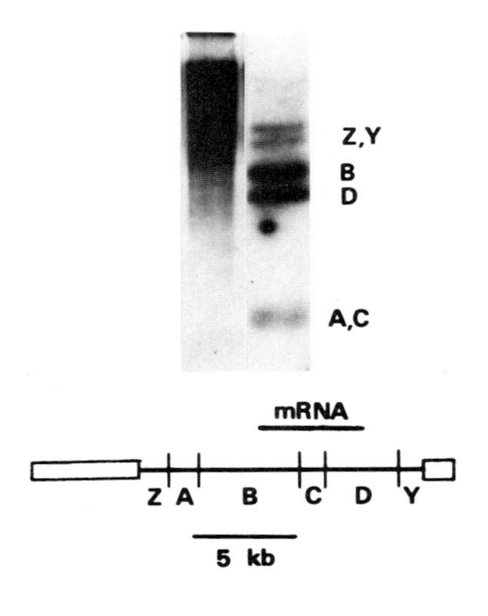

Fig. 3. Comparison of the BamI restriction fragments in cDm1020 with the BamI fragments from total D. melanogaster DNA that contain Dm1020 sequences. The map at the bottom of the figure shows the location of the sites in cDm1020 that are cleaved by BamI (vertical lines). The Dm1020 segment of the hybrid is indicated by the horizontal line connecting the two open blocks at both ends of the map, which represent the ColE1 DNA. The single SmaI cleavage site in the ColE1 DNA forms the ends of the map. The lengths of the A,B,C and D BamI fragments are 1.65, 4.7, 1.55 and 3.8 kb, respectively, and the Y and Z regions of Dm1020 are approximately 1.9 and 1.5 kb long.

D. melanogaster (Oregon R) embryonic nuclear DNA was digested with BamI and the resulting fragments fractionated by electrophoresis in a 1.0% agarose gel. Staining the fragments in the gel with ethidium bromide yields the left-hand panel in the upper part of the figure. The right-hand panel consists of the autoradiograph obtained when the fragments were hybridized with [^{32}P]cDm1020 DNA according to Southern's (31) method (see text). The autoradiographic bands labeled B,D and A,C comigrate with the respective BamI fragments from Dm1020. See the text for an explanation of the two bands labeled Y and Z.

blotting technique (31), they were hybridized with a ^{32}P-labeled DNA probe prepared by nick translation of cDm1020 (Materials and Methods). The result of this hybridization is given by the autoradiograph at the top of Figure 3, and is seen to conform to the single-copy prediction. In addition to the four internal fragments (A-D), two other fragments, labeled Y and Z, hybridize with the probe. These fragments, although longer than B (the largest internal fragment), hybridize less of the probe because the Y and Z regions of Dm1020 contain less of the Dm1020 DNA than does B. Clearly the 15 kb Dm1020 segment was formed by breakage at two points within a 26 kb block of chromosomal DNA consisting of all six fragments shown in the autoradiograph.

Since the level of the observed autoradiographic response is about that expected for fragments consisting of nonrepetitive sequences, the repetition of a significant fraction of the Dm1020 sequences (⩾0.5 kb) in some fragment other than the indicated six should have been detected. Consequently, repetition of Dm1020 sequences of length ⩾0.5 kb would require repetition of the entire 26 kb block. Even if such a repetition were tandemly arranged at the chromosomal site defined by *in situ* hybridization (Figure 2B), no more than two or three copies could be accommodated by the resolution and the intensity of that hybridization, given that the length of the block equals the haploid DNA content of the average polytene band.

Finally, we note that the sequence in the Dm1020 segment, or more precisely, in the B, C and D fragments of that segment, are not present in any of the other cDm plasmids that hybridize the polyA$^+$ cytoplasmic RNA. This was determined by colony hybridization of these cDm clones with [^{32}P]cRNA probes transcribed *in vitro* from cloned DNAs consisting of the B, C or D fragments inserted at the BamI site of the pSC105 plasmid (Materials and Methods). Since there is no sequence overlap between the ColE1 and pSC105 vectors, the colony hybridization will detect sequence homologies between the B, C or D fragments and the cloned Dm segments. However, only the colonies carrying cDm1020 exhibited a positive response to these hybridizations.

Taken together, the observations given in this section provide a strong argument that the Dm1020 sequences are nonrepetitive, although none, taken singly, eliminate the possibility that these sequences are repeated two or three times per haploid genome.

Dm1020 contains one copy of the sam gene which codes for a large mRNA. - The [^{32}P]polyA$^+$ cytoplasmic RNA was fractionated according to size by electrophoresis in a denaturing gel and the RNA in individual gel slices hybridized to saturating amounts of cDm1020 DNA fixed to nitrocellulose filters. The results, shown in Figure 4, indicate that the RNA molecules that hybridize with the Dm1020 segment are confined to a rather narrow size class with a modal value of 4-5 kb.

The sequences in the cDm1020 DNA that are responsible for this hybridization were located by restriction mapping. The BamI restriction fragments of cDm1020 were fractionated by agarose gel electrophoresis and then hybridized with [^{32}P]polyA$^+$ cytoplasmic RNA according to Southern's method (31). The resulting autoradiograph (not shown) indicated

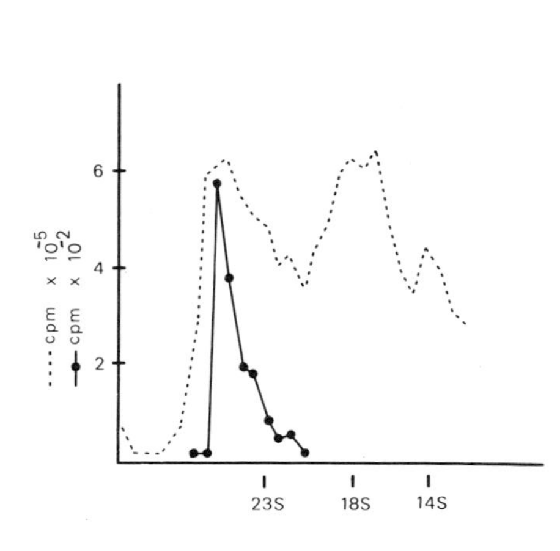

Fig. 4. Fractionation of the sam mRNA by electrophoresis in a denaturing gel. [^{32}P]polyA$^+$ cytoplasmic RNA (Materials and Methods) was applied to a gel consisting of polyacrylamide (3.5%) dissolved in 99% formamide. Following electrophoresis for 10 hr at 7 volts/cm, each 2-mm slice of the gel was sonicated in 0.5 ml of 5.6 x SSCP, 44% formamide so that the final concentration was 5 x SSCP, 50% formamide. The ^{32}P in the resulting solution was determined by Cerenkov counting (----). RNA in each fraction was then hybridized to saturating amounts (2 μg) of denatured cDm1020 DNA fixed to 13-mm nitrocellulose filters, as described in Materials and Methods (-●-). The portions of the gel to the right of the hybridization peak and for which no hybridization values are given on the graph, do not contain significant amounts of RNA that can hybridize with the cDm1020 DNA. The 14S, 18S and 23S marks on the abscissa indicate the positions of D. melanogaster 14S mitochondrial RNA (1.5 kb; D. Kemp, personal communication), D. melanogaster 18S rRNA (2.0 kb; 23, 33, 35), and E. coli 23S rRNA (3.1 kb; 32). These markers, along with the mRNA of the 412 gene (29; Rubin, Finnegan and Hogness, in preparation), were used to estimate a value of 4-5 kb for the RNA at the peak of the hybridization distribution.

that approximately equal amounts of the RNA hybridized to fragments B, C and D. Since the C fragment is 1.55 kb long, the sequences homologous to the RNA should be approximately three times this length, or about 4.6 kb of DNA. The correspondence between this value and the length of the hybridizable RNA molecules indicates that the homologous DNA sequences are contiguous and centered about the C fragment, as shown on the map in Figure 3. We therefore conclude that the Dm1020 segment contains a single copy of a gene that codes for an RNA that is 4-5 kb long. We call this gene sam and presume that it is a structural gene, although we have yet to show that its RNA can be translated to yield a polypeptide.

DISCUSSION

We have described a procedure whereby one can define, isolate and map a variety of structural genes without recourse to analysis of mutations or polypeptides, or to the prior isolation of specific mRNAs. We have illustrated the essential elements of the procedure by the isolation and mapping of the *sam* gene, and will conclude this paper by considering certain extensions of the procedure.

One obvious extension is the use of a cloned Dm segment for purification of a specific mRNA. As practiced in our laboratory, denatured cDm DNAs are covalently linked to cellulose (22), forming a DNA-cellulose matrix, which can then be hybridized with a heterogeneous population of mRNAs. The desired RNA is subsequently freed from the matrix by denaturating the RNA/DNA-cellulose hybrids, and it may be recycled through the same procedure if homogeneity is not obtained in a single passage.

The isolation of the *D. melanogaster* histone mRNAs provides a good illustration of this technique. All five of the histone genes are present in the same hybrid DNA molecule, cDm500, so that the product of the DNA-cellulose hybridization consists of a mixture of the five histone mRNAs. As it turns out, these can be easily separated from each other by electrophoresis in denaturing gels (Lifton, Goldberg and Hogness, in preparation). However, the point we wish to emphasize is that even if this separation were not feasible, it would still be possible to isolate the individual mRNAs by a second round of hybridization, if this time the DNA linked to the cellulose consists of cloned restriction fragments from cDm500 that contain sequences homologous to only one of the mRNAs. That is to say, the principle of cloning a DNA segment from a more complex DNA source and then using that segment to isolate specific RNA(s) from a heterogeneous population can be applied at whatever level is necessary to obtain the RNA that derives from a single gene.

Isolation of the mRNA for a given structural gene is important if one wishes to define the boundaries of that gene to a resolution of single base pairs. It is also important if one wishes to know the primary polypeptide product of the gene, since this can be determined by translation of the mRNA *in vitro*. However, it should be recognized that the amino acid sequence of the polypeptide can frequently be obtained from a determination of the sequence of base pairs in the structural gene, and that such determinations can now be accomplished with relative ease (20). For example, while it was possible to identify which of the five histones derived from each of the histone mRNAs by *in vitro* translation, that identification was less precise and more difficult to obtain than was the unambiguous identification which resulted from an analysis of the sequence of base pairs in part of each of the corresponding genes (Goldberg, Lifton and Hogness, in preparation). Of course, where the mRNA is quite long and where there is no prior knowledge about the polypeptide, as is the case for the *sam copia* and *412* genes, then translation of the isolated mRNA is the obvious first step in the identification process.

It should be emphasized, however, that identification of the polypeptide product of a structural gene is not a prerequisite for attacking many problems regarding its regulation. For example, there is no apparent need of this knowledge for a determination of the regulatory mechanisms of transcription and processing that lead to the formation of the gene's mRNA. By contrast, such a determination would be considerably aided by a map of the sequences in a cloned Dm segment that are homologous to the primary transcript. It should be possible to map these sequences by extending the intrasegmental mapping procedures that we have described for mRNAs to pulse-labeled nuclear RNAs. Indeed, by a modest extrapolation of these procedures, it may be possible to define the sequences corresponding to the promoters and terminators of transcription, and to the sites where the primary transcript is cleaved during processing.

MATERIALS AND METHODS

Preparation of DNAs and RNAs. - Plasmid DNAs were isolated as described by Finnegan, Rubin, Bower and Hogness (in preparation). Plasmid DNAs were labeled with ^{32}P by nick translation as described by Rigby *et al*. (27). [^{3}H] or [^{32}P]cRNAs to plasmid DNAs were prepared by transcription with *E. coli* RNA polymerase *in vitro*, as described by Wensink *et al*. (34). The [^{32}P]polyA^{+} cytoplasmic RNA was prepared from Eschalier's Kc line of *D. melanogaster* cells (7, 8) that have been adapted to grow in the absence of serum. They were obtained from W. Gehring as his Kco line. Growth and ^{32}P-labeling of these cells were carried out as described by Rubin and Hogness (28), using a 12-hr labeling period in mid-log phase. The [^{32}P]polyA^{+} cytoplasmic RNA was prepared from these cells according to a modification of the procedure of McKenzie, Henikoff and Meselson (21).

Cells were lysed at 4°C in 0.25 M KCl, 0.05 M $MgCl_2$, 25 mg heparin/ml, 0.2% Triton X-100, 0.05 M Tris-HCl, pH 7.4, and nuclei removed by centrifugation as in their procedure. RNA was isolated from the cytoplasmic fraction by combining it with an equal volume of SDS buffer (0.02 M NaCl, 0.04 M EDTA, 1.0% sodium lauryl sulfate (SDS), 0.02 M Tris-HCl, pH 7.5) at 90°C. After incubation at 90°C for 90 sec, the mixture was rapidly cooled to room temperature by immersing it in an ice bath. A $1/20^{th}$ volume of 10 mg proteinase K/ml (EM Laboratories, Inc.) was added and the mixture incubated for 10 min at room temperature. A $1/10^{th}$ volume each of 1 M Tris-HCl, pH 9 and of 5% SDS was then added and this solution mixed with an equal volume of phenol: chloroform:isoamyl alcohol (50:50:1). This mixture was stirred for 20 min at room temperature before the phases were separated by centrifugation in a Sorval HB-4 rotor for 15 min at 10,000 rpm. The aqueous phase was reextracted as above and the RNA precipitated from it by the addition of a $1/10^{th}$ volume of 2.0 M sodium acetate, pH 5.0 and 2 volumes of ethanol, the mixture being held at -20°C over-night. The polyA$^+$ molecules were isolated from this total cytoplasmic RNA by chromatography on oligo dT-cellulose (type T-3, Collaborative Research, Inc.) in 0.5 M KCl, 0.01 M Tris·HCl, pH 7.4 as described by Aviv and Leder (1) except that the intermediate salt wash was omitted. Prior to use as a probe for colony hybridization, this [^{32}P] polyA$^+$ cytoplasmic RNA was further purified by gel electro-phoresis as described in Figure 4, except that all fractions that contained appreciable amounts of migrated RNA were combined.

Colony hybridization. - Colony hybridization was carried out by a modification of the method described by Grunstein and Hogness (12). The bacterial colonies were formed on millipore HA filters in the same manner, but the bacteria were lysed and their DNA denatured by placing the millipore filters on top of a sheet of Whatman 3 MM paper soaked with 0.5 N NaOH, keeping the colonies uppermost. After 7 min, the filters were transferred to another sheet of 3 MM paper soaked in 1.0 M Tris-HCl, pH 7.6. This was repeated after 1 min, and the filters then transferred to a final sheet of 3 MM paper soaked with 1.5 M NaCl, 0.5 M Tris-HCl, pH 7.4. After 5 min the filters were vacuum washed with 60 ml of 95% ethanol to dry the DNA prints of the colonies and to fix them in place. Filters were then baked in a vacuum oven at 80°C for 2 hrs. Hybridization was carried out by wetting each filter (diam. = 9 cm) with 1 ml of a 5 x SSCP (SSCP is 0.12 M NaCl, 0.015 M sodium citrate, 0.02 M sodium phosphate, pH 6.8), 50% formamide solution containing 200 μg of polyA and approximately 5 x 10^5 cpm of the probe which consisted either of [^{32}P]polyA$^+$ cytoplasmic RNA from cultured cells, or of [^{32}P]cRNA transcribed *in vitro* from a cloned Dm segment. Filters were overlayed with mineral oil and incubated for 36 hr at 42°C to effect the hybridization. They were then washed for 30 min in 50% formamide, 5 x SSCP at 42°C. Washing was continued at room temperature with two changes of 2 x SSCP over a 30-min period, with a final 30-min wash in 2 x SSCP containing 20 μg/ml of pancreatic ribonuclease (Worthington). Excess liquid was removed by blotting and the filters covered with "Saran Wrap". Kodak XR-5 x-ray film was placed over the wrapped filters and a Kodak X-Omatic Regular intensifying screen placed over the film. The autoradiography was then carried out at -77°C.

Other hybridization procedures. - [^{32}P]polyA$^+$ cytoplasmic RNA was hybridized to 13-mm nitrocellulose filters containing 2 μg of plasmid DNAs containing Dm segments (e.g., cDm1020) according to the procedure described by Glover *et al*. (10), except that hybridization took place at 50°C for 18 hr under mineral oil. After washing twice for 30 min in the hybridization buffer (5 x SSCP, 50% formamide) at 50°C, the filters were incubated for 30 min at room temperature in 2 x SSCP containing 20 μg pancreatic ribonuclease/ml, rinsed in 2 x SSCP, dried and counted by scintillation.

Hybridization of [^{32}P]RNA or DNA to denatured restriction fragments that have been transferred from gels to nitrocellulose strips by Southern's blotting procedure (31) was carried out as follows. [^{32}P]RNA probes were hybridized under the conditions described in the preceding paragraph. In the case of [^{32}P]DNA probes, the nitrocellulose strip was soaked in 5 x SSCP, 50% formamide, 0.02% polyvinyl pyrrolidone, 0.02% ficoll and 0.02% bovine serum albumin (6) prior to hybridization which was carried out in this same solution. Both RNA/DNA and DNA/DNA hybridization was carried out at 42°C for 48 hr in minimal volumes contained in heat-sealed plastic bags. The strips were then washed as described above except that the temperature was 42°C and the RNase treatment was omitted for the DNA/DNA hybridization. The hybridization was then monitored by autoradiography as described for colony hybridization.

The method that was used to hybridize [^{3}H]cRNA probes to polytene chromosomes has been described (34).

Construction and cloning of hybrid plasmids. - The construction and cloning of all cDm plasmids is treated under Results. The B, C and D BamI fragments of cDm1020 (Figure 3) were inserted into the pSC105 plasmid (5) at its BamI site to yield the cloned pkDm1218, pkDm1172 and pkDm1180 hybrid plasmids, respectively. pSC105 consists of a 7.2 kb EcoR1 fragment determining kanamycin resistance (11), inserted into the pSC105 plasmid at its EcoR1 site (5). It contains the BamI cleavage site located in the *tet* gene of pSC101, and insertion at this site therefore inactivates the *tet* gene to yield plasmids that confer kanamycin but not tetracycline resistance to host cells (i.e., cells containing such pkDm plasmids are Km^R, Tc^S; 24). The BamI fragments were inserted by the polydA/polydT method of Wensink *et al*. (34) as modified by Brutlag *et al*. (4), where polydA tails were added to the fragments in a BamI digest of cDm1020, and polydT tails were added to a BamI digest of pSC105. Cloning was accomplished by transforming the HB101 strain of *E. coli* K12 (3) to Km^R, Tc^S with the annealed hybrid molecules. The specific BamI fragment of cDm1020 that is contained in each of a dozen independently cloned pkDm hybrid plasmids was identified as follows. The BamI fragments of cDm1020 were separated by electrophoresis on agarose gels (Figure 3), transferred to nitrocellulose strips by Southern's method (31) and hybridized with [^{32}P]pkDm DNA probes prepared from each hybrid by nick translation. Probes made from pkDm1218, pkDm1172 or pkDm1180 hybridize only to the BamI fragments B, C or D, respectively.

Enzymes. - *E. coli* DNA polymerase I (17), *E. coli* RNA polymerase (2), BamI (36) and SmaI (R. Green and C. Mulder, personal communication), and calf thymus terminal transferase (18) were the gifts of M. Goldberg, D. Brutlag, S. Goff, J. Lis and R.L. Ratliff, respectively, and were prepared according to the indicated references. The solvent for BamI digestions was 6mM β-mercaptoethanol, 6mM $MgCl_2$, 6mM Tris·HCl, pH 7.5, and for SmaI was 25mM KCl, 3mM $MgCl_2$, and 30mM Tris·HCl, pH 9.0. Digestions were carried out at 37°C. Pancreatic RNase was purchased from Worthington Biochemical.

REFERENCES

1. Aviv, H. and Leder, P. (1972) Proc. Nat. Acad. Sci. USA 69, 1408.

2. Berg, D., Barrett, K., and Chamberlin, M. (1971) In Methods in Enzymology, 21D, L. Grossman and K. Moldave, eds. (New York; Academic Press), 506.
3. Boyer, H.W. and Roulland-Dussoix, D. (1969) J. Mol. Biol. 41, 459.
4. Brutlag, D. Fry, K., Nelson, T., and Hung, P. (1977) Cell 10, 509.
5. Cohen, S.N., Chang, A.C.Y., Boyer, H.W., and Helling, R.B. (1973) Proc. Nat. Acad. Sci. USA 70, 3240.
6. Denhardt, D.T. (1966) Biochem. Biophys. Res. Comm. 23, 641.
7. Dolfini, S. (1971) Chromosoma 33, 196.
8. Eschalier,G. and Ohanessian, A. (1970) In Vitro 6, 162.
9. Glover, D.M. (1976) In New Techniques in Biophysics and Cell Biology, 3, R. Pain and B. Smith, eds. (London; John Wiley), in press.
10. Golver, D.M., White, R.L., Finnegan, D.J. and Hogness, D.S. (1975) Cell 5, 149.
11. Glover, D.M. and Hogness, D.S. (1977) Cell 10, 167.
12. Grunstein, M. and Hogness, D.S. (1975) Proc. Nat. Acad. Sci. USA 72, 3961.
13. Hershfield, V., Boyer, H.W., Yanofsky, C., Lovett, M.A., and Helinski, D.R. (1974) Proc. Nat. Acad. Sci. USA 71, 3455.
14. Jacob, F. and Monod, J. (1959) C. R. Acad. Sci. 249,1282.
15. Jacob, F. and Monod, J. (1961) J. Mol. Biol. 3, 318.
16. Jacob, F. and Monod, J. (1961) Cold Spring Harbor Symp. Quant. Biol. 26, 193.
17. Jovin, T.M., Englund, P.T., and Bertsch, L. (1968) J. Biol. Chem. 244, 2996.
18. Kato, K., Goncalves, J.M., Houts, G.E., and Bollum, F.J. (1967) J. Biol. Chem. 242, 2780.
19. Levy, W.B. and McCarthy, B. (1975) Biochemistry 14, 2440.
20. Maxam, A.M. and Gilbert, W. (1977) Proc. Nat. Acad. Sci. USA 74, 560.
21. McKenzie, S.L., Hinikoff, S., and Meselson, M. (1975) Proc. Nat. Acad. Sci. USA 72, 1117.
22. Noyes, B.E. and Stark, G.R. (1975) Cell 5, 301.
23. Pelegrini, M., Manning, J., and Davidson, N. (1977) Cell 10, 213.
24. Rambach, A. and Hogness, D.S. (1977) Proc. Nat. Acad. Sci. USA, in press.
25. Rasch, E.M., Barr, H.J. and Rasch, R.W. (1971) Chromosoma 33, 1.
26. Ritossa, F.M. (1976) In Organization and Expression of Chromosomes, V.G. Allfrey, E.K.F. Bautz, B.J. McCarthy, R.T. Schimke, A. Tissieres, eds. (Dahlem Konferenzen, Berlin),153.

27. Rigby, P.W.J., Dieckmann, M., Rhodes, C., and Berg, P. (1977) J. Mol. Biol., in press.
28. Rubin, G.M. and Hogness, D.S. (1975) Cell 6, 207.
29. Rubin, G.M., Finnegan, D.J., and Hogness, D.S. (1976) In Progress in Nucleic Acid Research and Molecular Biology, Vol. 19, W. Cohn ed. (Academic Press, New York), in press.
30. Rudkin, G.T. (1972) In Results and Problems in Cell Differentiation, Vol. 4, W. Beerman ed. (Springer-Verlag, New York), 59.
31. Southern, E.M. (1975) J. Mol. Biol. 98, 503.
32. Van Holde, K.E. and Hill, W.E. (1974) In Ribosomes, M. Nomura, A. Tissieres, P. Lengyel eds. (Cold Spring Harbor Laboratory), 53.
33. Wellauer, P.K. and Dawid, I.B. (1977) Cell 10, 193.
34. Wensink, P.C., Finnegan, D.J., Donelson, J.E., and Hogness, D.S. (1974) Cell 3, 315.
35. White, R.L. and Hogness, D.S. (1977) Cell 10, 177.
36. Wilson, G.A. and Young, F.E. (1975) J. Mol. Biol. 97, 123.

IDENTIFICATION OF A CIS-ACTING CONTROL ELEMENT IN *Drosophila melanogaster*[*]

Arthur Chovnick, Margaret McCarron and William Gelbart[†]

Genetics and Cell Biology Section, The Biological Sciences Group, University of Connecticut, Storrs, Connecticut 06268

ABSTRACT. The present report summarizes our progress in the genetic dissection of an elementary genetic unit in a higher organism. The rosy locus (*ry*:3-52.0) in *Drosophila melanogaster* codes for xanthine dehydrogenase, and is characterized by several classes of induced mutants as well as naturally occurring variants. Pursuing the hypothesis that the rosy locus includes a non-coding, control region as well as a structural element coding for the XDH peptide, experiments have been carried out which place genetic boundaries to the structural element in terms of a map of unambiguous structural element variants. Presently, our research is largely concerned with the elaboration of a cis-acting control element located adjacent to the structural element.

INTRODUCTION

Prokaryote studies provide a model of genetic organization which, in many respects, is directly applicable to higher organisms. However, recent evidence from several independent research directions suggests that the genetic material of higher organisms possesses unique organizational features, and the entire subject has become a major focus of much current research. Central to this topic is the issue of controlling elements and their relationship to the genes they control. For some years, major effort in this laboratory has been directed towards the development of a favorable experimental system for the investigation of such questions with *Drosophila melanogaster*. The early work has been the subject of extensive prior review (3, 9, 11). The present report emphasizes recent developments, and specifically, the identification of a cis-acting control element.

[*]This investigation was supported by research grant GM09886 from the Public Health Service and by research grant BMS74-19628 from the National Science Foundation.

[†]Present address: Cellular and Developmental Biology Group, Harvard University, 16 Divinity Avenue, Cambridge, MA 02138

THE GENETIC SYSTEM

The rosy locus in *Drosophila melanogaster* (*ry*: 3-52.0) is a genetic unit controlling xanthine dehydrogenase (XDH) activity, located on the right arm of chromosome 3 within polytene chromosome bands 87D8-12 (20). Originally defined by a set of brownish eye color mutants deficient in drosopterin pigment, such mutants were shown subsequently to exhibit no detectable XDH activity (17). Methods for the study of this enzyme are described elsewhere (5, 7, 22). Two observations place the coding information for XDH in or near rosy: (1) Variation in dosage of ry^{+} alleles, from 0 - 3 doses, appears to be the limiting factor in determining level of XDH activity/fly in otherwise wild-type flies, (2) the genetic basis for variation in electrophoretic mobility of XDH seen in wild-type strains maps to the immediate vicinity of the rosy locus (16, 18, 25). Both of these observations have been confirmed and extended in this laboratory (3, 22). Moreover, we now know that the rosy locus codes for a single peptide of approximately 160,000 daltons, and that the XDH molecule is a homodimer, possessing two copies of this peptide (1, 2, 14).

Recombination studies involving tests of rosy mutant heteroalleles are facilitated by the addition of purine to the culture medium which permits only ry^{+} recombinants to survive. Prior intensive fine structure analysis (4, 6) has provided a linear map of sites associated with XDH inactivation. Our attention now is directed towards a dissection of this genetic unit in terms of structural and control functions. To this end, the rosy locus map of XDH^{-} mutant sites provides a starting point (Fig. 1A). Recent experiments have been designed to locate the boundaries of the structural element by positioning sites of unambiguous structural variants on the existing map. Three classes of such variants are included in the recombination analysis: (1) XDH electrophoretic variants, (2) XDH^{-}, rosy eye color mutants which exhibit interallelic complementation, and (3) variants possessing low levels of XDH activity, which are purine sensitive like XDH^{-} mutants and which produce altered XDH molecules (13, 14).

Table 1 summarizes our current nomenclature relating wild-type isoalleles of the rosy locus to the electrophoretic mobility of the XDH that each produces. The XDH produced by ry^{+0} serves as a mobility standard, and is designated $XDH^{1.00}$. All variant XDHs that are slower are designated by relative mobilities < 1.00, while faster XDHs are designated by arabic superscripts > 1.00. In addition, the ry^{+4} allele is associated with sharply higher XDH activity than all

TABLE 1

WILD TYPE ISOALLELES OF ROSY

ry^+ alleles	Mobility	XDH Activity
+12, +13	0.90	N
+14	0.94	N
+10	0.97	L
+0, +6	1.00	N
+1, +11	1.02	N
+4	1.02	H
+2	1.03	N
+3, +5	1.05	N

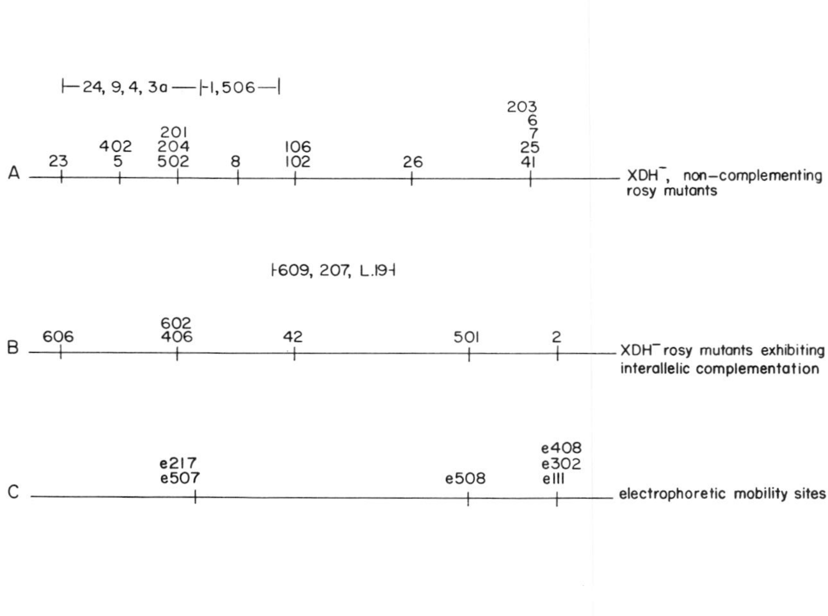

Fig. 1. Genetic fine structure maps of rosy locus sites. Map locations of unambiguous structural element variants (B, C, and D) are positioned relative to map of XDH^- non-complementing mutants (A).

others and is classified in Table 1 as high (H), while ry^{+10} exhibits much lower activity than all others and is classified as low (L). The remaining ry^{+} isoalleles exhibit intermediate levels of XDH activity which we presently classify as normal (N). Differences in intensity of XDH activity, like the mobility differences, are stable phenotypic characters which also map to the rosy locus (7).

Rosy variants are labeled with superscripts which identify the ry^{+} isoallele from which the variant site is derived. Thus the XDH^{-}, rosy eye color mutant, ry^{406} (Fig. 1B) is the sixth variant derived from ry^{+4}, while the electrophoretic site, ry^{e408}, (Fig. 1C) identifies the site that confers an increased mobility to the ry^{+4} product ($XDH^{1.02}$) over that seen with the ry^{+0} enzyme ($XDH^{1.00}$). The variant, ry^{ps214}, represents one of a class of induced purine sensitives (Fig. 1D). They have wild type eye color, but die on purine supplemented medium because of very low levels of XDH activity during larval development.

The relationship of the rosy locus map to the boundaries of the XDH structural element is summarized in Figure 1. The standard map of XDH^{-} mutants is presented in Figure 1A, while allele complementing mutant sites, electrophoretic sites, and purine sensitive structural variant sites are located on maps 1B, C, and D, respectively. At the right end, several electrophoretic sites, and the complementing mutant ry^{2} identify the right border of the structural element with no known rosy locus variants beyond them. The left end of the map is of greater interest for this report. Here, the complementing rosy eye color mutant, ry^{606}, is the leftmost unambiguous structural variant. On the basis of comparative recombination data, and the complete failure to recover recombinants in large scale test with ry^{606}, the leftmost member of the standard map, ry^{23}, also must be very close to the left border of the structural element.

The maps of Figure 1 position 39 sites to the right of our present left boundary of the XDH structural element. Moreover, an additional ten sites not indicated in Figure 1, in fact, map in the structural element. Admittedly, these boundaries are conditional. However, the extensive genetic data upon which we base the present left boundary suggests that, at least for this end of the structural element, we are close to a terminus.

VARIATION IN INTENSITY OF XDH ACTIVITY

Consider ry^{+4} and ry^{+10} which are associated with much greater and much less activity, respectively, than all of our other wild-type alleles (Table 1). These differences are readily classified in cuvette assays (spectrophotometry or fluorimetry) or upon gel electrophoresis. A detailed analysis of the basis for the ry^{+4} phenotype is presented elsewhere (7), and a report on the ry^{+10} character is now in preparation. Together, these studies provide the basis for the identification of a cis-acting control element located adjacent to the XDH structural element. The following sections outline the experimental basis for this conclusion, and focus upon the difference between the high level of activity of ry^{+4} and the normal lines. Analysis of the ry^{+10} line follows the same logic.

We have considered the possibility that there are structural differences between the XDH molecules produced by ry^{+4} and those produced by other ry^{+} isoalleles which are responsible for their differences in level of activity. Three sets of observations have failed to provide evidence in favor of a structural basis for the activity difference (7).

1. *There is no systematic relationship between level of XDH activity and electrophoretic differences.* Consider ry^{+0}, ry^{+1}, ry^{+2} and ry^{+11}, which exhibit normal levels of XDH activity. We know that ry^{+0} differs from ry^{+4} (which has a high level of activity) in one identified electrophoretic site (see earlier discussion), and ry^{+2} differs from ry^{+4} in two known sites, *e217* and *e408* (Fig. 1C). On the other hand, ry^{+1} and ry^{+11} produce XDH molecules of the same mobility as ry^{+4} (Table 1).

2. *There is no systematic relationship between XDH thermolability and level of activity.* One might suspect that the increased level of ry^{+4} enzyme activity is a reflection of a greater molecular stability resulting in increased numbers of XDH molecules. Such differences may be exposed by examination of XDH thermolability associated with the several ry^{+} isoalleles. Consider ry^{+0}, ry^{+2} and ry^{+11} which exhibit normal levels of XDH activity as compared to the high level associated with ry^{+4}. Heat inactivation experiments carried out over a range of temperatures with matched extracts of these lines has failed to identify a thermolability difference between them and ry^{+4}. We note that there are differences in thermolability of XDH molecules produced by other ry^{+} isoalleles. However, no association with level of XDH activity is apparent.

3. *There is no systematic relationship between level of XDH activity and substrate affinity of the XDH molecules.* Consider ry^{+4} which is associated with high levels of XDH activity and ry^{+11} which is characterized by standard or normal levels of activity. Matched, partially purified XDH containing extracts of these strains exhibit identical XDH electrophoretic mobilities, heat denaturation profiles, and

K_m. Similar results are seen in kinetic comparisons of ry^{+4}, on the one hand, and ry^{+0} and ry^{+2} on the other. That the K_m's are identical suggests that the difference in level of activity might be associated with variation in number of molecules of XDH/preparation.

Evidence from immunological experiments. Immunological studies provide additional support for the notion that the difference in activity reflects a difference in number of XDH molecules. One approach has used an antiserum titration experiment in which a set of dilutions of an anti-XDH serum is tested for ability to remove XDH activity from matched extracts of ry^{+4} and standard strains. The titration curves are very different for such preparations with the standard extracts being inactivated at antiserum dilusions that are incapable of removing more than 50% of the XDH activity from ry^{+4}. Still another approach has utilized the method of quantitative "rocket electrophoresis" (19, 24) to compare the relative number of molecules of XDH in matched extracts of ry^{+4} and normal strains. Such experiments also indicate that preparations from normal strains contain fewer XDH molecules than does ry^{+4} extract.

GENETIC FINE STRUCTURE EXPERIMENTS

Let us now turn to localization of the genetic basis for the high level of XDH activity associated with ry^{+4}. The first experiments involve recombination tests of null enzyme rosy mutants of the ry^{+4} allele tested against established site mutants within the XDH structural element. Five series of tests were carried out which sampled a total of 4.7 x 10^6 progeny, and yielded 35 crossovers. Figure 2 summarizes the results of these experiments which are divided into two classes in terms of the distribution of rosy mutant sites in the heterozygote. Heterozygote (A) of Figure 2 illustrates a cross which tests ry^x/ry^{40y} where ry^x is located to the right of ry^{40y}. Heterozygote (B) illustrates a cross involving ry^z/ry^{40q}, where ry^z is located to the left of ry^{40q}. In all of these crosses, there is an additional heterozygous site at the right end of the XDH structural element. This site, designated as e^S or e^F in Figure 2, is responsible for the electrophoretic mobility difference between the parental wild-type isoalleles ($XDH^{1.00}$ and $XDH^{1.02}$). Consider now the ry^+ recombinants that are associated with exchange for the flanking outside markers a and b. For heterozygote (A) each exchange product will recombine the left portion of the ry^x bearing strand with the right section of the ry^{40y} chromosome. As noted (Fig. 2) all of these exchanges are $a\ ry^+ +$, and upon electrophoresis, exhibit an $XDH^{1.02}$ with normal (N)

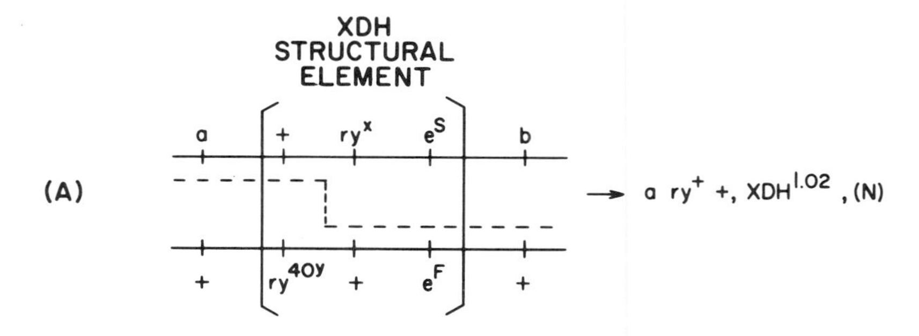

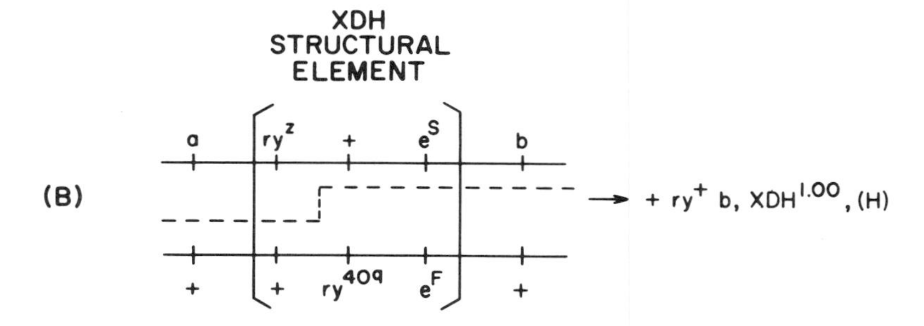

Fig. 2. Genetic fine structure analysis of crossovers within the XDH structural element.

levels of activity. Complementary results are seen in experiments involving heterozygote (B). Fourteen of the crossovers were recovered from tests of heterozygotes of type (A) while the remaining 21 were recovered from type (B) heterozygotes. It should be noted that in all of these experiments, as well as others to be described in this report, *no intermediate or non-parental classes* with respect to XDH activity level were recovered. Obviously, the site giving rise to the variation in XDH activity level is different and separable from the electrophoretic site. We designate this site as *i409* with ry^{+4} carrying *i409H* and other wild-type isoalleles as *i409N*. Thus, from this group of 35 crossovers representing exchange points across the XDH structural element, we note that the genetic basis for the difference in level of XDH activity (*i409*) fell to the left of all crossover points, while the electrophoretic site is located to the right of all exchange points. Another point should be noted. Two of the crosses, involving type (B) heterozygotes (Fig. 2), tested the leftmost known ry^{400} series mutant against ry^{z} mutants which represent markers for the left border of the XDH structural element. These recombination tests span a region which we estimate to represent the leftmost 1/10 of the XDH structural element map. Six of the crossovers described above emerged from these experiments. Thus, we are able to localize the site of *i409* to the left of all of these six exchange points. From this, we infer that *i409* is located either to the left of the structural element or just inside the left border, and very close indeed to the cluster of rosy mutant sites at that end of the map (Fig. 1).

One final point should be noted about the experiments described in Figure 2. Progeny survive if they carry a chromosome resulting from an event leading to the production of a ry^{+} recombinant. In addition to recombinants associated with exchange for flanking markers, numerous conversions of the XDH^{-} rosy mutants are recovered as well. These events provide additional information about the location of *i409*. In these experiments, conversions of $XDH^{-} \rightarrow XDH^{+}$ sites are selected events, while *i409* represents an unselected heterozygous marker. We may now ask about the frequency with which conversion of a given mutant allele is accompanied by co-conversion of the unselected *i409* site. The significance of the phenomenon of co-conversion was recognized in fungal studies as demonstrating that conversion events involve variable sized segments of DNA. Fogel, Hurst and Mortimer (12) have studied frequency of co-conversion as a function of distance between the sites, and have shown a linear relationship inversely proportional to distance. While the rosy locus data are not as extensive, our observations are entirely consistent with the fungal data (7, 8, 13, 14). Now, consider conversion observations on rosy mutants located at the left end of the XDH structural element map (Fig. 1) in which the recombination experiment provided opportunity to observe co-conversion for the *i409* site. If the *i409* site were located just inside the structural element left border, then frequent co-conversions of *i409* would be expected with conversion of these neighboring sites. In fact, a low frequency of co-conversion (2/14) is seen with the two rosy mutants which mark the present left border of the structural element (ry^{23}, ry^{606}), and there were no co-conversions seen among 19 conversions of mutant alleles just inside that boundary (ry^{ps214}, ry^{402}, ry^{5}, ry^{406}). Taken together, interpretation of both crossover and co-conversion data would place *i409* well to the left of the XDH structural element.

Confirmation of the position of i409. Figure 3 summarizes the previous discussion which locates *i409* relative to

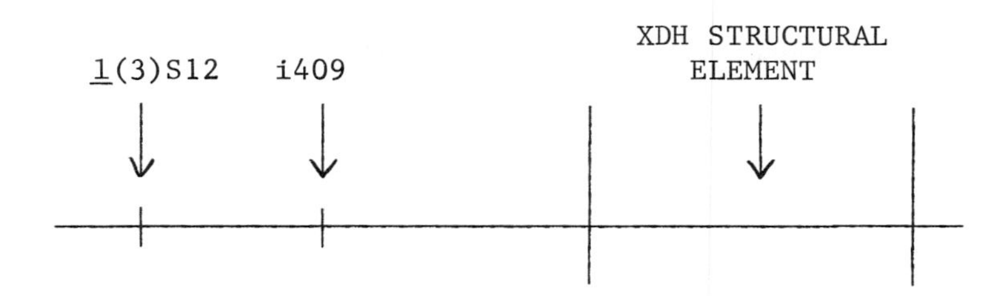

Fig. 3. Location of *i409* suggested by analysis of the data recovered from experiments outlined in Fig. 2.

the left end of the XDH structural element. Note also the relative position of *l(3)S12*, an X-ray induced recessive lethal mutant. Despite its proximity to the rosy locus (< 0.01 map units in preliminary experiments), there is no evidence to consider *l(3)S12* as an alteration within the rosy locus. Rather, we believe it to be a site mutant in the very next genetic unit to the left of rosy.

The next group of experiments were designed to test the localization of *i409* relative to *l(3)S12* and the XDH structural element. These experiments selected for survival crossovers between *l(3)S12* and an XDH^- rosy mutant located near the left border of the structural element. If *i409* lies to the left of the structural element, then the exchange events involved heterozygotes of the composition $\frac{l(3)S12 \quad i409N \quad +}{+ \quad i409H \; ry}$, which were mated in a selective system cross designed to kill all *l(3)S12* and/or *ry* bearing meiotic products of the heterozygote. Since recombination to generate meiotic products that are $l(3)S12^+$, ry^+ may take place on either side of the *i409* site, survivors should fall into two classes with respect to level of XDH activity. On the other hand, if *i409* were located within the XDH structural element (i.e., to the right of the rosy mutant marker), the test heterozygote would be $\frac{l(3)S12 \quad + \; i409N}{+ \quad ry \; i409H}$, and selection for $l(3)S12^+$, ry^+ recombinants would be expected to yield only *i409N* bearing products.

In fact, a total of 16 crossovers were recovered in 1.18 x 10^6 sampled progeny, and these fell into the two classes *i409N* (6) and *i409H* (10). These results, therefore, provide a very strong confirmation of the position of *i409* inferred in the previous section.

RELATIONSHIP BETWEEN *i409* AND THE STRUCTURAL ELEMENT

Might i409H represent a tandem duplication of the structural element? On this notion, the ry^{+4} allele would be considered to possess two functional XDH structural elements in tandem, presumably resulting from an unequal exchange event. Such a model is precluded on several counts:

1. EMS mutagenesis of ry^{+4} has produced rosy eye color mutations at a frequency that does not distinguish this allele from other ry^{+} isoalleles.

2. The ry^{+4} allele is associated with a single XDH electrophoretic band of the mobility class, $XDH^{1.02}$ (Table 1). XDH is a homodimer, and the presence of two electrophoretically distinct structural elements will produce individuals possessing three XDH moieties. The tandem duplication model then requires that the ry^{+4} allele possess two XDH structural elements whose peptide products are indistinguishable, and of the mobility class, $XDH^{1.02}$. Thus, *i409H* should be associated with an $XDH^{1.02}$. On this point, the tandem duplication model fails. In all experiments which recombine *i409H* with other electrophoretically distinct structural elements, there is no evidence of the production of an $XDH^{1.02}$ moiety.

3. Tandem duplications are characterized by instability in homozygotes due to increased incidence of unequal exchange events. The ry^{+4} stock has been quite stable. Moreover, fine structure recombination experiments involving tests of ry^{400} series mutants against other XDH^{-} mutants have been characterized by regular exchange events, and the complete absence of unequal crossing over.

4. Cytological examination of polytene chromosomes reveals no such tandem duplication.

Might i409 represent a regulatory variant? Let us consider the possibility that *i409* marks a genetic element that regulates XDH. On the basis of evidence thus far presented, a broad array of regulatory roles is possible. However, by a simple experiment, we are able to describe a key feature of this regulatory function. Under one class of regulatory roles, dominance-recessiveness, or incomplete dominance in heterozygotes of *i409* may be seen. In still another class of roles, the regulatory function of a specific *i409* element would be restricted to the specific XDH structural element adjoining it on the chromosome ("cis-acting" regulator).

Consider the heterozygote *i409N* ry^{+1}/*i409N* ry^{+12}. Such individuals should produce approximately equal quantities of the ry^{+1} and ry^{+12} peptides. Assume further that random union of monomers takes place to yield active XDH dimers.

Electrophoresis of extracts of such heterozygotes should reveal the presence of three XDH moieties, $XDH^{0.90}$ (ry^{+12} enzyme), $XDH^{0.96}$ (hybrid dimer), and $XDH^{1.02}$ (ry^{+1} enzyme) in a ratio of 1:2:1. Figure 4A is a photograph of an electropherogram illustrating the XDH molecular composition of extracts of *i409N* ry^{+1}/*i409N* ry^{+12}, and is entirely consistent with the expectation and its underlying assumptions.

Consider next the XDH composition of a matched extract of the genotype *i409H* ry^{+4}/*i409N* ry^{+12} (Fig. 4B). The structural element of the ry^{+4} allele produces an XDH possessing a mobility identical to that of ry^{+1} ($XDH^{1.02}$). Thus, in terms of XDH moieties, this genotype should possess the same three mobility classes as seen in Figure 4A. However,

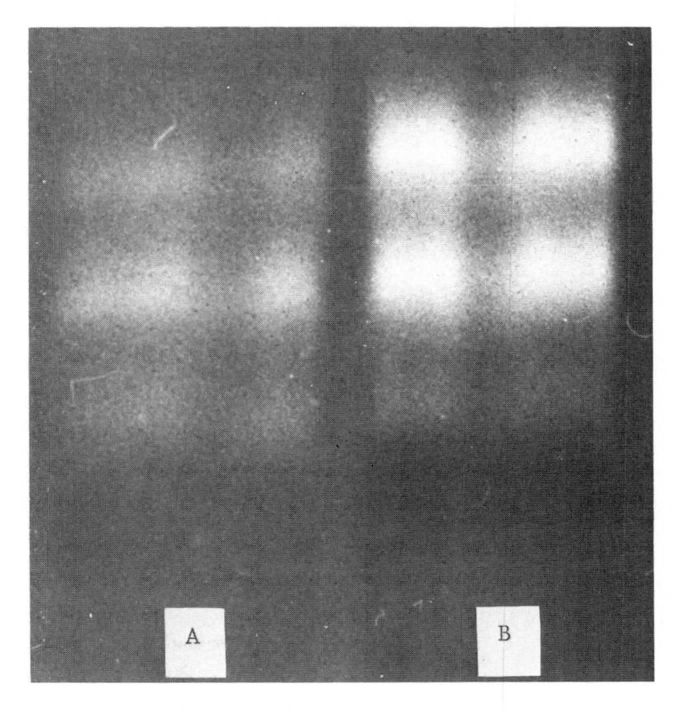

Fig. 4. XDH electropherogram indicating the relative amounts of $XDH^{0.90}$, $XDH^{0.96}$ and $XDH^{1.02}$ present in matched extracts of flies of the genotypes
(A) *i409N* ry^{+1}/*i409N* ry^{+12} and
(B) *i409H* ry^{+4}/*i409N* ry^{+12}.

this genotype differs from the former in that it is heterozygous, *i409H/i409N*. Under one class of regulatory models, dominance of one or the other form of these alternatives, or even incomplete dominance, may be expected. In such event, we should find the same 1:2:1 ratio of the three mobility classes. However, all would be produced in high, standard or intermediate quantities depending upon the specific dominance relationship that obtains. Alternatively, each *i409* element may operate to regulate only the specific XDH structural element (or product thereof) located adjacent to it on the same chromosome (i.e., a "cis-acting" regulator). Thus, for the heterozygote, $i409H\ ry^{+4}/i409N\ ry^{+12}$, this model would predict that greater quantities of the ry^{+4} polypeptide than ry^{+12} peptide would be produced. Random union of monomers would then result in increased quantities of $XDH^{1.02}$ (ry^{+4} enzyme) and $XDH^{0.96}$ (hybrid dimer) at the expense of $XDH^{0.90}$ (ry^{+12} enzyme). Examination of Figure 4B reveals that *i409* operates as a "cis-acting" regulator.

On the basis of this observation, we may eliminate all regulatory models which associate *i409* with the synthesis of a negative or positive acting diffusible regulatory molecule (10, 15). We are drawn, rather, to the possibility that *i409* marks the 5' control element of the rosy locus. Hence, we expect that variants of this element will exhibit alterations in DNA sequences which serve as binding sites for regulatory signal(s), sites for RNA polymerase binding and initiation of transcription, transcript processing sites, ribosome binding and initiation of translation.

Additional support for the existence of a non-coding region adjacent to the rosy structural information that functions as a "cis-acting" control element emerges from our recent work with the ry^{+10} allele (Table 1). This allele is associated with very low levels of XDH activity (approximately 1/8 that of ry^{+4}). In a parallel series of experiments (manuscript in preparation), the very low level of XDH activity associated with ry^{+10} has been shown to reflect number of molecules of XDH. Moreover, the genetic basis for this stable phenotypic character also maps to the interval between *l(3)S12* and the XDH structural element, and is separable from *i409*.

STRUCTURAL AND CONTROL ELEMENT SIZE ESTIMATES

The recombination map length of the XDH structural element is estimated at 0.005 map units (13). At present, we suggest that the distance from *i409* to the structural element (0.0034 map units) serve as a minimal estimate while the distance from *l(3)S12* to the XDH structural element (0.0054 map units) be taken as a maximum estimate of the control element size (7).

Translation of these estimates into DNA base lengths proceeds from the XDH peptide molecular weight of 160,000 daltons. Assuming an average amino acid molecular weight (adjusted for peptide linkage) to be 110, then the length of DNA in the XDH structural element responsible for such a peptide is approximately 4.4 kB (160,000 x 3/110). Then, from the recombination map length of the structural element (0.005 map units), we may relate map length to physical length (0.01 map unit = 8.8 kB), which should be directly applicable to the adjacent control element. Thus, the size of the control element is estimated to be 3.0 - 4.75 kB, and the total length of the rosy locus from 7.4 - 9.15 kB.

Another estimate of the length of DNA in the rosy locus may be derived directly from its genetic length (0.0084-0.010 map units). With 1.6×10^5 kB as the total genome DNA represented as single copy and middle repetitive sequences (21, 23), and 275 map units as the total map length, then 0.01 map unit represents 5.8 kB. By this method, the rosy locus DNA length is estimated at 4.87 - 5.8 kB. The reader should note that this method ignores regional differences in recombination frequency, and that the rosy locus is subject to reduced recombination due to its centromere proximal position (13). In this context, we view the former method to yield a better estimate.

REFERENCES

1. Andres, R.Y. (1976) *Eur. J. Biochem.* 62, 591.
2. Candido, E.P.M., Baillie, D.L. and Chovnick, A. (1974) *Genetics* 77, S9.
3. Chovnick, A. (1966) *Proc. Roy. Soc. London B* 164, 198.
4. Chovnick, A. (1973) *Genetics* 75, 123.
5. Chovnick, A., Ballantyne, G.H., Baillie, D.L. and Holm, D.G. (1970) *Genetics* 66, 315.
6. Chovnick, A., Ballantyne, G.H. and Holm, D.G. (1971) *Genetics* 69, 179.
7. Chovnick, A., Gelbart, W., McCarron, M., Osmond, B., Candido, E.P.M. and Baillie, D.L. (1976) *Genetics* 84, 233.
8. Chovnick, A., Gelbart, W., McCarron, M. and Pandey, J. (1974) *In* "Mechanisms in Recombination" (R.F. Grell, ed) pp. 351-363. Plenum, New York.
9. Dickinson, W.J. and Sullivan, D.T. (1975) *In* "Results and Problems in Cell Differentiation", Vol. 6 (W. Beermann, J. Reinert and H. Ursprung, eds), Springer-Verlag, New York.

10. Dickson, R.C., Abelson, J., Barnes, W.M. and Reznikoff, W.S. (1975) *Science* 187, 27.
11. Finnerty, V. (1976) *In* "The Genetics and Biology of *Drosophila*", Vol. 1b. (M. Ashburner and E. Novitski, eds), pp. 721-760. Academic Press, London.
12. Fogel, S., Hurst, D.D. and Mortimer, R.K. (1971). *In* "Stadler Genetics Symposia" (G. Kimber and G.P. Redei, eds), Vol. I and II, pp. 89-110. Agricultural Experiment Station, Univ. of Missouri, Columbia, Missouri.
13. Gelbart, W., McCarron, M. and Chovnick, A. (1976) *Genetics* 84, 211.
14. Gelbart, W.M., McCarron, M., Pandey, J. and Chovnick, A. (1974) *Genetics* 78, 869.
15. Gilbert, W. and Müller-Hill, B. (1970) *In* "The Lactose Operon" (J.R. Beckwith and D. Zipser, eds) pp. 93-109. Cold Spring Harbor Laboratory, Cold Spring Harbor, New York.
16. Glassman, E., Karam, J.D. and Keller, E.C. (1962) *Z. Vererb.* 93, 399.
17. Glassman, E. and Mitchell, H.K. (1959) *Genetics* 44, 153.
18. Grell, E.H. (1962) *Z. Vererb.* 93, 371.
19. Laurell, C.B. (1966) *Anal. Biochem.* 15, 45.
20. Lefevre, G., Jr. (1971) *Drosophila Inf. Serv.* 46, 40.
21. Manning, J.E., Schmid, C.W. and Davidson, N. (1975) *Cell* 4, 141.
22. McCarron, M., Gelbart, W. and Chovnick, A. (1974) *Genetics* 76, 289.
23. Rasch, E.M., Barr, H.J. and Rasch, R.W. (1971) *Chromosoma* 33, 1.
24. Weeke, B. (1973) *Scand. J. Immunol.* 2, (Suppl. 1) 37.
25. Yen, T.T.T. and Glassman, E. (1965) *Genetics* 52, 977.

EXPRESSION OF GENES ESSENTIAL FOR EARLY DEVELOPMENT IN THE NEMATODE, C. ELEGANS

David Hirsh, William B. Wood*, Ralph Hecht[+], Stephen Carr and Rebecca Vanderslice

Department of Molecular, Cellular and Developmental Biology
University of Colorado
Boulder, Colorado 80309

ABSTRACT. Genetic tests for maternal effects have been performed on 25 temperature-sensitive zygote-defective mutants of the nematode *Caenorhabditis elegans*. For most of the genes defined by these mutants (22 out of 25), maternal expression is sufficient for zygote survival, even if the gene is not expressed in the zygote. Twelve of these 22 genes *must* be expressed in the mother for zygote survival (strict maternals). For the remaining ten, expression either in the mother or in the zygote is sufficient for survival. One mutant shows a paternal effect in which wild-type sperm cytoplasm appears to rescue mutant zygotes. Maternal effect tests on mutants that block as late as the second larval stage after hatching indicate that in 3 of 11 mutants maternal contributions still can rescue mutant progeny.

Temperature shift experiments on the zygote-defective embryos show that all but one of the strict maternal mutants are temperature sensitive only before gastrulation. One of the mutants that can be rescued by gene expression in the zygote is temperature sensitive prior to gastrulation, suggesting that some zygote genes can function in early embryogenesis.

INTRODUCTION

Many classical studies indicate that much of the information for early embryogenesis in a variety of organisms is contained in the egg, and that expression of the embryonic genome after fertilization is not essential at least until the onset of gastrulation (reviewed in 1). We have begun to

*Present address: Biology Division
California Institute of Technology
Pasadena, CA 91125

[+]Present address: Department of Biophysical Sciences
University of Houston, Houston, TX 77004

investigate early embryogenesis in the free-living soil nematode *Caenorhabditis elegans*, whose short life cycle, convenience for growth in the laboratory, and ease of genetic manipulation (2) promise to make it a favorable organism for research into genetic control of development. In this paper we report studies on a set of temperature-sensitive (*ts*) mutants that are zygote-defective, that is, the progeny eggs of mutant animals fail to hatch at the nonpermissive temperature due to a block in zygote development (3). Temperature-sensitive mutants have two advantages: their lethality is conditional, allowing growth of homozygous mutant stocks at the permissive temperature, and the expression of the genes they define can be manipulated experimentally by changing the growth temperature. We have asked two questions about the essential gene functions defined by a set of 25 *ts* zygote-defective mutants of *C. elegans*. First, how many of these genes must be expressed during oogenesis in the maternal parent, and how many after fertilization in the zygote? This question has been investigated using three genetic tests for maternal effects. Second, we have asked when during embryogenesis the products of these genes must act, using temperature shift experiments. The results define at least four classes of genes essential for early development.

DESCRIPTION OF THE ORGANISM

C. elegans is a self-fertilizing hermaphrodite with a pair of sex chromosomes (XX) and five pairs of autosomes. Males (XO), which are morphologically distinguishable from hermaphrodites, appear in the population at a frequency of about 1 in 700 animals. Males mate with hermaphrodites; hermaphrodites do not mate with each other. A mated hermaphrodite produces both self and outcross progeny. The genetics of *C. elegans* (2) and the development, anatomy, and function of the reproductive system in both hermaphrodites (4) and males (5) have been described previously.

The life cycle of *C. elegans* is diagrammed in Figure 1. By about 10 min after fertilization the zygote has formed a vitelline membrane and a chitinous shell and become autonomous, in that it can continue normal development if removed from the mother. Normally, however, embryos are laid about 3 hours after fertilization at a 30-cell stage that corresponds approximately to the time of gastrulation. Hatching occurs at about 14 hours. The newly hatched first stage larva (L1), about 250 μm in length, contains 546 somatic nuclei and 4 primordial gonadial nuclei (6). During the next

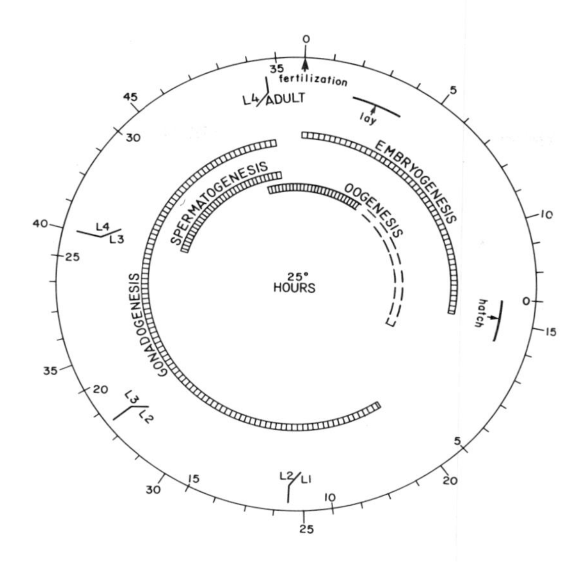

Fig. 1. Life cycle of *Caenorhabditis elegans*. Numbers on the outside of the circle indicate hours after ferterilization at 25°C, and numbers on the inside indicate hours after hatching. L1, L2, L3, and L4 refer to the larval stages, which are separated by molts at the times shown. Oogenesis continues for up to 8 days after the last molt.

35 hours the animal passes through 3 more larval stages (L2-L4), separated by 4 molts, and grows to a 1 mm adult with 808 somatic nuclei. The 4 primordial gonadial nuclei begin to divide a few hours after hatching and proliferate to form the gonad during larval development. In the hermaphrodite, spermatogenesis takes place during the L4 stage to produce about 150 mature sperm in each of the two spermathecae. Mature oocytes appear just prior to the fourth molt, and oogenesis continues for up to 8 days in the adult (4).

TEMPERATURE-SENSITIVE MUTANTS

A set of 223 temperature-sensitive developmental mutants of *C. elegans* has been isolated by screening approximately 7500 F2 clones derived from parents mutagenized with ethyl methane sulfonate (3). The permissive and nonpermissive temperatures for these mutants are 16°C and 25°C, respectively. About 25 of the mutants, designated *zyg* mutants, are zygote-defective. At the nonpermissive temperature these mutants lay embryos that fail to hatch due to a block in early development. About 48 of the mutants, designated _acc_ mutants, are larval-defective. At the nonpermissive temperature these mutants produce progeny that accumulate at a particular larval stage due to a later developmental block. Most of the experiments to be described here were carried out on _zyg_ mutants, all of which were backcrossed at least twice to wild type. All the _zyg_ mutations are recessive or partially dominant; none are dominant.

MATERNAL EFFECTS

The presence and nature of maternal effects among _zyg_ mutants have been investigated using three genetic tests, designated the selfing or S-test, the rescue by wild-type male or R-test, and the rescue by heterozygous mutant male or H-test. The procedures for and results of these tests are summarized in the following paragraphs; details of the experiments will be published elsewhere. In the descriptions that follow, wild-type alleles are designated by "+", and mutant (_ts_) alleles by "m".

The S-test ascertains whether a maternal + allele is sufficient to allow survival of a homozygous mutant (m/m) zygote. Heterozygous (m/+) hermaphrodites are obtained from a cross between a homozygous mutant (m/m) hermaphrodite reared at 16°C and a wild-type (+/+) male. The hermaphrodite parent carries a recessive morphological marker so that self (m/m) and outcross (m/+) progeny can be distinguished. Heterozygous hermaphrodites are reared and allowed to self-fertilize at 25°. The zygotes produced are of 3 genotypes: m/m, m/+, and +/+ in a 1:2:1 ratio; however, all zygotes have wild-type ooplasm contributed by the m/+ parent. Survival of the m/m zygotes is assayed by picking about 100 progeny larvae onto individual plates at 25°C and determining how many display the mutant phenotype as they mature. If about 25% of the progeny do so, then the m/m zygotes must have successfully completed embryogenesis as a result of the maternal effect of the heterozygous mother's + allele, and the mutant is scored maternal (M). If none of the progeny exhibit the mutant phenotype as adults, it is assumed that the m/m zygotes failed to hatch, and the mutant is scored nonmaternal (N). This result is checked by ascertaining that about 25% of the eggs laid by the m/+ hermaphrodite fail to hatch. When the set of 25 _zyg_ mutants was S-tested, 22 were scored M and 3 were scored N. This result indicates that for most of the genes defined by these mutants, maternal expression is sufficient for zygote survival, even if the gene is not expressed in the zygote.

The R-test ascertains whether fertilization with sperm from a wild-type male is sufficient to allow survival of m/+ zygotes produced in a homozygous mutant (m/m) hermaphrodite. Mutant hermaphrodites carrying a morphological marker are hatched at 16°C and shifted to 25°C as third or fourth stage larvae. When they have reached adulthood these hermaphrodites, which can produce only nonviable zygotes by self-fertilization, are mated with wild-type males. Since almost all of the _zyg_ mutants are reversible, proof of mating can be obtained by removing the hermaphrodites from the mating plates after 24 hours, shifting these animals to 16°C, and determining that outcross progeny are produced. If mated hermaphrodites produce no viable outcross progeny zygotes at 25°C, then the mutant is scored M, indicating that the presence of the + allele in the zygote is _not_ sufficient to allow m/+ progeny of the m/m hermaphrodite to survive. If viable m/+ zygotes are produced then the mutant is scored N, indicating _either_ that zygotic expression of the + allele is sufficient for survival, _or_ that a sperm function or factor needed for early embryogenesis can be supplied paternally. These alternative interpretations of an N result can be distinguished by the H-test, as described below. When the set of 25 _zyg_ mutants was R-tested, 13 were scored M, and 12 were scored N.

The combined results of the S- and R-tests are summarized in Figure 2. Twelve of the 25 mutants are strict maternals (M,M). The genes defined by these mutants _must_ be expressed in the hermaphrodite parent for the zygote to survive. Ten of the mutants are M,N; expression of these genes _either_ in the hermaphrodite parent _or_ in the zygote is sufficient for zygote survival. Two of the mutants are N,N; these genes _must_ be expressed in the zygote. Finally, one of the mutants is N,M; this gene must be expressed _both_ in the hermaphrodite parent _and_ in the zygote for the zygote to survive.

The H-test is designed to ascertain for M,N mutants whether male rescue in the R-test is due to expression of the paternally derived + allele in the zygote or to some other factor or function contributed by the sperm. The H-test procedure is the same as for the R-test, except that the homozygous mutant (m/m) hermaphrodite is mated with a male that is heterozygous (m/+) for the same mutation, so that the outcross progeny zygotes will be of two genotypes, m/m and m/+, in a 1:1 ratio. If viable zygotes of only m/+ genotype are found, then rescue must depend on the presence of the + allele in the zygote, and the mutant

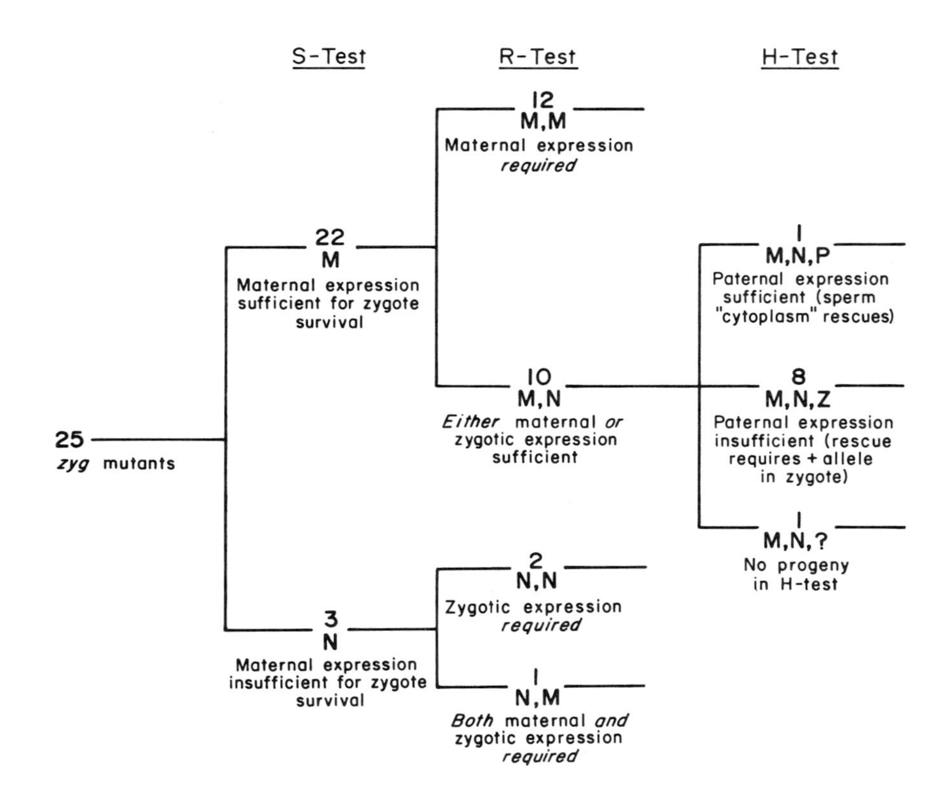

Fig. 2. Summary of maternal-effect test results for zyg mutants.

is scored zygotic (Z). However, if viable zygotes of both m/m and m/+ genotypes are produced in approximately equal numbers, then rescue must be due to a paternal effect resulting from expression of the + allele in the male parent but independent of sperm genotype. Such mutants are scored P. When the set of 10 M,N mutants was H-tested, 8 of them were scored Z, and 1 produced equal numbers of m/m and m/+ progeny and was scored P. Some sperm factor or function appears to be sufficient to allow zygote survival in this mutant, even when the gene it defines is expressed neither in the maternal parent nor in the zygote. This mutant displays only weak rescue in the R-test (~12%), suggesting that the sperm component responsible for rescue may be near a threshold level. Whether this component also can be transmitted through the egg is not known, because unfortunately m/m males of the mutant are infertile, so that the reciprocal test cannot be carried out. The one remaining M,N mutant produced no progeny in the H-test, although m/+ males were shown to be fertile. This mutant is being investigated further.

To determine whether maternal contributions can affect mutational blocks beyond hatching, S- and R-tests also were carried out on a set of 11 ts acc mutants that arrest during the L2 stage, which extends from about 26-32 hours after fertilization. Eight of these mutants scored N,N, indicating that development beyond the L2 stage is determined solely by the expression or lack of expression of the corresponding genes in the zygote. However, the remaining 3 of the mutants scored M,N, indicating that maternally supplied factors can allow bypass of mutational blocks as late as the L2 stage even when the corresponding genes are not expressed in the embryo.

CRITICAL TIMES OF TEMPERATURE SENSITIVITY

Experiments on the effects of shifting embryos from permissive to nonpermissive temperature or vice versa at various stages of embryogenesis have been performed with 19 of the 25 zyg mutants. Two-cell embryos dissected from hermaphrodites reared at one temperature are incubated for various times at the same temperature and then shifted to the other temperature and scored for zygote survival. Alternatively, gravid hermaphrodites reared at one temperature are shifted to the other and 2-cell embryos dissected out at various times thereafter are scored for zygote survival. Thus the critical period of temperature sensitivity can be defined relative to the 2-cell embryonic stage. The interpretation of temperature shift experiments to define critical periods of temperature sensitivity has been discussed by Hirsh and Vanderslice (3). In general, shift-down experiments are used to define the beginning of the temperature sensitive period, and shift-up experiments to define the end.

Almost all of the strict maternal (M/M) mutants are ts during oogenesis and/or early embryogenesis, but are no longer ts by the time of gastrulation. Only one of these mutants is ts after gastrulation until close to the time of hatching. The M,N mutants appear to exhibit two periods of sensitivity, one before fertilization and the other at gastrulation, as might have been expected from the results of the maternal effect tests. However, one of these mutants is clearly temperature-sensitive between fertilization and gastrulation, suggesting that the gene it defines is expressed exceptionally early in the zygote. The N,N mutants are ts late in embryogenesis beginning at least two hours after gastrulation, and the M,N mutant, as expected, is ts throughout oogenesis and embryogenesis. Details of these experiments will be published elsewhere.

DISCUSSION

Our results are consistent with the view that in *C. elegans* as in many other organisms early embryogenesis is dependent on maternal gene functions. The 25 mutationally identified genes that we have studied were selected only as being essential for zygote development prior to hatching based on the lethal phenotype of the corresponding ts mutations. For 12 of these genes maternal expression is required for zygote survival, and for 10 more maternal expression is sufficient. The temperature shift experiments, although impossible to interpret unequivocally, suggest that for the 12 maternally required genes, about half of the corresponding proteins are produced or utilized before fertilization, and the other half are produced or utilized between fertilization and gastrulation. The products of the latter genes could be supplied to the zygote as maternal mRNA's. Only one of the maternally required proteins clearly is utilized beyond gastrulation.

Only 5 of the 10 maternally sufficient genes have been tested by temperature shift; in all cases the corresponding proteins appear to be produced or utilized after fertilization as expected, because zygotic function of these genes also is sufficient for survival. The products of these genes may be supplied to the zygote as mRNA's, but these messages also must be transcribed from the zygote genome before the times when the corresponding proteins are required in embryogenesis. For one of these genes zygotic transcription and translation appear to occur before gastrulation.

Only 2 of the 25 *zyg* genes studied show no maternal effect, and as expected the temperature-sensitive periods of the corresponding mutants are relatively late in embryogenesis, well past gastrulation. For only one of the 25 genes are both maternal and zygotic expression necessary for zygote survival. This category might have been expected to include many genes that code for essential metabolic functions. The finding of only one suggests that essential metabolic proteins in general may be supplied maternally in large excess.

The paternal effect seen with the one M,N,P mutant represents to our knowledge a unique example of a "cytoplasmic" component essential for early development that can be transmitted through the sperm. The defect in this mutant cannot be in fertilization _per se_, because zygotes are produced by a homozygous mutant hermaphrodite at 25°C, and these embryos can be rescued by shifting them to 16°C at any stage prior to gastrulation. The sperm of _C. elegans_ are not flagellated but ameboid, with about 1/300 the volume of the mature egg. The mechanism of fertilization is not known, but the paternal effect described here would be consistent with a fusion of sperm and egg cytoplasm.

The functional nature of the 25 mutationally identified gene products remains an intriguing question, but some clues can be obtained from the gross mutant phenotypes so far observed. Nineteen of the 25 _zyg_ mutants display other phenotypes in addition to the zygote defect if placed at nonpermissive temperature shortly after the end of the temperature sensitive period in the zygote. For example, many show gonadogenesis defects and do not produce zygotes if shifted to 25°C at the first larval stage. Others grow very slowly at the nonpermissive temperature or arrest at specific larval stages. Such multiple phenotypes suggest that these mutants may carry defects in common metabolic functions. In contrast, 5 of the strict maternal-effect (M,M) mutants as well as the paternal-effect (M,N,P) mutant display no other phenotypes at the nonpermissive temperature, have normal growth rates at both 16°C and 25°C, and produce normal numbers of progeny at the permissive temperature. Therefore these 6 mutants may define functions that are unique to, or at least particularly important for early embryogenesis. More detailed phenotypic characterization of these mutants is now in progress with the goal of further defining the corresponding gene functions.

ACKNOWLEDGMENTS

We thank Kimberly Johnson for assistance with the temperature shift experiments. This work was supported by USPHS grant No. GM19851 and by USPHS Career Development Award No. GM70465 to D.H. W.B.W. was the recipient of a Guggenheim Fellowship, and R.H. was supported by a USPHS postdoctoral fellowship from NICHD.

REFERENCES

1. Davidson, E.H. (1977) Gene Activity in Early Development. Academic Press, New York.
2. Brenner, S. (1974) Genetics 77, 71.
3. Hirsh, D., and Vanderslice, R. (1976) Dev. Biol. 49, 220.
4. Hirsh, D., Oppenheim, D., and Klass, M. (1976) Devel. Biol. 49, 200.
5. Klass, M., Wolf, N., and Hirsh, D. (1976) Devel. Biol. 52, 1.
6. Sulston, J. E. and Horvitz, R. (1977) Devel. Biol. 56, 110.

Summary of Workshop on Nematodes

W.B. Wood, Convener

The workshop on nematodes presented current research from four laboratories on the development and physiology of *C. elegans*.

1) Dr. Fred Schachat, in collaboration with H. E. Harris, R. L. Garcia, J. A. W. La Pointe and H. F. Epstein (Stanford University Medical Center) showed that native myosin purified from wild-type and mutant nematodes is composed principally of two kinds of molecules. The normal molecular weight of body wall myosin heavy chains is 210,000, but in the mutant *unc-54*^{e675} there is an altered myosin heavy chain with a molecular weight of 203,000 as well. Myosins homogeneous for either class of heavy chains can be resolved by hydroxyapatite chromatography. These myosins yield distinct peptide maps when cleaved with cyanogen bromide. Similar results are obtained with wild-type myosins, which also can be fractionated by hydroxyapatite chromatography into species with distinct cyanogen bromide maps. Antigenic differences exist between the two myosins in the mutant, and antibodies specific for the class that contains the *unc-54* gene product is being purified in order to study the location and synthesis of this myosin during development. (Reference: Schachat *et al.*, 1977.)

2) Dr. R. K. Herman, in collaboration with P. M. Meneely (University of Minnesota) described two approaches to isolation and maintenance of recessive lethal mutations. In the first an X-ray induced translocation of an X chromosome fragment to chromosome V has been used as a balancer for EMS-induced lethals that fall into a small region of X. So far, sixteen independent X-linked lethals with various defective phenotypes have been isolated and shown to fall into eleven complementation groups. In the second approach an X-ray induced crossover suppressor on chromosome II has been isolated and used to balance several dozen recessive lethal mutations linked to the *unc-4* gene on chromosome II. (Reference: Herman *et al.*, 1976.)

3) Dr. Kenneth K. Lew, in collaboration with S. Ward (Harvard Medical School) has determined the cell lineage of the intestinal tract, which arises from a single precursor cell in the eight-cell embryo and is composed of 32 to 34 cells in the mature adult. Biochemical differentiation of gut cells was demonstrated as early as the eight-cell stage of the developing gut primordium, by taking advantage of the

mutant *flu-1*^{e1002}, which carries a defect in tryptophan catabolism that causes gut cells to fluoresce. The fluorescence technique was used to examine gut development in several double mutants that carried e1002 and a *ts* zygote-defective mutation. In one of these mutants at the nonpermissive temperature gut cell division was arrested at the two-cell stage and all of the cells in the embryo became fluorescent. In other mutants gut development was arrested at the six- to eight-cell and twenty-cell stages. (Reference: Lew and Ward, 1977.)

4) Dr. Richard L. Russell*, in collaboration with C. D. Johnson, J. B. Rand*, S. Scherer, and M. S. Zwass (California Institute of Technology) has developed rapid assays for several neurotransmitter-related enzymes, including acetylcholine esterase and choline acetyl transferase. These assays have been used to screen large numbers of mutants for enzyme defects. The worm has four distinguishable acetylcholine esterase activities; one mutant has been found that lacks two of these species. This mutant is partially paralyzed, and behavioral tests suggest that the defect primarily affects body and not head movement. The mutation maps near the right end of the X chromosome and may be an allele of the *unc-3* gene. Two additional mutants have been found to be severely deficient in choline acetyl transferase activity. These two mutations give rise to different phenotypes; one is severely paralyzed, whereas the other shows only aberrant defecation behavior. (References: Russell *et al*., 1977; Johnson, C. D. and Russell, R. L., submitted for publication.)

REFERENCES

Herman, R. K., Albertson, D. G., and Brenner, S. (1976) *Genetics* 83, 91.

Lew, K. K. and Ward, S. (1977) *Proc. Nat. Acad. Sci.* in press.

Russell, R. L., Johnson, C. D., Rand, J. B., Scherer, S., and Zwass, M. S. (1977) This volume.

Schachat, F. H., Harris, H. E., and Epstein, H. F. (1977) *Cell* 10, 721.

*Present affiliation, University of Pittsburgh.

MUTANTS OF ACETYLCHOLINE METABOLISM IN THE NEMATODE *CAENORHABDITIS ELEGANS*

Richard L. Russell*, Carl D. Johnson†, James B. Rand*, Stewart Scherer† and Maurice S. Zwass†

*Department of Life Sciences, University of Pittsburgh, Pittsburgh, PA 15260; †Division of Biology, California Institute of Technology, Pasadena, CA 91125

ABSTRACT. Radiochemical assays based on the selective extraction of either substrate or product from an aqueous reaction volume into an organic scintillator have been developed for acetylcholinesterase and choline acetyltransferase. These rapid, convenient assays have made it possible to screen large numbers of mutant lines for potential enzymatic defects. One mutant with a partial acetylcholinesterase defect and two more with choline acetyltransferase defects have been identified. The acetylcholinesterase defective mutant lacks two of the four isozymic forms of acetylcholinesterase found in wild type *C. elegans*. Behaviorally, it is selectively defective in the propagation of contractile waves in the body region. Of the two mutants with choline acetyltransferase defects, one is remarkably paralyzed and uncoordinated, while the other is behaviorally nearly normal.

INTRODUCTION

As the work reported elsewhere at this conference amply illustrates, the small soil nematode *Caenorhabditis elegans* has advantages of genetic manipulability and cellular simplicity which recommend it for genetic approaches to a wide range of problems in metazoan organization (1). Amongst these problems, one of particular interest for the organization of the nervous system is that of synapse formation, the process by which specific contacts are formed between cells to give the nervous system its functional capacities. One genetic approach to this problem is to obtain mutants with behavioral anomalies of varying specificity in the expectation that some of these will harbor synaptic disruptions of interest (2,3,7). This approach has been and is being pursued actively, but will not be the subject of this paper. Instead we will be concerned here with a second genetic approach, one which aims at obtaining mutants affected in known components of synaptic transmission. This approach has been begun by the isolation of mutants affected in enzymes involved in the metabolism of acetylcholine, a common neurotransmitter.

The successful pursuit of mutations affecting neurotransmitter enzymes seemed to us *a priori* to have three requirements, 1) the availability of rapid easy assays for the enzymes in question, 2) a sufficient familiarity with the wild type enzymes to facilitate screening for mutant defects, and 3) a ready source of mutants or clones to be screened. Below we describe first a general principle (liquid extraction) which has proven adaptable to the assay of several neurotransmitter enzymes with ease and convenience; then we describe some properties of the wild type *C. elegans* acetylcholinesterase revealed using an assay based on this general principle, and finally we describe the properties of an interesting mutant with a partial acetylcholinesterase defect, isolated by a screening procedure based on the wild type enzyme properties.

METHODS

We have found it convenient and reliable to assay several enzymes of neurotransmitter metabolism using the same general principle, i.e. the use of radiochemically labeled substrates followed by the use of liquid extraction between an organic and aqueous phase to separate substrate and product (8). The convenience is clearly exemplified by our assay for acetylcholinesterase (11). We carry out the enzymatic reaction in an aqueous reaction volume of 100 μl in a 1 dram shell vial. The substrate is [^{3}H-acetyl]choline, and the radiochemical product is ^{3}H-acetate. The enzymatic hydrolysis is carried out for a short time (often five minutes is adequate) and the reaction is then stopped by the addition of an additional 100 μl of strong pH 3 buffer. Under these conditions, the ^{3}H-acetate is protonated, and therefore available for organic extraction; we add directly to the vial 4 ml of a mixture composed of 9 parts standard toluene-based scintillation fluid and 1 part n-butanol. The vial is then capped, shaken briefly to achieve equilibrium partition, allowed to settle briefly, and then counted directly in a scintillation counter. The ^{3}H-acetate in the organic phase counts efficiently, but the ^{3}H-acetylcholine, which remains in the aqueous phase, does not, since the very weak β particles from its decay do not reach the scintillator. Thus, the convenience of the assay arises because no elaborate post-assay manipulations are necessary to achieve selective counting of the radiochemical product.

Several possible modifications of this general procedure are available to facilitate assay of other enzymes. To cite just two, ion pairing extractants can be added to the organic phase to facilitate the selective extraction of charged substrates or products, and substrates and products that do not differ sufficiently for selective extraction can be made to differ by post-assay enzymatic modification. Both of these modifications are used in our assay for choline acetyltransferase (8). Here, the substrates are ^{3}H-choline and unlabeled acetyl CoA, and the radiochemical product is ^{3}H-acetylcholine. At the end of the choline acetyltransferase assay period, excess choline kinase from yeast is added to convert all the unreacted ^{3}H-choline to ^{3}H-choline phosphate, and the ^{3}H-acetylcholine is then extracted into the organic phase by the addition of sodium tetraphenylboron. Still other modifications have allowed us to develop quick and convenient assays for 5 other enzymes concerned with the metabolism of other transmitters, and no doubt the general method can be used for a wide range of different enzymes.

WILD TYPE *C. ELEGANS* ACETYLCHOLINESTERASE

Early attempts to solubilize *C. elegans* acetylcholinesterase revealed that even under the best of conditions (0.05M borate buffer, pH 8.8) no more than 50% of the acetylcholinesterase activity of the crude homogenate could be solubilized (12). Attempts to solubilize the remaining activity with detergents showed that ionic detergents were necessary to effect complete solubilization, but that the solubilized enzymes were unstable in the presence of the detergent. Therefore, a sequential detergent method was used to effect complete solubilization; crude extracts were treated for 1 minute with 0.2% desoxycholate, and then centrifuged rapidly into a sucrose gradient containing 0.1% Tween 80. The results of such a gradient are shown in Fig. 1 which contains two clear peaks and indications of some others. In fact, by selective isolation and resedimentation it has been possible to discern four discrete activity peaks composing the original pattern; the sedimentation profiles of the separated (but otherwise unpurified) peaks are shown in Fig. 2. In order of increasing size, these have been given the labels I, II, III, and IV, and their sedimentation coefficients are respectively 5.3S, 7.1S, 11.4S, and 13.0S.

Although these four forms have not been further purified, they have been sufficiently well separated to allow separate study of their kinetic properties. Fig. 3 shows that the four forms can be distinguished on the basis of substrate affinity; forms I and II have apparent Michaelis contants 4-5 times higher than those of forms III and IV.

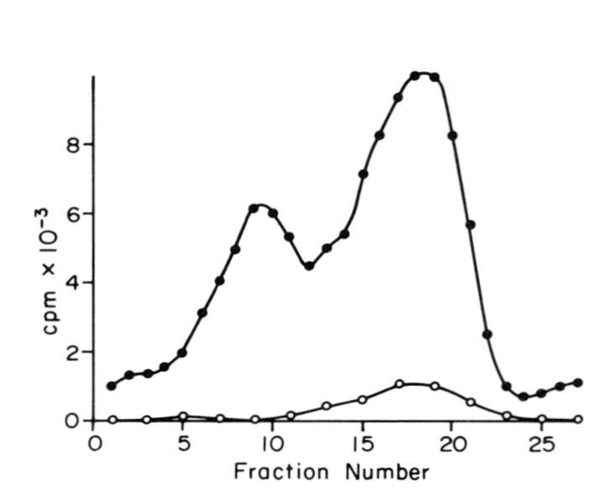

Fig. 1. Velocity sedimentation after extraction with desoxycholate. One ml of frozen nematodes were freeze powdered and extracted with 0.8 ml of 0.1M borate, pH 8.8 and 0.03 ml 10% desoxycholate at 4°C. After 2 min a 1500 G 1 min supernatant was prepared and 100 µl aliquots were loaded onto 5.0 ml, 5-20% sucrose gradients in 0.05M borate, pH 8.8, 0.1% Tween 80. After centrifugation at 65,000 rpm for 4.5 hours, fractions were collected and 10 µl of each was assayed radiometrically for acetylcholinesterase.

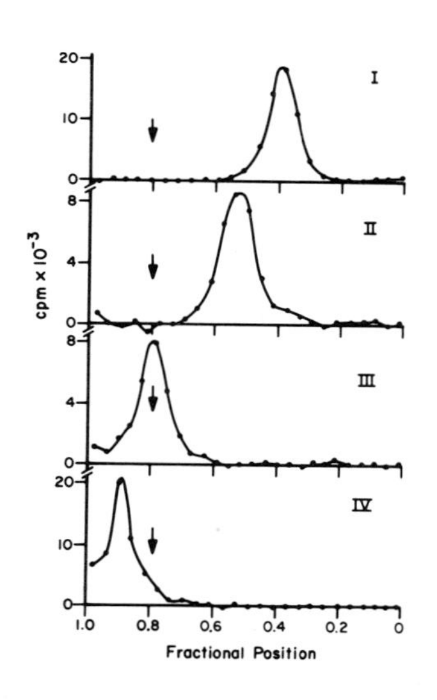

Fig. 2. Velocity sedimentation of separated acetylcholinesterase forms. Selected fractions from an initial sucrose gradient separation of a crude homogenate were pooled and resedimented on sucrose gradients. Peak fractions from the resedimentation were then loaded separately onto additional sucrose gradients (5-20%) and sedimented for 7 hrs at 65,000 rpm. Fractions were collected and assayed radiometrically for cholinesterase.

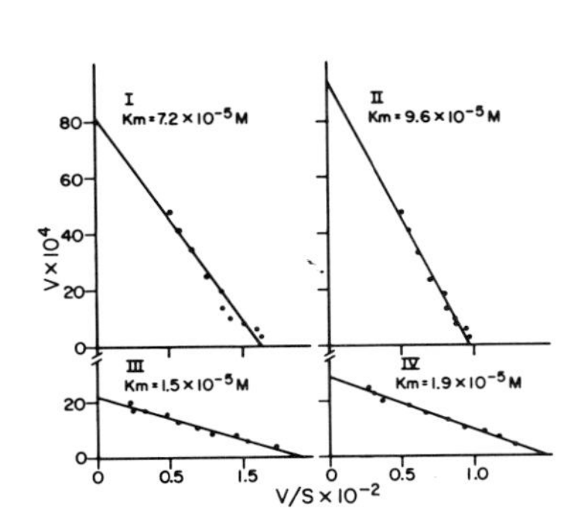

Fig. 3. Substrate affinities of separated forms. The separated forms I-IV were assayed with 10 concentrations of acetylcholine ranging from 2 x 10^{-6}M to 10^{-4}M. Velocity (V in units/ml) was plotted against V/S, an Eadie-Hofstee plot. The apparent Km (the negative of the slope) was calculated with the linear regression program on an HP 65 calculator. The standard deviation of the slopes were: form I, ± 0.6 x 10^{-5}M; form II, ± 0.5 x 10^{-5}M; form III, ± 0.1 x 10^{-5}M; and form IV, ± 0.1 x 10^{-5}M.

The same functional division is suggested by studies of substrate specificity, as shown in Table 1; forms III and IV are considerably less specific in their preference for acetylthiocholine, relative to other acylthiocholines, as a substrate. (As with other invertebrate cholinesterases, the _C. elegans_ cholinesterases do not resemble either the "true" or the "pseudo" cholinesterases of vertebrates.)

Evidence from inhibitor studies and thermal inactivation studies (data not shown) further substantiates the functional division of the four cholinesterase forms into two classes (I + II and III + IV). Forms III and IV are considerably more sensitive than forms I and II to inhibition by a variety of inhibitors ranging from neostigmine to DFP, but on the other hand, forms III and IV are considerably more resistant than forms I and II to thermal inactivation.

A final, and useful, difference between the two functional classes is revealed by studies of detergent sensitivity. As Fig. 4 shows, the low molecular weight forms (I and II), are considerably more sensitive to inactivation by 0.5% desoxycholate than are the higher molecular weight forms (III and IV). This difference, because it could be easily studied on large numbers of samples at one time, turned out to be very convenient for screening for possible mutant defects.

TABLE 1

ACYL-THIOCHOLINE SUBSTRATE SPECIFICITY OF SEPARATED FORMS

Substrate	Electric Eel AChE	Horse Serum ChE	Activity of Nematode ChE Forms			
			I	II	III	IV
Acetyl-thiocholine	100	100	100	100	100	100
Acetyl-β-methyl-thiocholine	68	37	45	47	66	78
Propionyl-thiocholine	61	189	54	59	104	103
Propionyl-β-methyl thiocholine	36	92	35	34	94	102
Butyrylthiocholine	0.3	181	26	25	64	71

Hydrolysis of acyl-thiocholine substrates was followed spectrophotometrically. Initial substrate concentrations were 1 mM. All of the enzymes had about the same activity with acetylthiocholine as substrate. Activity (change in OD_{412}/min) is recorded as % of the activity measured with acetylthiocholine.

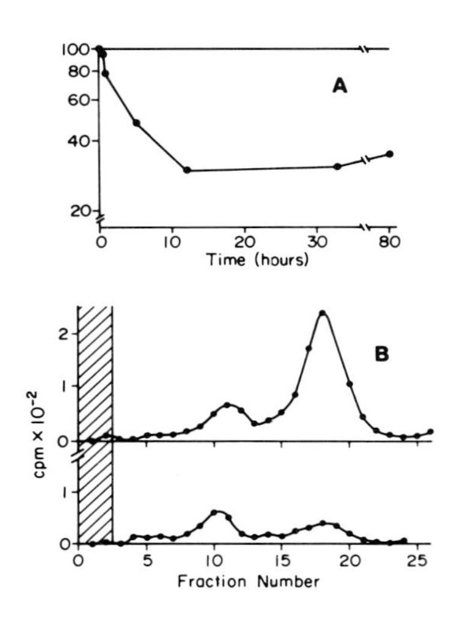

Fig. 4. Inactivation of homogenate acetylcholinesterase by 0.5% desoxycholate. A. Nematode homogenate was diluted to 0.05M of borate, pH 8.8; 0.5% desoxycholate and stirred at 4°C. At specific times, an aliquot was assayed with 5.0 x 10^{-5}M acetylcholine. Cholinesterase activity was plotted as a % of that measured 5 min after dilution. B. Homogenate was diluted as in A and after 30 min (top gradient) or after 8 hrs (lower gradient) at 4°C, 100 µl aliquots were loaded onto 4.5 ml 5-20% sucrose gradients (0.05M borate; pH 8.8; 0.1% Tween 80) with 0.5 ml renografin cushions. After centrifugation for 4 hrs at 65,000 rpm, fractions were collected and 10 µl of each was assayed for acetylcholinesterase activity.

A PARTIALLY CHOLINESTERASE DEFECTIVE MUTANT

Because of the qualitative differences found between the various forms of wild type *C. elegans* cholinesterase, it seemed quite possible that more than one structural gene for cholinesterase might exist in *C. elegans*. Because this possibility would preclude the isolation of a completely defective mutant (which might be lethal anyway) we sought a method which would allow detection of a mutant defective in one or the other of the two recognized functional classes. For this purpose we chose a detergent survival method like that depicted in Fig. 4 and used it to study extracts of about 200 mutants previously isolated. These mutants were of a variety of types, but many were of the generally behaviorally deficient category known as "uncoordinated" (1). Most of these mutants resembled the wild type (N2), as shown in Fig. 5. However, one mutant, BC46, was quite noticeably different (13); it showed little or no activity resistant to 0.5% desoxycholate. The implication that BC46 might lack the detergent resistant forms III and IV was confirmed by sucrose gradient sedimentation. As Fig. 6 shows, the usual sequential detergent extraction method revealed no evidence of forms III and IV in BC46, nor could any evidence for these

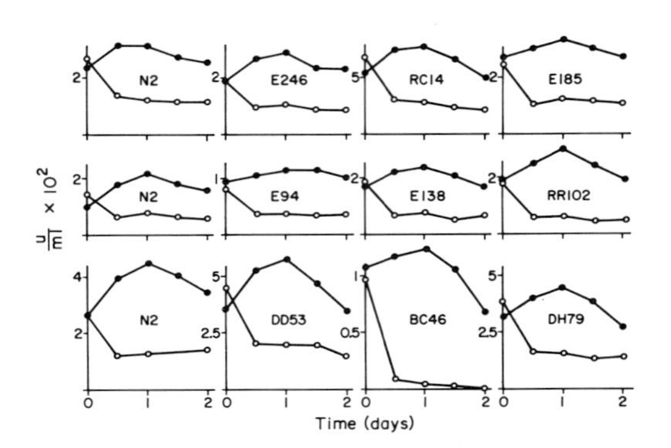

Fig. 5. Desoxycholate inactivation of mutant extracts. Ten plates of animals for each strain to be assayed were eluted, washed with distilled water and freeze-powdered in 1 ml. One hundred µl aliquots of homogenate were diluted with an equal volume of 0.1M borate, pH 8.8. Separate aliquots received 20 µl of 10% Tween 80 or 10 µl of 10% desoxycholate. Ten µl aliquots of the extracts were assayed for 5 min at about 12 hr intervals. Activity is plotted versus time of assay. The desoxycholate extract of one mutant (BC46) was devoid of activity after about 24 hrs.

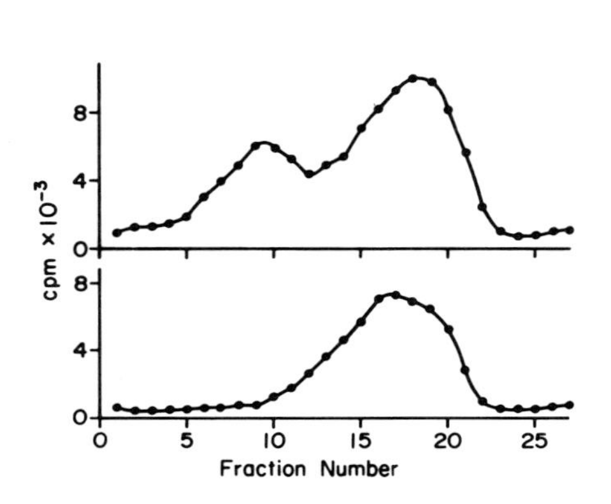

Fig. 6. Velocity sedimentation of detergent extracted N2 and BC46. One ml of homogenate from N2 (top gradient) or BC46 (lower gradient) was diluted with 0.8 ml of 0.1M borate, pH 8.8. At 4°C, desoxycholate was added to 0.2%, and after 2 min, a 1500 G 1 min supernatant was prepared and 100 µl aliquots were loaded onto 5 ml 5-20% sucrose gradients in 0.05M borate, pH 8.8, 0.1% Tween 80.

forms be derived by extraction in the absence of detergent, where small amounts of these forms are ordinarily seen. Forms I and II, however, could be extracted from BC46 and were indistinguishable from the corresponding wild type forms with respect to substrate affinity, substrate specificity, inhibitor sensitivity, thermal inactivation and detergent inactivation (data not shown). Biochemically, then, BC46 appears to be quite selectively defective in the two high molecular weight forms of C. elegans acetylcholinesterase, forms III and IV.

Behaviorally, BC46 is an interesting mutant. Its overall mobility is quite low, but not because of any generalized weakened paralysis like that which characterizes mutants defective in muscle-structural proteins (4,5). Instead, BC46 is quite active; its head shows wave like motions which are comparable in extent and frequency with those of the wild type. However, the waves generated at the head are apparently incapable of propagating into the body proper, and the same is apparently true of the waves, associated with backward movement, generated at the tail. Interestingly, the inactive muscles receive their innervation from the ventral nerve cord (18), but the muscles of the head, which remain active, receive their innervation instead from the nerve ring (17).

The specificity of the BC46 behavioral deficit is shown by studies of its sensory behavior. Although the generalized motion defect makes it difficult to assess BC46's capability to respond to sensory stimuli, the tests which have been so far carried out indicate that all the known normal sensory functions are intact. For instance, BC46, although taking a great deal longer than the wild type to execute the response, nonetheless shows orientation and accumulation in radial gradients of sodium chloride, an attractant (16). Similarly, BC46 retains the wild type's ability to avoid conditions of extremely high osmotic strength (2). And finally, BC46 retains many aspects of the normal response to mechanical stimuli (13). As Table 2 shows, when tapped lightly on the snout, wild type C. elegans moves backward for a period of about 4.5 seconds, during which it executes 2-3 backward waves, and during this response the low amplitude searching movements of the head which characterize forward movements are eliminated. In response to the same stimulus, BC46 does not move backward, apparently because of its wave propagation deficit, but the low amplitude searching movements cease and this response persists for very nearly the same length of time as in the wild type. Thus, both the initial sensation and the subsequent timing appear unaffected by the BC46 defect. Moreover, tail touch sensitivity appears intact in BC46 as well. When wild

TABLE 2

MECHANICAL RESPONSES OF N2 AND BC46

	N2	
Frequency of waves (sec^{-1})	0.59 ± 0.07 (n=10)	0.52 ± 0.14 (n=10) 0.60 ± 0.12 (n=10)
Duration of head-tap-response (htr) (sec)	4.4 ± 1.6 (n=30)	4.7 ± 2.1 (n=27) 5.8 ± 2.0 (n=10)
Frequency of stopping htr with tail-tap (htr^{-1})	0.85 (n=20)	0.90 (n=10)

Frequencies of head waving were observed for the indicated numbers of animals over a period of 30 sec each. For the head tap response, animals were touched on the snout with either an eye-lash or a fine disposable syringe needle and observed thereafter for either backward movement (N2) or a period of cessation of head searching movements (BC46). Results for two independent observers are given for this and for the previous response. For the tail tap response, animals were touched on the snout as in the head tap response, but then were touched within 1 sec on the tail. The recorded value is the probability of reversing the original head tap response.

type _C. elegans_ is touched on the tail shortly after having been started backward by a touch on the snout, it reverses direction and heads off forward immediately. Under the same conditions, BC46 touched on the tail immediately resumes the low amplitude searching movements of the head. Thus, despite the malfunction of the body proper for the propagation of contractile waves, information from tail receptors can apparently traverse this region of the animal to produce an effect in the head.

The selective biochemical and behavioral deficiencies of BC46 suggest that forms III and IV of _C. elegans_ acetylcholinesterase may normally be localized to synapses concerned with the propagation of contractile waves in the body proper. This possibility is currently being examined by light microscope histochemistry and by electron microscopy of the body neuromuscular junctions in BC46. Attempts are also being made to determine whether BC46 represents a structural gene mutation for acetylcholinesterase or whether, as an alternative, the selective acetylcholinesterase deficiency is secondary to some other primary defect.

CHOLINE ACETYLTRANSFERASE DEFECTIVE MUTANTS

Recently (and in the absence of much information on the wild type enzyme), many of the same mutants screened for acetylcholinesterase deficiencies were screened for possible deficiencies in the biosynthetic enzyme for acetylcholine, choline acetyltransferase. Although it was anticipated that absolute deficiencies of this enzyme might well prove to be lethal, two mutants were indeed found to have very low levels of this enzyme. The first of these is tcfr-1, a mutant resistant to the drug trichlorfon, isolated in the laboratory of David Hirsh at Boulder (10). Extracts of this mutant contain at most 2% of the wild type level of choline acetyltransferase activity, and the mutants themselves are extremely paralyzed and uncoordinated. However, whether the relationship between the biochemical and behavioral deficiencies of tcfr-1 is causal remains to be seen, particularly in the light of the second mutant, DH103. This mutant was originally isolated as a partial defective in defecation behavior (6) but is otherwise very nearly normal. Nonetheless, extracts of DH103 contain a little less than 5% of the wild type level of choline acetyltransferase, suggesting either that choline acetyltransferase is not in fact an important synthetic enzyme, or, perhaps more likely, that the differences between DH103 and tcfr-1 in residual choline acetyltransferase levels may be significant. Further work to examine these possibilities is in progress.

DISCUSSION

The mutants described above represent the first steps toward a genetic analysis of the synthesis, degradation, and possible functional significance in C. elegans of one potential neural transmitter, acetylcholine. Their isolation has depended on the availability of convenient assays, of information on the relevant wild type enzymes, and of a pool of screenable mutants. These same conditions can be met for other potential neurotransmitters, and we anticipate future work in this direction.

The properties of BC46 illustrate the kind of valuable, unexpected results that the genetic approach can produce. In the absence of a selectively defective mutant, it is unlikely that specific functional roles would have been inferred for the multiple forms of C. elegans acetylcholinesterase. However, the selective biochemical and behavioral deficiencies of BC46 suggest rather strongly that the two larger molecular weight forms (III and IV) are selectively involved in the propagation of contractile waves in the body proper. This suggestion is now being checked by selective histochemistry and by anatomical fractionation experiments, and if it proves true, may have interesting developmental implications; many of the neuromuscular synapses which drive the body muscles are formed post-embryonically, after the involved motor neurons have differentiated as the result of a series of stereotyped lineage divisions from neuroblast precursor cells (15). It may be that these late forming synapses incorporate forms of acetylcholinesterase different from those incorporated in earlier synapses.

BC46 might be either a mutation of an acetylcholinesterase structural gene or alternatively a mutation producing an acetylcholinesterase deficiency secondarily. For resolving these alternatives, a fortunate feature of BC46 is its genetic map position. BC46 maps to the right arm of the X chromosome, in a portion for which a stable translocated duplication has been isolated (9). This fortunate circumstance should permit the use of gene dosage experiments to examine the nature of BC46.

Less is known so far about the two choline acetyltransferase deficient mutants, and much information still remains to be gathered on the wild type enzyme as well. However, the behavioral differences of the two mutants already pose some interesting problems, and it will be of particular interest to ask how levels of acetylcholine are affected by the two mutations. For the future, we would like very much to isolate temperature sensitive alleles of one or both of these mutants, in the hope of adjusting internal acetylcholine levels at will.

ACKNOWLEDGMENTS

We are grateful to Margot Szalay for technical assistance, to Carol Vermaes for materials, and to Ed Hedgecock for useful discussions. The work reported here was supported by a U.S. Public Health Service Training Grant (GM02031) to C.D.J., by an American Cancer Society Postdoctoral Fellowship to J.B.R., and by a Sloan Foundation Neurosciences Grant and a U.S. Public Health Service Research Grant (NS09654) to R.L.R.

REFERENCES

1. Brenner, S. (1974) *Genetics* 77,71.
2. Culotti, J. and Russell, R.L. (1977) submitted for publication.
3. Dusenbery, D., Sheridan, R.E. and Russell, R.L. (1975) *Genetics* 80, 297.
4. Epstein, H.F., Waterston, R.H. and Brenner, S. (1974) *J. Mol. Biol.* 90, 291.
5. Epstein, H.F. and Thomson, N. (1974) *Nature* 250, 579.
6. Hall, D., Personal Communication.
7. Hedgecock, E.M. and Russell, R.L. (1975) *Proc. Nat. Acad. Sci. U.S.* 72, 4061.
8. Hedgecock, E.M., Johnson, C.D., Scherer, S., Rand, J.B. and Russell, R.L. (1977) submitted for publication.
9. Herman, R.K., Albertson, D.G. and Brenner, S. (1976) *Genetics* 83, 91.
10. Hirsh, D., Personal Communication.
11. Johnson, C.D. and Russell, R.L. (1975) *Anal. Biochem.* 64, 229.
12. Johnson, C.D. and Russell, R.L. (1977a) submitted for publication.
13. Johnson, C.D. and Russell, R.L. (1977b) submitted for publication.
14. Lewis, J.A. and Hodgkin, J.A. (1977) submitted for publication.
15. Sulston, J. (1976) *Phil. Trans. Roy. Soc.* (B) 275, 287.
16. Ward, S. (1973) *Proc. Nat. Acad. Sci. U.S.* 70, 817.
17. Ware, R.W., Clark, D., Crossland, K. and Russell, R.L. (1975) *J. Comp. Neurol.* 162, 71.
18. White, J.G., Southgate, E., Thomson, N. and Brenner, S. (1976) *Phil. Trans. Roy. Soc. London* (B) 275, 327.

STUDIES ON TWO BODY-WALL MYOSINS IN WILD TYPE AND MUTANT NEMATODES

Frederick H. Schachat, Harriet E. Harris*, Robert L. Garcea, Janice W. LaPointe and Henry F. Epstein

Department of Pharmacology, Stanford University Medical School
Stanford, California 94305

ABSTRACT. Native myosin purified from the wild-type, N2, and a body-wall defective mutant, E675, of the nematode contains two myosins, each homogeneous for different heavy chains. These myosins can be resolved from one another on hydroxyapatite and, when cleavaged with CNBr, they yield different peptide-fragments.

In E190, one of the homogeneous myosins is absent. e190 and e675 are alleles of the same gene, unc-54. The myosin lacking in E190 is the same one affected in E675. This suggests that unc-54 is the structural gene for a myosin heavy chain.

In order to determine the role of these different myosins, we plan to use antibodies to locate the myosins on thick filaments from body-wall muscle. Additionally, we are studying the patterns of synthesis and degradation of the two myosins in the wild-type and muscle-defective mutants in order to discover how the observed stoichiometry is maintained.

Introduction

In 1974, Epstein, Waterston and Brenner (2) proposed that there were two structural genes for the heavy chains of body-wall muscle myosin in the nematode, *Caenorhabditis elegans*. They were led to this suggestion by the finding that E675, a mutant defective in body-wall muscle structure had two different myosin heavy chains of 210,000 daltons and 203,000 daltons. This contrasted with the wild type, N2, which had only 210,000 dalton heavy chains.

In studies from this laboratory (4), we have confirmed that proposal and have demonstrated the existence of two myosin species from E675 each homogeneous for one of the different heavy chains. Analysis of CNBr-fragments from these two species shows that each myosin species has heavy chains coded for by a different structural gene.

* Present address: MRC Laboratory of Molecular Biology,
Hills Road,
Cambridge, CB2 2QH
England

Our ability to characterize these different myosin species by hydroxyapatite chromatography and CNBr-fragments enabled us to characterize myosin from both the wild type and E190 (another muscle-defective mutant in the same gene as E675) for the presence of the two myosins. Myosin from the wild type, N2, yielded a complex chromatogram. CNBr-peptide from different regions of the chromatogram demonstrated the presence of both homogeneous myosin species. In contrast to both E675 and N2 myosin, myosin from E190 is composed of only one of the two species. Its chromatography on hydroxyapatite yields a single chromatographic species and peptide maps indicate that it is lacking the CNBr-peptides characteristic of the 203,000 dalton heavy chain present in E675.

Also, we have prepared antibodies against myosin from the wild type nematode, N2. Our understanding of the nature of the mutations in E675 and E190 have enabled us to demonstrate antigenic differences between the two myosins.

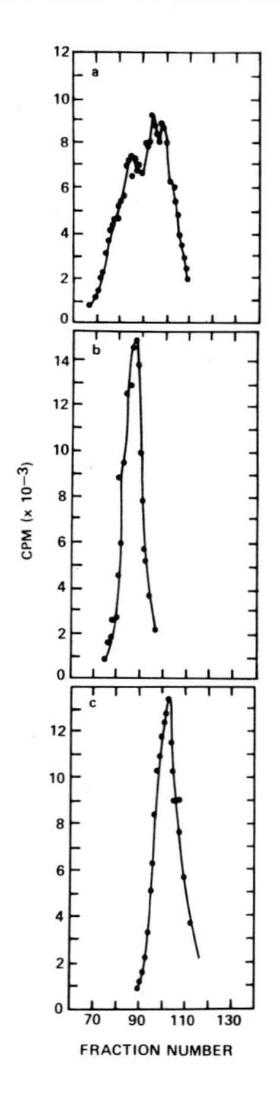

Fig. 1. Hydroxyapatite chromatography and rechromatography of ^{35}S-labelled myosin from E675. a) fractionation of E675 myosin. b) re-chromatography of fractions 73 to 81 of 1a. c) rechromatography of fractions 98 to 110 of 1a. Chromatography is described in Schachat *et al.* (4).

Two Homogeneous Myosins in E675

As can be seen in fig. 1a, when myosin from E675 is chromatographed on hydroxyapatite it yields a complex profile. There is a leading peak (fractions 70 to 89), a central region (fractions 92 to 96) and a trailing peak (fractions 98 to 110). When myosin from the pooled fractions 73 to 81 of region I is rechromatographed (fig. 1b) a chromatographically homogeneous myosin species is recovered. Analysis of the heavy chains of this material by 4.5% polyacrylamide-SDS gel electrophoresis shows that only the 210,000 dalton heavy chain is present. Similarly, when fractions 101 to 108 of region III are rechromatographed (fig. 1c) a chromatographically homogeneous myosin species which contains only the 203,000 dalton heavy chain is recovered.

The material in the central region appears to be composed primarily, perhaps exclusively, of complexes of the two homogeneous myosins and not a heterogeneous myosin species (myosin with one of each of the different molecular weight heavy chains).

Behavior of Wild-Type and E190 Myosin on Hydroxyapatite

When myosin from the wild type *C. elegans* is chromatographed on hydroxyapatite, it yields a complex profile similar to myosin from E675. However, in contrast to both N2 and E675 myosin, E190 myosin elutes as a single chromatographic species (fig. 2). That species behaves like the one recovered by rechromatography of myosin from region I of the E675 hydroxyapatite chromatogram.

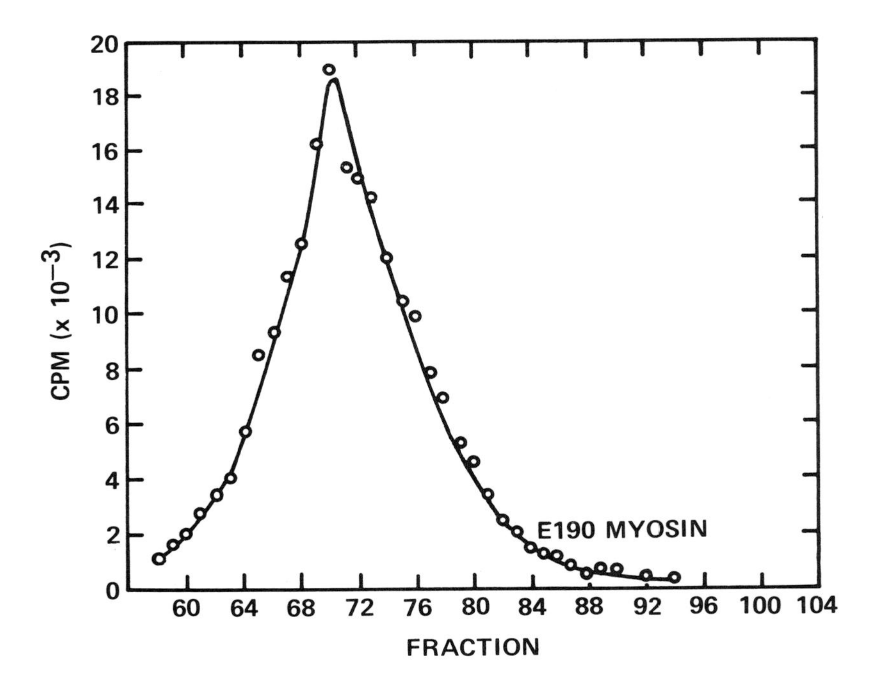

Fig. 2. Hydroxyapatite chromatography of ^{35}S-labelled myosin from E190. Chromatography is described in Schachat *et al.* (4).

CNBr-peptides of Myosin from Wild Type, E675 and E190 Show There are Two Homogeneous Myosins

When E675 myosin was fractionated preparatively on hydroxyapatite and CNBr-peptides generated from myosin eluting in regions I and III were compared, several characteristic differences were observed (4). Fig. 3a and b shows the patterns of peptides

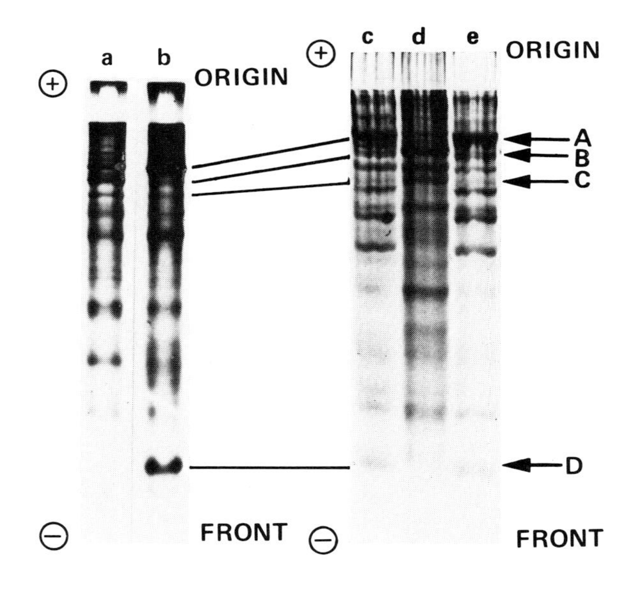

Fig. 3. 12% Polyacrylamide-8M Urea-SDS gel electrophoresis of CNBr-fragments of a) region I, and b) region III of a preparative hydroxyapatite fractionation of E675 myosin, c) N2 myosin, d) E190 myosin, e) E675 myosin. CNBr-peptides were prepared as in Schachat et al. (4). Gels run as described in Epstein and Wolff (1).

following 12% polyacrylamide-8M urea-SDS gel electrophoresis. The peptides A and D are characteristic of the 203,000 dalton homogeneous myosin, B and C of the 210,000 dalton homogeneous myosin. Several other peptides were also characteristic of the 203,000 dalton heavy chain (4). These differences are explicable most simply if the 203,000 dalton and 210,000 dalton myosin heavy chains of E675 are the products of different structural genes.

Analysis of the CNBr-peptides of myosin from the wild-type and E190 explains their chromatographic properties on hydroxyapatite. Preparative fractionation of wild type myosin and analysis of CNBr-peptides from areas corresponding to regions I and III of the E675 myosin chromatogram yields the same characteristic peptides seen from E675 (4). On the other hand, E190 myosin doesn't have the characteristic peptides A and D, (fig. 3d). Since those peptides are characteristic of the 203,000 dalton homogeneous myosin of E675, it follows that both the *e675* and *e190* mutations affect the same myosin heavy chain. The peptides from wild type myosin (fig. 3c) show that that heavy chain is present and since there are only 210,000 dalton myosin heavy chains in the wild type it must have a molecular weight of 210,000 in the wild type. The effect of e675 is to produce a lower molecular weight heavy chain and *e190* causes it to be absent.

Antibody to *C. elegans* Myosin

Antiserum to wild type myosin have been generated in this laboratory (Schachat and Garcea, manuscript in preparation). Myosin purified by the procedure of Harris and Epstein (3) was injected into rabbits and after several boosts or injections, serum was prepared. We have shown by both direct and indirect precipitation that this antiserum reacts with purified myosin. Further, as can be seen in fig. 4, we have shown that it is specific to myosin when reacted with a radio-labelled actomyosin extract in the presence of ATP and the antibody precipitate analyzed by electrophoresis on 10% polyacrylamide-SDS gels.

Antigenic Differences Between the Two Myosins

Our knowledge of the effects of the *e675* and *e190* mutations have allowed us to ask whether there are antigenic differences between the two homogeneous myosins. In E675, heavy chains from the different myosins can be distinguished by polyacrylamide-SDS gel electrophoresis. The heavy chain present in E190 is the same as the 210,000 dalton heavy chain in the E675 210,000 dalton homogeneous myosin. If the 210,000 dalton homogeneous myosin in E675 is antigenically distinct from the 203,000 dalton homogeneous myosin in E675, then addition of cold E190 myosin to mixtures with constant amounts of antimyosin sera and ^{35}S-E675 myosin should with excess E190 myosin result in the competition of radio-labelled 210,000 dalton E675 heavy chain from the precipitate. If all antigens are

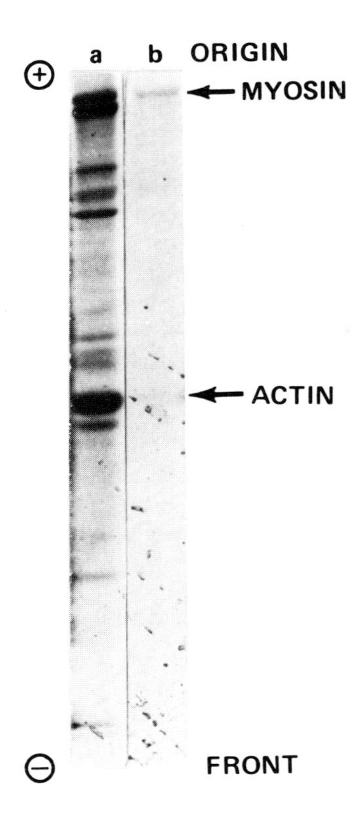

Fig. 4. Specificity of antimyosin sera in crude extract. Autoradiogram of 10% polyacrylamide-SDS gel of ^{35}S-labelled a) actomyosin extract of wild type animals, b) indirect precipitate of antimyosin sera with actomyosin extract. The indirect precipitate was collected by a modification of the procedure of Schimke et al. (5) (Schachat, Garcea and Epstein, manuscript in preparation). The preparation of the homogenate and actomyosin extract are described in Harris and Epstein (3).

common between the two heavy chains, then excess E190 myosin should compete equally with both radiolabelled heavy chains from the precipitate. As figure 5 shows, the 210,000 dalton heavy chain of E675 myosin can be competed completely, but the 203,000 dalton heavy chain of E675 myosin is only partially competed by E190 myosin. This indicates that the two homogeneous myosins both share some common antigenic sites and possess unique sites of their own.

The results of this preliminary study suggest the feasibility of isolating antibody specific for the myosin affected by both the e675 and e190 mutations, and work along these lines is in progress.

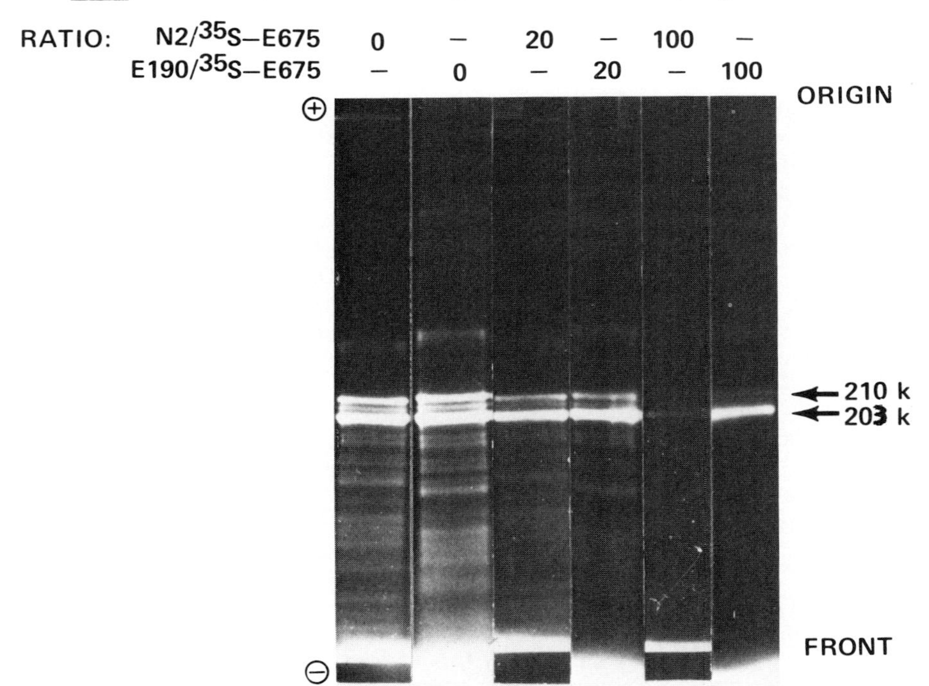

Fig. 5. Immunocompetition of ^{35}S-E675 myosin. Positive of autoradiogram of 4.5% polyacrylamide-SDS gel of competition of ^{35}S-E675 myosin in labelled actomyosin extract by unlabelled wild type, N2, actomyosin extract or E190 actomyosin extract. With constant volume, ^{35}S-E675 actomyosin extract, and antimyosin sera increasing amounts of wild type extract compete out the binding of both E675 myosin heavy chains while increasing amounts of E190 extract compete out only one of the heavy chains.

Discussion

The studies reviewed and presented here demonstrate:

1. The existence of two different structural genes for myosin heavy chains from body-wall muscle of *Caenorhabditis elegans*.

2. That like heavy chains can associate in the formation of native myosin to produce two different homogeneous myosins.

3. That two different mutations in the unc-54 gene, E675 and E190 affect the same myosin heavy chain.

4. That there are antigenic differences between the two homogeneous myosins.

All these studies have been done with native myosin (3) and we have shown that the hydroxyapatite fractionation yields myosin which is capable of polymerization, ATPase activity, and still has its light chains associated (4).

In order to determine the role of these different myosins, we plan to use antibodies to locate the myosins on thick filaments from body-wall muscle. Additionally, we are studying the patterns of synthesis and degradation of the two myosins in the wild-type and muscle-defective mutants in order to discover how the observed stoichiometry is maintained.

Acknowledgements

The work was performed during the tenure of Muscular Dystrophy Association Fellowships to F.H. Schachat and R.L. Garcea, a McCormick Fellowship through Stanford University School of Medicine to H.E. Harris, and a Mellon Foundation Fellowship and NICHD Career Development Award to H.F. Epstein. The research was supported by grants from the National Institute of Aging, from the American Heart Assocaition, Inc. through the Santa Clara County Chapter, and from the Muscular Dystrophy Associations of America, Inc.

References

1. Epstein, H.F. and Wolff, J.A. (1976). Anal. Biochem., 76, 157.

2. Epstein, H.F., Waterston, R.H. and Brenner, S. (1974). J. Molec. Biol., 90, 291.

3. Harris, H.E. and Epstein, H.F. (1977). Cell, in press.

4. Schachat, F.H., Harris, H.E. and Epstein, H.F. (1977). Cell, in press.

5. Schimke, R.T., Rhoades, R.E. and McKnight, G.S. (1974). Methods in Enzymology. Moldave and Grossman (Eds). Academic Press, New York, 30, 694.

THE ISOLATION OF A SUPPRESSIBLE NONSENSE MUTANT IN MAMMALIAN CELLS

M. R. Capecchi, R. A. Vonder Haar, N. E. Capecchi and M. M. Sveda

Department of Biology, University of Utah, Salt Lake City, Utah 84112

ABSTRACT. An HGPRT$^-$ cell line, derived from mouse L cells, has been shown to have the following properties: (1) It is CRM$^+$. (2) The defective HGPRT molecules are altered in the carboxyterminal peptide. (3) The mutant cells regain HGPRT activity when ochre-suppressor tRNA is microinjected into them, but not when amber-suppressor, or wild-type tRNAs are injected. We conclude from these properties that this mutant cell line contains an ochre nonsense mutation (UAA) in the structural gene for HGPRT.

INTRODUCTION

The isolation and characterization of nonsense mutants and nonsense suppressors in mammalian cells should provide a valuable new tool for genetic analysis of mammalian cells and their viruses. For example, the classification of a mutation as suppressible by a nonsense suppressor provides a strong criterion that the mutation is in a structural gene. This inference can be made because of the very nature of nonsense mutants and their suppressors. A nonsense mutation generates an in-phase, polypeptide-chain-termination codon (UAA, UAG, or UGA) in the interior portion of a structural gene. As a consequence of this mutation an aminoterminal polypeptide fragment, rather than the completed polypeptide chain is synthesized (Sarabhai _et al_., '64). In bacteria and yeast, suppressors of nonsense mutations arise from mutations in tRNA genes which permit the mutant tRNA to translate a termination codon as an amino acid codon (Capecchi and Gussin, '65; Engelhardt _et al_., '65; Goodman _et al_., '68; Capecchi, Hughes and Wahl, '75; Gesteland _et al_., '76). Thus, in the presence of a suppressor tRNA, the completed polypeptide product of a nonsense mutant can be synthesized. If the amino acid inserted at the site of the mutation does not markedly alter the protein structure, then an active gene product may be restored.

Nonsense mutants are clearly conditional lethal mutations since the physiological effect of the nonsense mutation can be studied in the presence and absence of the suppressor. It is of practical importance that this should be a very stringent conditional lethal system in mammalian cells. The expectation of stringency is based on the following considerations: a) Most NH_2-terminal polypeptide fragments should not exhibit biological activity. b) Such fragments are probably rapidly degraded in mammalian cells (Capecchi et al., '74), and c) cell-free protein synthesis experiments indicate that the level of "read-through" of a nonsense mutation in the non-permissive system is less in mammalian cell-free extracts than in the comparable bacterial extracts (Capecchi, Hughes and Wahl, '75).

The question of stringency of this potential conditional-lethal system is an important one since application of the other common conditional-lethal system, temperature sensitive mutations, to mammalian cell studies has met with some technical difficulty. The versatility of temperature-sensitive mutants has been limited because mammalian cells cannot be grown over as wide a temperature range as bacteria or yeast. As a result of this narrow range many temperature-sensitive mutants isolated in mammalian cells and their viruses are "leaky", that is, they exhibit measurable activity at the non-permissive temperature.

A further important property of the nonsense/nonsense-suppressor system is that the affected mutant gene product can often be identified since one can usually distinguish between the mutant polypeptide fragment and the suppressed, complete, polypeptide chain.

The strategy which we adopted to search for nonsense mutants and their suppressors in mammalian cells is to start by isolating a large number of mouse L cell lines deficient for the non-essential enzyme hypoxanthine-guanine phosphoribosyl transferase (HGPRT; E.C.2.4.2.8). This enzyme catalyzes the conversion of the purines hypoxanthine and guanine to their respective nucleotides IMP and GMP. It was selected as the target enzyme for these studies for a number of reasons. The enzyme can be readily assayed in cell-free extracts and in intact cells. Selective methods for isolating cell lines having lost or regained enzymatic activity had been developed (Szybalski, Szybalska and Ragni, '62; Littlefield, '63). Purine analogues, such as 8-azaguanine and 6-thioguanine, kill cells containing HGPRT and select for resistant cell lines which have reduced HGPRT activity. Such mutants remain viable because purine nucleotides can be synthesized by de novo pathways in the absence of HGPRT. Revertants, which have regained HGPRT activity, can be selected by blocking de novo purine biosynthesis with methotrexate and simultaneously supplying hypoxanthine. Such treatment renders the cells dependent on HGPRT for synthesis of purine nucleotides from hypoxanthine.

Because it seemed likely that most nonsense mutants would exhibit very little enzymatic activity, the purine analogue concentration (8-azaguanine plus 6-thioguanine) used for isolating the HGPRT$^-$ cell lines was chosen so that all of the surviving clones contained less than 0.1% of the parental HGPRT activity.

These mutant cell lines were then divided into two classes, those which had lost enzymatic activity but still retained protein which cross-reacted with antisera prepared against purified HGPRT (CRM$^+$) and those which had lost both enzymatic activity and immunological crossreactivity (CRM$^-$). The crossreacting material from the CRM$^+$ mutants could be analyzed by standard methods of protein chemistry to determine whether the CRM exhibited smaller subunit molecular weights relative to the parental HGPRT molecules. If certain CRM's appeared to be fragments, one could further ask whether the alteration occurred at the carboxyterminal end of the polypeptide chain as predicted for nonsense mutants. The CRM$^-$ cell lines, which are expected to contain most of the HGPRT$^-$ nonsense mutants, are more difficult to analyze. Two separate approaches are being used. First, the CRM$^-$ cell lines are being tested for sensitivity to phenotypic suppression (to be described). Second, assuming that the collection of CRM$^-$ mutants contains suppressible nonsense mutants, revertants from each of the CRM$^-$ mutant cell lines have been isolated and are being tested *in vitro* for suppressor tRNA activity.

RESULTS

Properties of the HGPRT$^-$ cell lines.

The HGPRT$^-$ mutants used in this study were selected, after mutagenesis with nitrosoguanidine, for resistance to the purine analogues 8-azaguanine and 6-thioguanine (Sharp, Capecchi and Capecchi, '73). As previously mentioned, each of these cell lines contains less than 0.1% of the HGPRT activity present in the parental mouse L cells. A specific antibody directed against highly purified mouse-liver HGPRT was used to detect CRM in these HGPRT$^-$ clones. Two methods for detecting CRM activity were used: (1) a standard precipitation inhibition assay (Suskind, '57) and (2) a radioimmune precipitation assay (Wahl, Hughes and Capecchi, '75). The latter assay proved to be much more sensitive for detection of altered HGPRT molecules. By these methods, 40% of the HGPRT$^-$ cell lines were shown to contain CRM. Since the amount of CRM varies among cell lines, the distinction of CRM$^+$ from CRM$^-$ lines must be an operational one, based on some arbitrarily chosen limit of detectable CRM. We have designated a cell line as CRM^+ if it contains greater than 1% of the amount of CRM present in the parental cell line in an equivalent assay. Examination of the physical properties of CRM from different mutant cell lines indicated that they arose from mutations at many different loci within the HGPRT structural gene (Wahl *et al.*, '75). The vast majority of these independently-isolated CRM's did, however, have subunit molecular weights indistinguishable from wild-type HGPRT. In retrospect this is not a very surprising result. We have examined the rates of degradation of several missense mutants of HGPRT and found that they are selectively degraded twenty to one hundred-fold faster than the wild type protein (Capecchi *et al.*, '74). These studies showed that even a small change in a protein, such as a missense mutation that leaves the protein still immunologically detectable, is sufficient to cause rapid selective degradation. Most nonsense mutants seem likely to be degraded even faster. Following this line of reasoning, the only nonsense mutants we might expect to escape rapid degradation and be detected as CRM^+ might be those resulting from the introduction of a nonsense codon very near the carboxyterminal end of the coding region. In fact, as shown in Fig. 1, the $HGPRT^-$ CRM^+ mutant which we will discuss in this manuscript is not resolved from wild-type HGPRT by electrophoresis in S.D.S-urea polyacrylamide gels. Nevertheless, we will demonstrate that this mutant contains an altered carboxyterminal tryptic peptide and that the chromatographic behavior of this peptide is consistent with its being shorter than the corresponding wild-type peptide. For the experiment shown in Fig. 1., mutant and wild type cells were labeled with $[^{35}S]$- and $[^{3}H]$- methionine respectively, extracts were prepared and immune precipitated with anti-HGPRT sera. The immune precipitates were then mixed and electrophoresed on SDS-urea polyacrylamide gels. The resolving power of this gel is approximately $\pm$ 700 daltons (i.e., $\pm$ six amino acids).

Tryptic peptide analysis of wild-type and mutant HGPRT.

The ability to resolve differences between mutant and wild-type HGPRT is greatly enhanced by examination of their tryptic peptides. Further, an analysis of the tryptic peptides should afford the opportunity to identify the carboxyterminal peptide. Since nonsense mutants generate NH_2-terminal polypeptide fragments, it is clearly important to look for changes in the carboxyterminal region of the protein.

The processing of the mutant and wild type HGPRT for tryptic peptide analysis includes labeling the proteins *in vivo* with the desired radioactive amino acids, preparing extracts, immunoprecipitating the HGPRT molecules and purifying

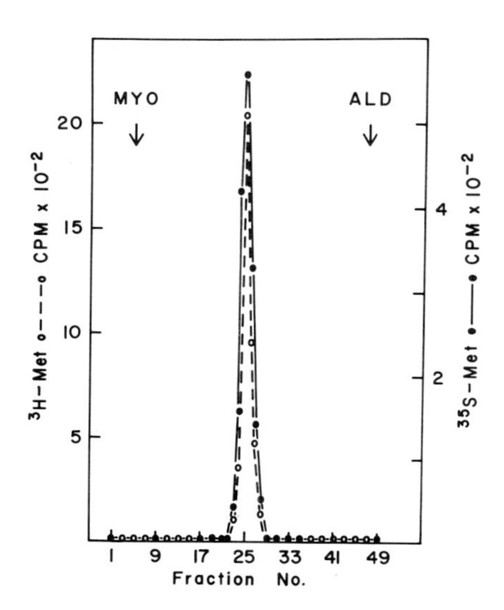

Fig. 1. Analysis of HGPRT isolated from L$^+$ and mutant (HGPRT$^-$, CRM$^+$) cell extracts by radioimmune precipitation and S.D.S.-urea polyacrylamide gel electrophoresis. The L$^+$ and mutant cell lines were labeled with [^{3}H]- and [^{35}S]-methionine respectively, the cell extracts were prepared and treated with antiserum against HGPRT and antibody against rabbit IgG. The radioimmune precipitates were then mixed and subjected to electrophoresis in a SDS-urea polyacrylamide gel. The gel was internally standardized with fluoresceinated myoglobin (MYO) and aldolase (ALD).

the immune precipitates by SDS-urea polyacrylamide gel electrophoresis. The mutant and wild-type HGPRT are then eluted from their respective gels, mixed and trypsinized with enzyme that has been treated with the chymotrypsin inhibitor TPCK. The tryptic peptides are separated by cation-exchange chromatography using a procedure similar to that developed by Milman, Krauss, and Olsen for analyzing the tryptic peptides of human HGPRT.

In Fig. 2 we illustrate such an experiment in which L$^+$ cells were labeled with [^{35}S]- methionine. The profile

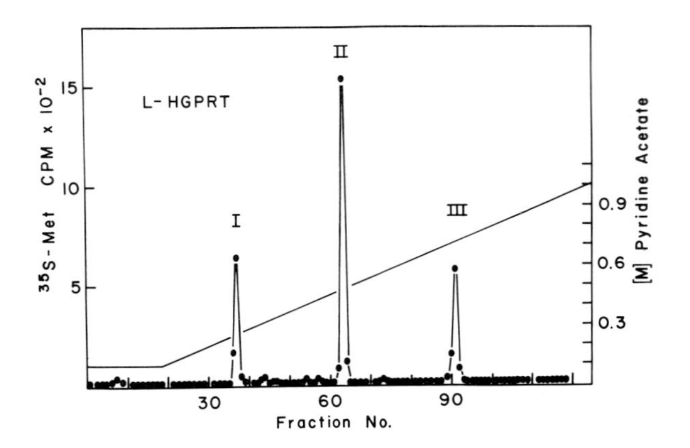

Fig. 2. Analysis of the methionine-containing tryptic peptides of HGPRT by cation exchange chromatography. L^+ cells were labeled with [^{35}S]- methionine. The labeled HGPRT was purified from the cell extract by immune precipitation and SDS-urea polyacrylamide gel electrophoresis. The HGPRT molecules eluted from the gel were digested with TPCK-treated trypsin and the peptides were separated by chromatography in Biorad-aminex 5 resin using a pyridine acetate gradient.

of the methionine-containing tryptic peptides of HGPRT is observed to be relatively simple. Peptides I and III contain one methionyl residue whereas peptide II appears to contain two. There is a fourth methionine containing peptide (illustrated in Fig. 5 and 6) which requires strong base to be eluted from the column. These results are consistent with the amino acid composition analysis of purified mouse-liver HGPRT which indicated that the protein contains five methionyl residues per 27,000 daltons of protein (unpublished results).

In Fig. 3, we compare the elution profile of the methionine containing tryptic peptides of HGPRT isolated from the mutant, labeled with [^{3}H]- methionine and wild type cells, labeled with [^{35}S]- methionine. The mutant and wild type HGPRT molecules were mixed prior to digestion with trypsin and then chromatographed on the cation exchange column. Mutant peptide I is observed to elute from the column before the corresponding wild type peptide from which we conclude

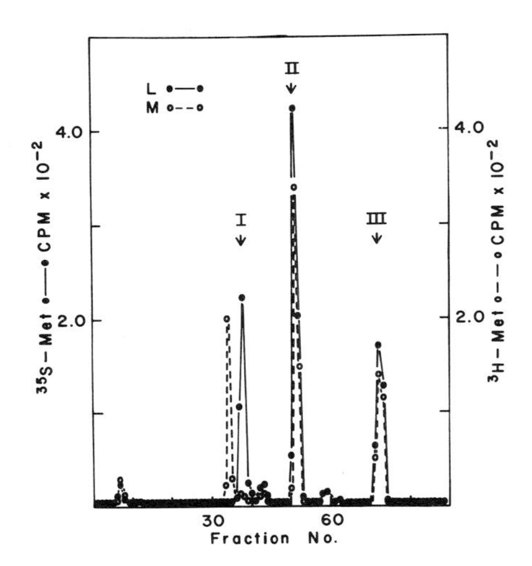

Fig. 3. Analysis of the methionine-containing tryptic peptides of HGPRT isolated from L^+ and mutant (M) cell extracts. L^+ and mutant cells were labeled with [^{35}S]- and [^3H]- methionine respectively. The wild type and mutant HGPRT were purified from cell extracts by immune precipitation and SDS-urea polyacrylamide gel electrophoresis. The purified protein eluted from the gels were mixed, digested with TPCK-treated trypsin, and analyzed as described in Fig. 2.

that the mutant peptide is either shorter than, or more negatively charged than the wild type peptide, or both.

The tryptic peptide profile shown in Fig. 4 is from an experiment similar to that shown in Fig. 3, except that the radioactive labels were reversed. Again it is observed that the mutant peptide I elutes from the column ahead of the wild-type peptide.

If the mutant in question is a nonsense mutant, the prediction is that peptide I is the carboxyterminal tryptic peptide of HGPRT. That this is so, is shown in Figs. 5 and 6. Fig. 5 shows that the methionine-containing peptides II

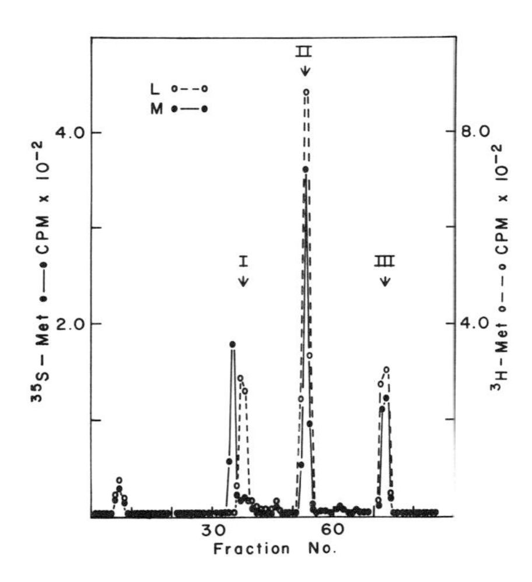

Fig. 4. Further analysis of the methionine-containing tryptic peptides of HGPRT isolated from L^+ and mutant (M) cell extracts. The experiment depicted in this figure was as described in Fig. 3 except that the radioactive labels were reversed.

and IV are lysyl tryptic peptides. Fig. 6 shows that peptide III contains arginine. However, peptide I contains neither lysine nor arginine. Since trypsin specifically cleaves proteins after lysine and arginine and peptide I contains neither, peptide I must be the carboxyterminal tryptic peptide of HGPRT.

These results are consistent with this CRM^+ mutant being a nonsense mutant, however, a word of caution should be entered at this point. Showing that a mutant protein has a smaller subunit molecular weight than the wild type protein and even showing that the alteration occurs at the carboxyterminal end of the protein is consistent with but not a proof that the mutation is a nonsense mutation. Alternative explanations include that the mutation is the result of a small internal deletion at the carboxyterminal end, a frameshift mutation, or an error in post-synthetic

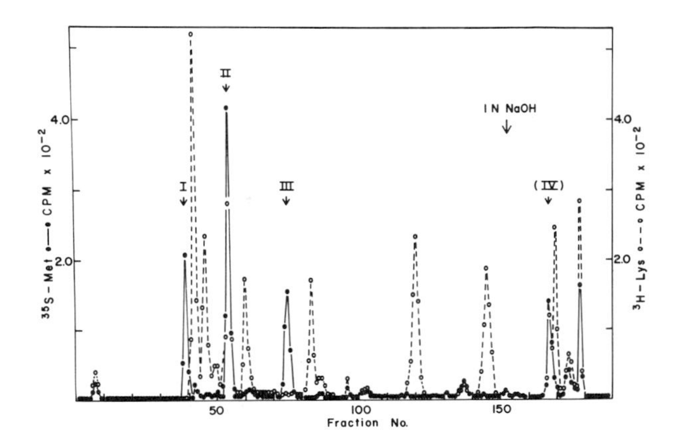

Fig. 5. Analysis of the lysine- and methionine-containing tryptic peptides of HGPRT. Mouse L^+ *cells were separately labeled with* [^{35}S]- *methionine and* [^3H]- *lysine. The labeled HGPRT was purified, trypsinized and the peptides were analyzed as described in Fig. 2. The methionine-containing peptides II and IV also còntain lysine. A non-methionine-containing lysyl peptide elutes just after peptide IV. The reason for believing that peptide IV is a legitimate tryptic peptide of HGPRT is a) it is always found to be present in a ratio of one to one relative to peptides I and III, and b) the ratio of methionine to lysine in peptide IV is one and not less than one. The radioactive material beyong peptide IV, which requires 1N NaOH for elution, may represent more complex tryptic products (i.e., unresolved peptides or polypeptides which have not been completely hydrolyzed by trypsin).*

processing. Indeed, we believe that HGPRT is processed from a larger precursor. This is of interest since HGPRT does not appear to be a protein destined for export. Thus processing of mammalian proteins may be a much more general phenomenon than previously thought.

<u>Processing of HGPRT?</u>

As previously mentioned, the majority of $HGPRT^-$ CRM^+ mutants have subunit molecular weights indistinguishable from wild type protein. However, we have isolated mutants with altered subunit molecular weights, and while some are

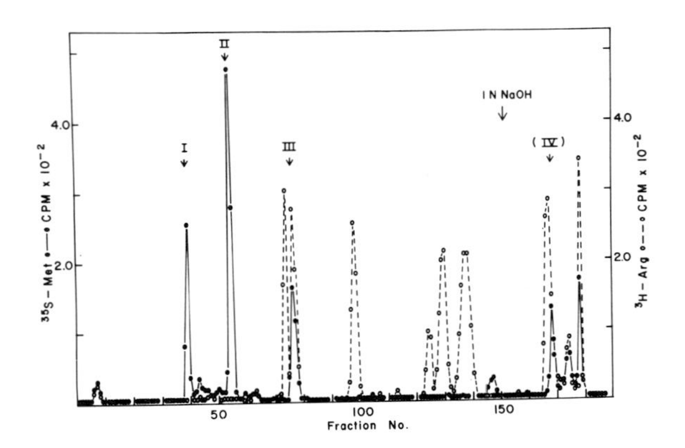

Fig. 6. Analysis of the arginine and methionine containing tryptic peptides of HGPRT. The analysis is as described in Fig. 5 except mouse L^+ cells were labeled with [^3H]- arginine instead of [^3H]- lysine. A non-methionine-containing arginyl tryptic peptide elutes just ahead of peptide IV. Peptide III is observed to contain arginine.

smaller, others are <u>larger</u> than wild type HGPRT. Examples of such mutants are illustrated in Fig. 7. The CRM's shown migrate respectively faster than, the same as, and slower than wild-type HGPRT on calibrated SDS-urea polyacrylamide gels. The altered migration of the mutants could reflect altered levels of post-synthetic chemical modification of HGPRT (carbohydrate addition, phosphorylation, adenylation, etc.). To date, however, we have failed to generate any data which would support the idea that HGPRT is post-synthetically chemically modified. Therefore, we believe that the altered mobilities of the CRM's in S.D.S-urea polyacrylamide gels reflect altered subunit molecular weights of the mutant proteins. We believe that most of these mutants are not nonsense mutants for a number of reasons. First, some are clearly larger than wild type HGPRT. Second, some of the revertants of such mutants retain their altered subunit molecular weight. Third, they are not sensitive to phenotypic suppression. Rather, we believe that many of these mutants with altered subunit size represent structural gene mutants, which as a result of the mutation, have acquired alternate processing sites. Proof of this hypothesis

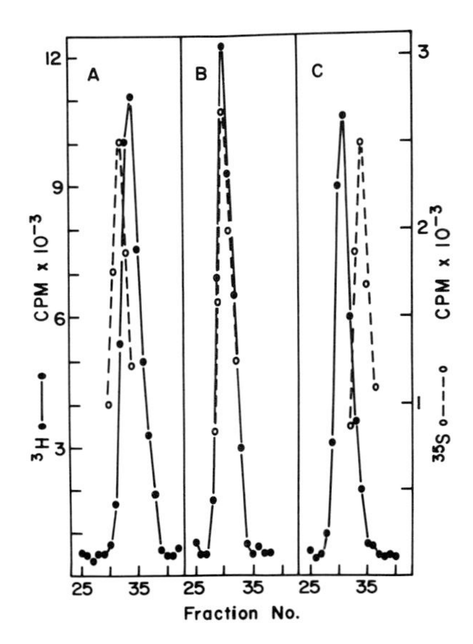

Fig. 7. Analysis of the electrophoretic mobility of three HGPRT$^-$, CRM$^+$ mutants in SDS-urea polyacrylamide gels. Mouse L$^+$ cells were labeled with [^{3}H]- lysine and HGPRT$^-$ CRM$^+$ mutant cells with [^{35}S]- methionine. The wild type and mutant HGPRT molecules were purified and subjected to electrophoresis in SDS-urea polyacrylamide gels as described in Fig. 1.

will require primary structural analysis of the affected proteins.

One of the reasons for introducing the topic of processing of HGPRT is to point out the dangers of relying solely on altered subunit molecular weight as a criterion for a nonsense mutation and to emphasize the need to develop additional criteria. The approach which we have used to provide a second criterion is to introduce by microinjection, suppressor tRNA isolated from bacteria and yeast into the mutant cell line and to look for restoration of HGPRT activity (i.e., a type of phenotypic suppression).

Microinjection experiments

Drs. Schlegel and Rechsteiner, as well as Dr. Loyter and his colleagues, have developed a method for microinjection of macromolecules into mammalian cells in culture (Schlegel and Rechsteiner, '75; Loyter, Zabai and Kulka, '75).

The method involves fusing red blood cells preloaded with macromolecules to the mammalian cells using U.V.-inactivated Sendai virus. The loading of the red blood cells with macromolecules is accomplished by hypotonic hemolysis of the red blood cells in the presence of the desired macromolecules. The details of loading and fusing of the red blood cells differ as described by the two groups. We have more closely followed the procedures of Schlegel and Rechsteiner, '75.

To be certain that microinjection was achieved under our experimental conditions, we have injected HGPRT molecules into the mutant cell line. The macromolecules within the red blood cells were removed by a prelysis step; the cells were then resealed, loaded with HGPRT, and fused to the $HGPRT^-$ cell line (Fig. 8B). For comparison red blood cells not loaded with HGPRT were also fused with the same cell line (Fig. 8A). HGPRT activity present in the two populations of mouse L cells after fusion was determined by

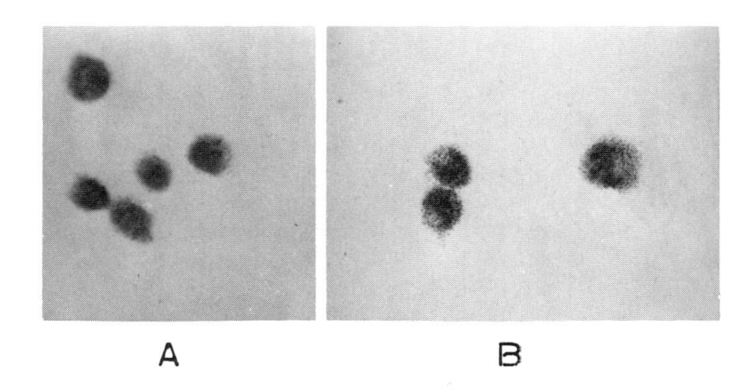

Fig. 8. Autoradiographic analysis of the HGPRT activity present in $HGPRT^-$ mouse L cells after fusion with red blood cells containing (B) and not containing (A) HGPRT molecules. Human red blood cells were hypotonically lysed in the presence of a 20-fold excess of buffer to remove internal macromolecules, resealed and loaded with purified HGPRT (B) and bovine serum albumin (A). The loaded red blood cells were fused to the $HGPRT^-$ cell line using U.V.-inactivated Sendai virus. After fusion, the cells were plated onto glass cover slips and incubated at 37°C for 36 hours. HGPRT activity was then measured by incubating the cells with [^3H]- hypoxanthine, fixing the cells with methanol, washing the cells with 10% TCA, and overlaying the cells with autoradiographic emulsion. The autoradiograms were exposed in the dark for 72 hours and developed.

incubating the cells with [^{3}H]- hypoxanthine, fixing the cells, washing the fixed cells with 10% TCA to remove the unincorporated hypoxanthine, and analyzing the level of incorporation by autoradiography. The cells which were fused with HGPRT-loaded red blood cells have approximately 10-fold greater density of exposed silver grains overlying them. The grain density over the control cells (Fig. 8A) is typical for this cell line under comparable incubation and exposure conditions whether or not the cells have been fused with unloaded red blood cells.

Assuming a Poisson distribution, an analysis of the number of null cells, those with low grain density suggesting they had not successfully fused with any HGPRT-loaded red blood cells, indicated that on the average we were fusing one loaded red blood cell per fibroblast cell.

In Table I a quantitative analysis of the HGPRT microinjection experiment is given. The HGPRT activity loaded

HGPRT Loaded into Red Blood Cells	^{3}H-Hx Inc. by HGPRT$^-$ Fibroblast After Fusion (CPM)	% of L-Cell Activity
0	844	0.08
lx	9,940	1.0
2x	20,720	2.0

Table I. Microinjection of HGPRT into the HGPRT$^-$ mouse L cell line. Human red blood cells were hypotonically lysed in the presence of a large excess of buffer to remove the internal macromolecules and resealed. The red blood cells were then loaded with three concentrations of purified HGPRT (0, 1X and 2X). The units of HGPRT are given as the concentration present in L$^+$ cells. The HGPRT activity transferred to the mutant L cell line was determined by measuring in culture the incorporation of [^{3}H]- hypoxanthine into TCA precipitable product.

into the red blood cells is given in units of concentration (act/unit Vol) relative to the activity in wild type mouse L cells. It is observed that the amount of HGPRT activity transferred to the mutant cell line is proportional to the amount of HGPRT loaded into the red blood cells. The volume of a red blood cell is approximately 1% of the volume of a mouse fibroblast cell. Thus, it is also seen that the amount of HGPRT activity transferred is consistent with the fusion, on the average, of one red blood cell per fibroblast cell.

Confident that we could microinject HGPRT into the mutant cell line, we could now test the effect of microinjecting suppressor tRNA's into our mutant cell line. In Fig. 9 the results of microinjecting partially purified tRNA from three isogenic strains of *E. coli* into the mutant cell line is given. The strains were S26, the suppressor-minus parental strain, S26R1E, which contains an SuI amber suppressor, and PS2 which contains the corresponding ochre suppressor. The tRNA's were purified using benzoylated-DEAE-cellulose chromatography as previously described (Capecchi, Hughes and Wahl, '75). HGPRT activity was measured 36 hours after fusion with the loaded red blood cells. It is observed that microinjection of ochre suppressor tRNA into the $HGPRT^-$ cell line specifically restores HGPRT activity.

In Fig. 10 we show a similar experiment except that the sources of tRNA's were three closely related strains of yeast, *S. cerevesiae*. Again it was observed that only the yeast ochre suppressor tRNA restores HGPRT activity to the mutant cell line.

DISCUSSION

We have described an $HGPRT^-$ mutant cell line derived from mouse L cells with the following properties:

1. It contains approximately 0.08% of the HGPRT activity present in the parental cell line.
2. It is CRM^+.
3. It has an altered carboxyterminal tryptic peptide. We will argue that its chromatographic behavior is consistent with this peptide being shorter than the corresponding wild-type peptide.
4. The mutant can be phenotypically corrected by microinjecting into the cells ochre-suppressor tRNA from either bacteria or yeast, but not by injecting amber-suppressor or wild-type tRNA from the same sources.

From these properties we conclude that this mutant contains an ochre nonsense mutation in the HGPRT structural gene.

The obvious next step is to analyze a series of revertants of this mutant for suppressor tRNA activity. For this

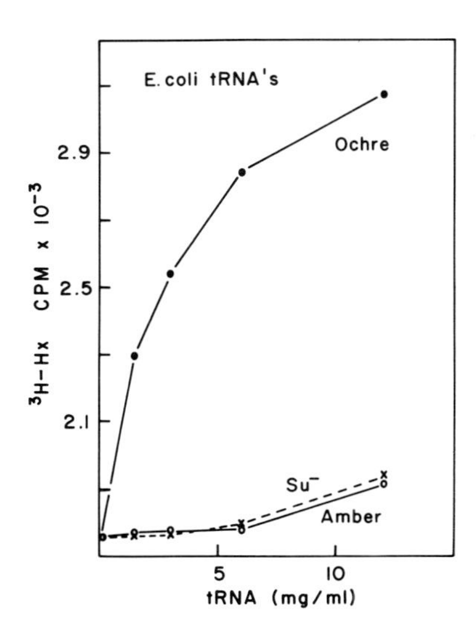

Fig. 9. Restoration of HGPRT activity to the HGPRT⁻ cell line by microinjection of E. coli ochre suppressor tRNA. Human red blood cells, loaded with increasing concentrations of tRNA, were fused to the mutant L cell line using U.V.-inactivated Sendai virus. The tRNA was isolated from three strains of E. coli: S26, S26R1E and PS2 which contain no suppressor, an amber suppressor and an ochre suppressor respectively. The units of tRNA along the abscissa are given as the concentration of tRNA loaded into the red blood cells. HGPRT activity after fusion was measured in culture by determining the level of [^{3}H]- hypoxanthine incorporation into TCA precipitable product.

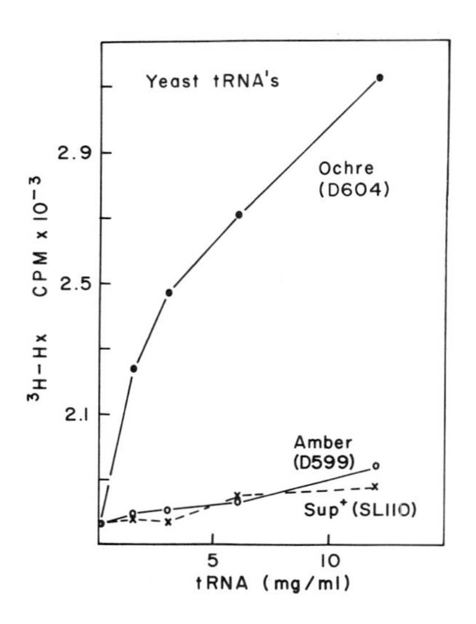

Fig. 10. Restoration of HGPRT activity to the HGPRT⁻ cell line by microinjection of yeast ochre suppressor tRNA. The experimental details were as described in Fig. 9, except the sources of tRNA were three related strains of yeast: SL110-4D, D599-4B and D604-7D containing no nonsense suppressor, an amber suppressor and an ochre suppressor respectively.

purpose we have developed in vitro suppression assays utilizing mammalian protein-synthesizing components (Capecchi, Hughes and Wahl, '75; and unpublished results). The mutant does revert to HGPRT$^+$ at a frequency, after mutagenesis, of 10^{-7}.

To date, the data indicating that the mutant peptide is shorter than the corresponding wild-type peptide is not complete. The argument favoring this interpretation, however, is the following. We have shown that the mutant peptide is shorter, more negatively charged or both. The pH of the column at the elution position is sufficiently low that the peptide could not be more negatively charged by having acquired an acidic amino acid residue. Further, it could not be more negatively charged by having lost a basic residue since the wild type peptide does not contain lysine or arginine. It could, however, have become more negatively charged by losing a histidine residue. This possibility is currently being tested. If the latter possibility proves not to be the case, then by elimination, one is left with the hypothesis that the mutant peptide is shorter than the wild-type one. Direct testing of this hypothesis will have to await sequence analysis.

A number of investigators have reported on shortened mutant gene products in higher eucaryotes and their viruses. Epstein *et al.* have reported on a mutant nematode which contains a shortened myosin peptide. Adetugbo, Milstein and Secher have described a mutant myeloma cell line which synthesizes a shortened immunoglobulin H chain, and Summers *et al.* have reported on herpes simplex viral mutants containing shortened thymidine kinase peptides. Some of these mutants have subsequently been shown to have been generated by internal deletions or frameshift mutations. Most recently, Dr. Gesteland and his co-workers have isolated nonsense mutants in an SV40-adeno virus hybrid (personal communication). These workers can identify the cell-free translation product of the affected gene and suppress it *in vitro* by the addition of the appropriate suppressor tRNA.

ACKNOWLEDGMENTS

We should like to thank Dr. M. Rechsteiner for introducing us to the methodology of microinjection and Mr. T. Dayhuff for expert technical assistance. We should also like to acknowledge past members of our laboratory, Drs. J. D. Sharp, S. H. Hughes and G. M. Wahl, who have contributed to different phases of this project. This investigation has been supported by a grant from the USPHS #GM 21168. This work was done while R.V.H. was post doctoral fellow of the American Cancer Society and USPH. M.R.C. is supported by a faculty research award from the American Cancer Society.

REFERENCES

Adetugbo, K., Milstein, C., and Secher, D. S., (1977) *Nature*, 265, 299.

Capecchi, M. R., and Gussin, G., (1965) *Science*, 149, 417.

Capecchi, M. R., Capecchi, N. E., Hughes, S. H., and Wahl, G. M., (1974) *Proc. Nat. Acad. Sci. USA*, 71, 4732.

Capecchi, M. R., Hughes, S. H., and Wahl, G. M., (1975) *Cell*, 6, 269.

Engelhardt, D. L., Webster, R. E., Wilhelm, R. C., and Zinder, N. D., (1965) *Proc. Nat. Acad. Sci. USA*, 54, 1791.

Epstein, H. F., Waterson, R. H., and Brenner, S., (1974) *J. Mol. Biol.*, 90, 291.
Gesteland, R. F., Wolfner, P., Grisafi, ., Fink, G., Botstein, D., and Roth, J. R., (1976) *Cell*, 7, 381.
Goodman, H. M., Abelson, J., Landy, A., Brenner, S., and Smith, J. D., (1968) *Nature*, 217, 1019.
Littlefield, J. W., (1963) *Proc. Nat. Acad. Sci. USA*, 50, 568.
Loyter, A., Zakai; N., and Kulka, R., (1975) *J. Cell Biol.*, 66, 292.
Milman, G., Krauss, S. W., and Olsen, A. S., (1977) *Proc. Nat. Acad. Sci., USA*, 74, 926.
Sarabhai, A. S., Stretton, A., Brenner, S., and Bolle, A., (1964) *Nature*, 201, 13.
Schlegel, R., and Rechsteiner, M., (1975) *Cell*, 5, 371.
Sharp, J. D., Capecchi, N. E. and Capecchi, M. R., (1973) *Proc. Nat. Acad. Sci. USA*, 70, 3175.
Summers, W. P., Wagner, M., and Summers, W. C., (1975) *Proc. Nat. Acad. Sci. USA*, 72, 4081.
Szybalski, W., Szybalska, E. H., and Ragni, G., (1962) *Nat. Cancer Inst. Monogr.*, 7, 75.
Wahl, G. M., Hughes, S. H., and Capecchi, M. R., (1975) *J. Cell Phys.*, 85, 307.

CYCLIC AMP, MICROTUBULES, MICROFIBRILS, AND CANCER

Theodore T. Puck

Eleanor Roosevelt Institute for Cancer Research
4200 E. 9th Ave., Denver, Colorado 80262

ABSTRACT. Transformed cells *in vitro* can acquire morphological characteristics, growth behavior, and an organized microtubular structure resembling those of the normal fibroblast, when treated with agents which raise the cellular level of cyclic AMP. The process has been called Reverse Transformation. The transformed state of the CHO K1 cell possesses microtubules which are present in sparse numbers and random distribution within the cell cytoplasm. Normal fibroblasts and the fibroblast-like forms developing after Reverse Transformation possess a highly organized microtubular system in which the microtubules traverse the length of the cell in parallel to its long axis. Immunofluorescence experiments indicate that the microfibrils participate in the organized structure displayed by the microtubules in the normal cell, and its disorganization in the transformed cell. Colcemide and cytochalasin B which respectively disrupt microtubules and inactivate microfibrils can re-convert cells in the normal or Reverse Transformed state to that resembling the transformed cell. It is postulated that malignancy can result from disruption of the organized arrangement of microtubules and microfibrils in the normal fibroblast, and that the organized form of this system permits interaction with the genome so as to limit reproduction to normal body levels. Other kinds of malignancy may also exist. Evidence is presented suggesting that specific patterns of the microtubular-microfibrillar apparatus may be associated with particular differentiation states. The state of organization of the microtubular-microfibrillar pattern of the CHO K1 cell is associated with changes in membrane agitation or tranquility, activity of cell surface antigens, and effectiveness of active transport of C^{14}-α amino butyrate. Aspects of cell behavior such as the nature of violently oscillating membranous knobs of the transformed cell are explained by means of these considerations.

INTRODUCTION

Normal fibroblasts transformed to the malignant state usually undergo characteristic changes in morphology and behavior. The cells become compact and pleomorphic instead of exhibiting the elongated, dipolar, spindle-shaped form of the fibroblast; the outside of the cell is frequently studded with knobs or blebs instead of presenting a reasonably smooth surface; the cells grow in random fashion so that colonies arising from single cells exhibit a roughly circular outline instead of displaying the characteristic whorls associated with the "sheaves of wheat" (1) pattern of fibroblastic colonies; and single cells in the transformed state form colonies with an efficiency in the neighborhood of 100% either when attached to solid surfaces or when grown in agar suspension, whereas normal fibroblasts grow only on surfaces and not at all in suspension.

The Chinese hamster ovary cell (CHO-K1), a permanent cell line which has been under study in these laboratories for many years, exhibits all of these transformation stigmata when grown in standard growth medium. In a series of studies, it was demonstrated that agents which increase the level of adenosine cyclic 3':5'-monophosphate (cAMP) in these cells cause loss of the transformation characteristics listed above so that the cell acquires morphology and behavior much more closely approximating those of the normal fibroblast. This cAMP-mediated process was named Reverse Transformation (2) (3) (4) (5) (6). (Fig. 1).

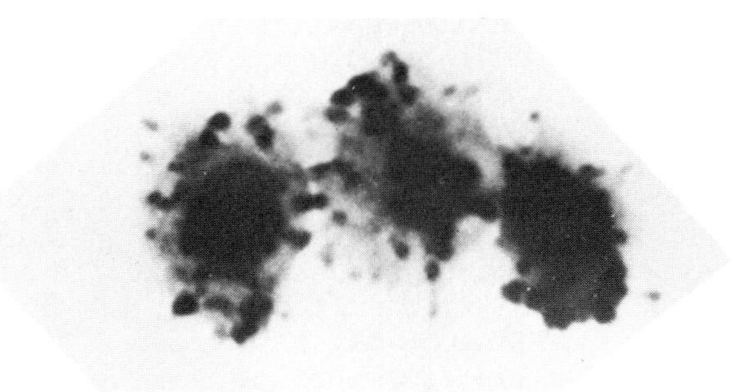

1A

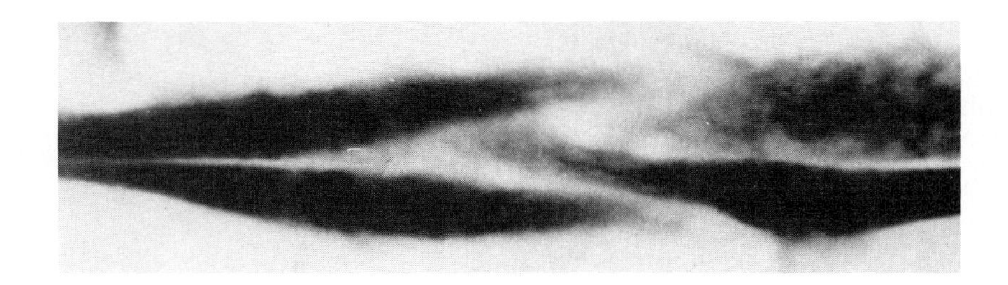

1B

Fig. 1A. CHO-K1 cells in normal growth medium. The cells are compact, pleomorphic, heavily knobbed, and display little tendency to associate.
Fig. 1B. The same cells as in 1A treated with DBcAMP plus testololactone as explained in the text. The cells are elongated in fibroblastic form, the knobs have disappeared leaving a smooth surface, and the cells tend to associate in parallel to their long axis.

The study has continued in an effort to understand the nature of these changes and their implications for the malignant process.

RESULTS AND DISCUSSION

Reverse Transformation can be effected by 10^{-3} M dibutyryl cyclic AMP (DBcAMP) alone or by $1x10^{-4}$ M DBcAMP plus a synergizing agent like testosterone (5), testololactone (7), or prostaglandin E (5). Cyclic AMP alone is ineffective but becomes highly so in the presence of an inhibitor of phosphodiesterase action like theophylline. The 8-Bromo derivative of cAMP which is resistant to the hydrolytic action of phosphodiasterase is highly effective by itself. Experiments like these demonstrate that Reverse Transformation is not due to the release from dibutyryl cyclic AMP of butyrate, a substance which by itself has some effect on cell morphology, but which displays differences from those of Reverse Transformation even though certain common elements may be involved (8). The action of the synergistic reagents here employed can be explained by their inhibition of phosphodiesterase action although the experimental proof is not yet complete. Cyclic GMP had no observable effect of its own nor did it antagonize the Reverse Transformation reaction brought about by cyclic AMP compounds (7).

Reverse Transformation has been shown to occur in a variety of transformed cells so it is not a phenomenon peculiar to the CHO culture. The morphological changes produced suggested that under the stimulus of an increased cyclic AMP concentration, skeletal structures inside the cell become organized so as to produce the resulting elongation. The obvious candidates for such an action would appear to be the rigid cellular microtubules. This supposition was supported by experiments which demonstrated that the Reverse Transformation could be prevented by the addition of colcemide or other agents which specifically disrupt microtubules (2). Further demonstration of the microtubular involvement in this process was carried out by means of transmission electron microscopy. It was found that in the native state the CHO-K1 cell exhibits cytoplasmic microtubules which are sparse in number and oriented randomly with respect to each other. When DBcAMP in appropriate concentration is added to the medium, a highly organized and dense array of microtubules arises, in which the individual tubules tend to become aligned in parallel to each other and to the long axis of the cell. The presence of such a highly ordered pattern in the cell with fibroblast-like morphology as opposed to the random condition of the transformed cell suggests that loss of the organized pattern is associated with acquisition of malignancy.

The assumption of a patterned, parallel arrangement of microtubular strands was confirmed by Dr. Brinkley, using immunofluorescence with an antiserum produced against tubulin (9).

It seemed reasonable to expect that the contractile actin-containing microfibrils in the cells might also participate in Reverse Transformation since the action produced appears to involve contractile-like motions. Consequently, experiments were carried out with cytochalasin-B, an agent known to interfere with microfibrillar functions to determine whether microfibrillar action is involved in this effect. It was found that cytochalasin-B, like colcemide, does indeed also antagonize Reverse Transformation and hence Reverse Transformation was interpreted as involving a common system involving both microtubules and microfibrils (5). Further confirmation of microfilament involvement has been obtained by immunofluorescent studies with antibodies produced by the rabbit against chick-actin (kindly supplied by Dr. Lazarides). These demonstrated that in the Reverse Transformation reaction, a change in the distribution of actin in the cells also occurs. The actin is converted from a relatively disorganized distribution into an orderly system of parallel strands resembling that observed with the microtubules. The conclusion appears substantiated then that microfibrils are also involved in the Reverse Transformation reaction and participate with the microtubules in production of the organized structure which is assumed by the cell under the influence of cAMP.

Experiments demonstrated that no new protein synthesis is required to bring about the Reverse Transformation reaction (10); therefore, the action of cyclic AMP brings about organization of tubulin and actin-containing protein already synthesized inside the cells. The nature and constitution of the knobs distributed over the entire surface of the cell in its transformed state but absent (or confined to the terminal polar regions) in the normal fibroblast-like form are of interest. Similar knobs or blebs have been described in a number of malignant cells. Experiments were undertaken with time-lapse cine photomicrography to examine the dynamics of disappearance and reappearance of the knob as the cells were taken in and out of the Reverse Transformed state.

A surprising result of these studies was that these knobs are not static but exist in a state of continuous agitation, executing oscillatory movements in the cell membrane (5). As a consequence, the membrane presents a violently boiling aspect. While the individual knob movements display fairly large variations, their amplitude lies roughly in the neighborhood of 1/5-1/10 of the cell diameter, and the time of a complete back and forth oscillation involves approximately 15-50 seconds (6). Consequently, the cell in its native transformed state exhibits a violently agitated membrane in contrast to the relatively tranquilized state of the Reverse Transformed cell. The action of cytochalasin B is not identical to that of colcemide, but tends even more than does the latter compound to produce violently agitating knobs in place of smooth cell surfaces. Indeed, cytochalasin B will frequently produce such knobs on cells which have retained their elongated fibroblast-like morphology, a phenomenon not observed with colcemide. This observation ruled out an explanation of the knob appearance or disappearance respectively as being due to the presence in the membrane of the transformed cell of loose pockets of membranous material which would stretch out when the cell assumed the elongated fibroblast-like state. This interpretation was also clearly made untenable by the time-lapse photographic studies. If disappearance of the knobs during Reverse Transformation is due to stretching of the cell membrane which takes up the slack constituting the knobs, one should expect disappearance of the knobs to proceed either simultaneously with or after the cell elongates. The movies, however, demonstrate clearly that knob disappearance is the <u>first</u> phenomenon which occurs after addition of the Reverse Transformation reagents. The violently oscillating knobs disappear, giving rise to smooth membranes before any other change in cell morphology can be observed. Consequently, instead of being a result of the stretching reaction, knob disappearance must reflect some earlier step in the Reverse Transformation sequence.

A clue to the dynamics of knobbing was afforded by immunofluorescence experiments. When cells in the transformed condition are treated either with anti-tubulin or antiactin sera, and then with a fluorescent antiserum to the first antibody, an increased intensity of fluorescence is observable within the knobs themselves. The effect seems even more pronounced when antiactin is used as the first antibody.

Knobs can even be induced in profusion in normal fibroblasts by the addition of 1 μg/ml of cytochalasin B. If such cells are treated with antiactin serum followed by fluorescent antibody to the latter, the knobbed regions become highly luminous indicating the intense concentration of actin in these bodies. Almost all of the discernable cell actin is confined to the knobs. (Fig. 2)

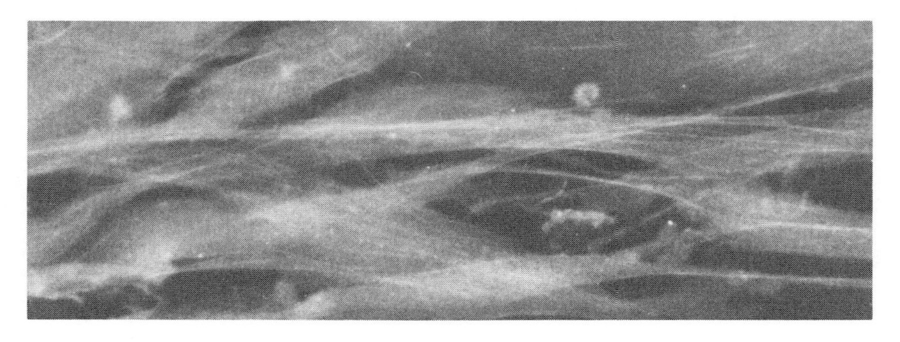

2A

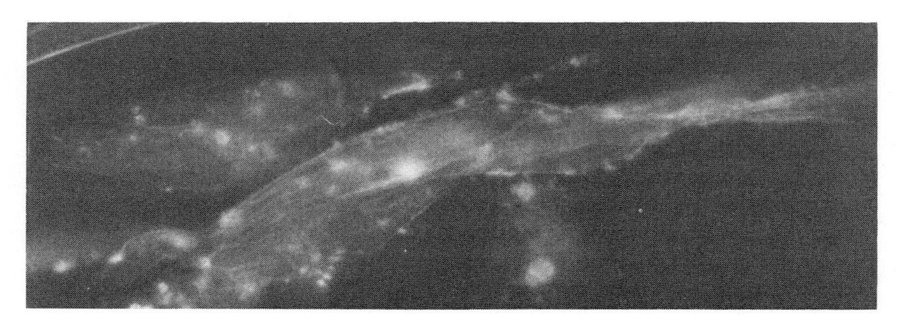

2B

Fig. 2A. Normal Chinese Hamster fibroblast culture (CHO-III) treated with rabbit-antiactin serum and then a fluorescent antibody to rabbit gamma globulin and illuminated and photographed by means of its own fluorescence. Few knobs are visible and large amounts of actin-containing strands traverse the entire cell length.
Fig. 2B. The same culture as in 2A but which has been treated for 4 hours with 1 μg/ml of cytochalasin B. While a few actin-containing strands are still visible, most of the actin fluorescence is now confined to discrete knobby deposits.

The conclusion appears warranted that actin and tubulin are concentrated in the knobbed bodies of the cells in the transformed condition.

These results suggest an explanation for the knobbing phenomenon. In the normal fibroblast, rigid microtubules and contractile actin-containing filaments are joined together to form a flexible, dynamic structure dispersed throughout the cell. Since this structure is disrupted in transformed cells, we postulate it to be fundamental to maintenance of the control of cell reproduction which is lost in cancer cells.

In the normal fibroblast we postulate that microtubules and microfibrils are arranged in a common, highly ordered network containing both rigid and contractile elements. It extends throughout the cell; it is essential for proper control of cell reproduction; and it requires the action of cyclic AMP for achievement of its organized form. If this structure deteriorates, microfilamentous and microtubular elements can collect together at points along the cell surface.These regions containing deposits of contractile protein are able to continue their flexing motions, producing the pulsing action that is observed in the knobbed structures. When the conditions which promote Reverse Transformation are established, the microtubules and microfibrils begin to reestablish their connection in the organized structure. As this process begins, the deposits of contractile protein in the knobs begin to disappear. Later, when the organized network arrangement nears completion, the cells undergo their characteristic elongation to produce the fibroblastic morphology. This picture, developed as a working hypothesis to explain the phenomena described in this paper, also explains other observations. For example, since transformed cells multiply readily, disruption of the microtubular-microfibrillar pattern should not interfere with the normal progression of the cells through G1, S, and G2. This is exactly the behavior which has been found when colcemide is added to cells in culture. The time required for passage of normal cells treated with colcemide through G1, S, and G2 is exactly the same as in untreated cultures (11). However, under the influence of colcemide, cells accumulate in anaphase. Thus, disruption of the microtubular structure in interphase does not halt any phase of the reproductive cycle, whereas disruption of the microtubular structure in metaphase prevents completion of mitosis for which the spindle is essential.

If the proposed dynamics underlying the formation and oscillatory character of the knobs are correct, one might expect to find the appearance of knob activity even in a normal or Reverse Transformed cell at one particular point in its life cycle. During mitosis a large fraction of the cellular microtubules and microfibrils are reorganized into the spindle which distributes the chromosomes to each of the newly-forming daughter cells. When mitosis is completed and the cell is ready to embark on a new Gl phase, the spindle is decomposed and its constituent tubulin and actin-containing protein are released, eventually to become reformed into their interphase arrangement. During the transition period, the microtubules and microfibrils are presumably unorganized much as might be expected from the addition of colcemide or cytochalasin to the interphase cell; therefore, an outburst of knob activity might be expected. Such a phenomenon is indeed observed (5). Time-lapse studies reveal that in the beginning of Gl, an outbreak of knob activity occurs even in the presence of Reverse Transformation conditions. This surface agitation rapidly subsides and the cells proceed to assume their stretched fibroblast-like condition.

Cyclic AMP causes organization of an intracellular network containing microfibrillar and microtubular elements which we have postulated is involved in loss of the control over cellular reproduction which underlies the difference between normal fibroblasts and transformed cells. The following examples demonstrate that the cell surface membrane is also involved in the phenomenon of Reverse Transformation. As already indicated, the presence of the oscillating knobs in the Reverse-Transformed state as opposed to the predominantly smooth, tranquilized membranes of the Transformed state itself constitutes a profound surface change achieved when the microtubular-microfibrillar (MT-MF) complex changes from the disorganized to the patterned array. Addition of the Reverse Transformation reagents to CHO-Kl cells can change the activity of cell surface structures responsible for cell agglutination and for cell cytolysis by specific antisera. An example of the latter effect is shown in Table I. Reverse Transformation also renders the cells less susceptible to the rounding reaction produced by lectins and inhibits the capping phenomenon (4) (12). Cyclic AMP plus testololactone increase the effectiveness of attachment of the cells to plastic surfaces and to themselves (13). Finally, the Reverse Transformation condition is associated with change in the capacity of the cell to carry out certain active transport processes. An increase in active transport of ^{14}C-α amino butyric acid develops as a result of the addition of cyclic AMP plus testololactone. It is of interest that this stimulation of active transport can readily be prevented by cytochalasin-B but not by colcemide. The result implies the possibility that some features of the alter-

TABLE 1

Demonstration that dibytyryl cyclic AMP protects CHO-K1 cells from killing by otherwise lethal concentrations of specific antiserum and complement. Antiserum made in the rabbit to CHO-K1 cells was used in each dish at a concentration of 0.01%, using 1.5% normal rabbit serum in all dishes as a source of complement.

10^{-3}M DBcAMP + 1.0 μg/ml of Testololactone	Antiserum	% Relative Plating Efficiency
absent	absent	100.
absent	present	1.3%
present	absent	100.
present	present	90.5%

ed cell behavior associated with the Reverse Transformation condition may depend more on the activities of microfibrillar elements than on those of the microtubules. The need for continued structural and functional analysis of the elements of this system is obvious.

The nature of the pattern assumed by the microtubular and microfibrillar elements under the influence of cyclic AMP has been defined for fibroblasts. If this pattern is indeed actively concerned with regulatory control of cell reproduction and possibly other metabolic behavior as well, the pattern adopted may vary for cells of different differentiation states. Other reports have demonstrated that the addition of cyclic AMP causes neuroblastoma cells in culture to extend dendrite-like structures (14). We have embarked on a series of experiments in which a standard CHO-K1 cell is hybridized with other Chinese hamster cells taken directly from biopsies. One such clone obtained by hybridization of CHO-K1 cell with cells obtained from a biopsy of Chinese hamster brain was secured. A stable cell culture was produced whose general morphology and growth behavior in standard nutritional medium is not greatly dissimilar from those of the CHO-K1 cell itself. However, when cyclic AMP and testololactone are added to such cells, each cell extends dendrite-like structures which eventually unite to form a network. (Fig. 3)

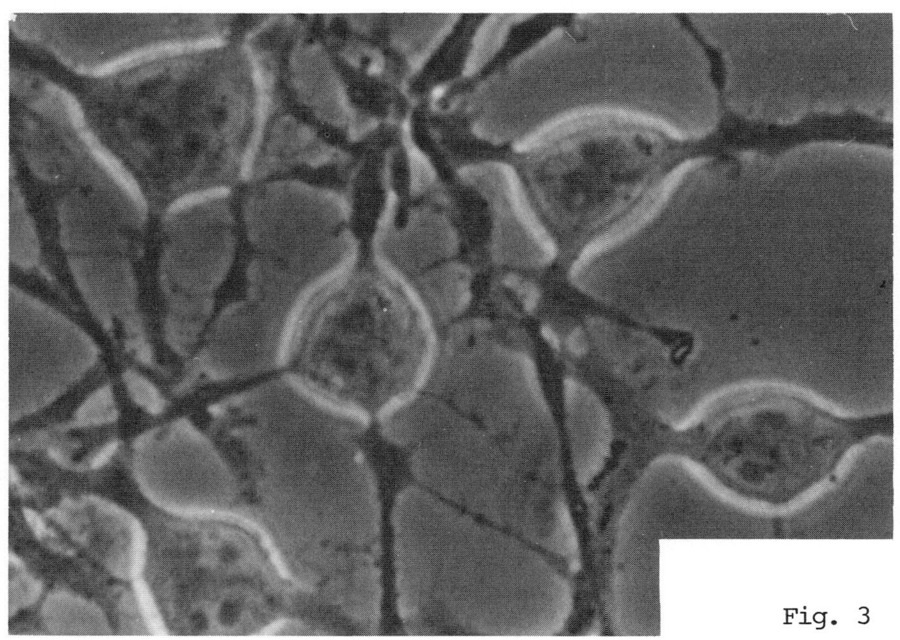

Fig. 3

Fig. 3. Hybrid cells of the CHO-K1 strain fused with trypsinized cells taken directly from a Chinese Hamster brain biopsy. In normal growth medium, the cells possess a morphology resembling that of the CHO-K1 cell. When DBcAMP plus testololactone is added, each cell extends dendrite-like processes which form a network as shown in the figure. When normal growth medium is restored, the dendritic processes are resorbed and the morphology again resumes that of an undifferentiated tissue culture cell.

When the drugs are removed, the dendritic structures are resorbed into each of their cells of origin, and the cells return to their original morphology. This experiment appears to have a number of implications. The one important for present purposes is that different cells appear to have different commitments to a particular kind of pattern of MT-MF elements which is activated in the presence of elevated concentrations of cyclic AMP. It appears tenable that this fundamental pattern may be an intrinsic characteristic of the differentiation state.

Presumably, any regulation of cellular metabolic behavior due to a particular pattern of microtubular and microfibrillar structure must sooner or later involve communication with the genome. How is this communication effected? We have proposed the speculation that just as microtubules become linked up to the chromosomes in mitosis, so also do they become attached to specific chromosomal regions during interphase. The attachment may involve participation of various classes of chromosomal proteins. This formulation suggests a role for regions of highly reiterated chromosomal DNA which characterize the mammalian genome. Just as the microtubules of the spindle become attached in mitosis to the centromeric regions which are rich in reiterated DNA so too might the microtubular elements become attached to specific reiterated DNA regions of the chromosomes in interphase. Thus, the microtubular and microfibrillar elements of the structure which we have been discussing might well be involved in the process of exposure or sequestration of particular regions of the genome in particular states of differentiation. The theory would demand that such specific connections be demonstrable. While such a clear demonstration is not yet available and may be difficult to secure because of the instability of at least some groups of chromosomal proteins, we have demonstrated by means of immunofluorescence that cells in interphase display ordered patterns of tubulin deposits around and possibly inside the nucleus.

The considerations in the foregoing pages demonstrate a fundamental involvement of cyclic AMP in organization of microtubules and microfibrils in a patterned array which can be different for cells committed to different differentiation states. It should be emphasized that breakdown of this mechanism need not be the only one by which malignancy could be produced since any break-down in the transfer of information from the cell surface to the genome in a way which would distort the control of cell reproduction could produce a malignant condition. However, the disruption of the MT-MF organization producing hyperactive cell membranes associated with the appearance of oscillating knob-like structures and a decreased tendency of such cells to attach to solid surfaces and to each other may illuminate the process of metastasis formation. The agitated membranes of such transformed cells may more readily cause dislodgement from their normal site of attachment and recolonization of new malignant loci.

ACKNOWLEDGEMENTS

This paper presents a summary of data that have been collected during the last several years plus the working hypotheses which are being entertained about their cellular significance. The participants in these studies have been Drs. B. Brinkley, A. Hsie, C. Jones, F.T. Kao, A. Kauvar, S. Keesee, D. Kelley, E. Lazarides, D. Patterson, K. Porter, C. Waldren, L. Wenger, and the author. These investigations were aided by grant #VC81G from the American Cancer Society and by grants #CA-15794 and CA-20810 of the National Cancer Institute.

REFERENCES

1. Puck, T.T., Cieciura, S.J. and Fisher, H.W. (1957) J. Exp. Med. 106, 145. Puck, T.T., Morkovin, D., Marcus, P.I. and Cieciura, S.J. (1957) J. Exp. Med. 106, 485.

2. Hsie, A.W. and Puck, T.T. (1971) Proc. Nat. Acad. Sci. USA 68, 358.

3. Johnson, G.E., Friedman, R.M. and Pastan I. (1971) P.N.A.S. 68, 425.

4. Hsie, A.W., Jones, C. and Puck, T.T. (1971) P.N.A.S. 68, 1648.

5. Puck, T.T., Waldren, C.A. and Hsie, A.W. (1972) P.N.A.S. 69, 1943.

6. Porter, K., Puck, T.T., Hsie, A.W. and Kelley, D. (1974) Cell 2, 145.

7. Puck, T.T., and Wenger L. (1974) J. of Cell Biol. 63, 275a. Also unpublished data.

8. Storrie, B., Puck, T.T., Wenger L. Manuscript in preparation.

9. Brinkley, B.R., Fuller, G.M., and Highfield, D.P.(1975) in Microtubules and Microtubule Inhibitors, edited by M. Borgers and M. deBrabander. p 297, America Elsevier Publishing Co., New York.

10. Patterson, D. and Waldren, C.A. (1973) Biochem. Biophys. Res. Comm. 50, 566.

11. Puck, T.T. and Steffen, J. (1963) Biophysical Journal, 3, 379.

12. Storrie, B. (1974) J. Cell Biology. 62, 247. Storrie, B. (1975) J. Cell Biol. 66, 392.

13. Kauvar, A.J. Wenger, L., and Puck, T.T. (1973) ICRS. FDA 73.

14. Prasad, K.N. and Hsie, A. (1971) Nature New Biol. 233, 141. Furmanski, P., Phillips, P.G., and Lubin, M. (1971) Nature New Biol. 233, 413.

GENE TRANSFER IN SOMATIC CELL POPULATIONS

Frank H. Ruddle

Department of Biology, Kline Biology Tower, Yale University
New Haven, Connecticut 06520

ABSTRACT. The transfer of genomes or parts of genomes between somatic cell populations is briefly reviewed and its relationship to the general problems of gene mapping and gene regulation in man is considered. The methods used have been: (1) cell hybridization, (2) microcell mediated chromosome transfer, and (3) phagocytosis mediated subchromosome transfer. Methods are compared with respect to efficiency of production and stability of a suitable partial genome of one parent and degree of resolution attainable in mapping. Common modes of analysis in all methods for establishment of synteny and gene assignment are outlined.

INTRODUCTION. Somatic cell genetic systems make use of somatic cells for the extraction of genetic information. For a recent review article, see Ruddle and Creagan (1975). The techniques of somatic cell genetics have been applied mainly to mammalian species, and within this group to the primates and the rodents. Considerable attention has been devoted to man, since in the case of man, somatic cell procedures possess certain advantages over Mendelian approaches. The primary mode of analysis has involved the transfer of genomes or parts of genomes between cells. This has been accomplished in three ways: (1) cell hybridization, (2) microcell mediated chromosome transfer, and (3) phagocytosis mediated subchromosome transfer. Each of these procedures and their relationships to gene mapping in man will be considered.

CELL HYBRIDIZATION

Cell hybridization involves the fusion of the cell membranes of individual cells and results in the transfer of whole parental genomes into a single resulting cell, generally termed the heterokaryon. Later in the life of the heterokaryon, the parental chromosome sets are located within a single nucleus. Such a cell is termed a hybrid, and frequently, the hybrid may possess the capacity for unlimited cell division. In cases where the hybrids are formed between cells of different specific origins, there is an elimination of chromosomes from one of the parental sets. The loss of chromosomes from one of the parents is sometimes referred to as segregation, since the process is in a sense anologous to chromosome segregation in meiosis. In this report, for the sake of convenience, the eliminated chromosome set will be called the donor set, whereas the non-eliminated, residual set will be called the recipient or host chromosome set. This type of terminology is useful since it is consistent with the terminology applicable to other gene transfer systems to be described later.

Human gene mapping by somatic cell genetics has been carried out extensively through the use of human-rodent cell hybrids. In these cell fusions, the human cells are usually derived from primary cell lines. The rodent cells are generally long-term mouse or Chinese hamster cell lines which have been adapted to grow in tissue culture. Such lines possess abnormal chromosome constitutions, but have a finite life expectancy. Hybrids of this type invariably lose the human chromosomes. The elimination of the human chromosomes is usually incomplete, so that one may obtain hybrid clones which retain partial sets of the human genome.

Gene mapping is established by correlating the presence or absence of particular human phenotypes or chromosomes with the presence or absence of other human phenotypes which may be expressed in the hybrid cells. The synteny test involves correlations between donor phenotypes, and when positive suggests that the corresponding donor genes are located on a common chromosome. The assignment test involves correlations between donor phenotypes and specific donor chromosomes, and when positive provides evidence for the assignment of the corresponding gene to a particular donor chromosome. The regional assignment test accomplishes the same thing for segments of chromosomes. In this instance, the segments are defined either by translocation breakpoints or be deletion breakpoints in the donor chromosome set, and may exist either in the donor parental cell prior to fusion, or may arise in the donor chromosome set at some point following fusion in the lifetime of the hybrid cell.

Cell hybridization techniques have been instrumental in mapping a large number of genes in the human genome. More than 200 genes have been mapped in man according to a recent review on the subject (McKusick and Ruddle, 1977). These assignments also include linkage data obtained by family study methods. It should be emphasized that somatic cell genetics and Mendelian family methods can interact synergistically, and together can provide a powerful tool for the human genome mapping. Presently, a genetic marker is mapped to each chromosome, and regional linkage data are available for many of the individual chromosomes.

MICROCELL HYBRIDS

Microcell-mediated transfer of chromosomes represents a second method of gene transfer in mammalian cell systems. This method was pioneered by Ege and Ringertz (1974), but the first viable microcell hybrids were produced by Fournier and Ruddle (1977). The donor cells are treated with Colcemid in order to arrest the cells in mitosis. The mitotic block is imposed for a number of days, and under these conditions of the prolonged Colcemid treatment, the mitotic cells will leak through the block giving rise to polykaryocytes. Thus, the resulting multinucleated cells have their chromosomes distributed into a number of micronuclei which contain only one or a few chromosomes. The treatment of the polykaryocytes with cytochalasin, together with centrifugation, permits one to isolate microcells. Microcells represent a micronucleus together with a rim of cytoplasm plus an intact plasma membrane. The resulting mixture of intact cells and microcells of varying sizes may be separated on a stayput gradient in order to purify the small microcell component of the mixture. Microcells may then be fused to host cells in order to form microcell heterokaryons, and ultimately, these will give rise to viable long term microcell hybrids. The microcell hybrids are unique in that from the outset they possess only one or several donor chromosomes. Thus, the direction of a chromosome elimination is controlled by the choice of the donor parent. To date, using this type of protocol, we have transferred mouse donor chromosomes into mouse, Chinese hamster, and human host cells.

The microcell procedure developed by Ege and Ringertz is not suitable for the preparation for human microcells because of the sensitivity of human cells to Colcemid. However, a second method applicable to human cells has been developed by Johnson and co-workers (Johnson, _et al_, 1975). According to this procedure, the human donor cells are arrested mitotically with nitrous oxide at 5 atmospheres pressure. The arrested cells are then plated out at 4°C. This treatment results in the formation of polykaryocytes which can then be sheared to form microcells. We have successfully prepared long term microcell hybrids using human cells as a donor, using a modification of the Johnson procedure. In this type of experiment performed by Mr. Klobutcher in our laboratory, it has been possible to transfer human chromosome 17 to mouse L strain derivative, deficient in the cytosol form of thymidine kinase. Under conditions of HAT selection (Littlefield, 1966), it was possible to recover a microcell hybrid which possessed chromosome 17, and no other detectable human chromosome or fragment.

ISOLATED CHROMOSOME TRANSFER

A third method of genome transfer involves the phagocytotic uptake of donor chromosomes (Ruddle and McBride, 1977). In this procedure, first developed by McBride and Ozer (1973), the donor cells are arrested mitotically and then broken to release their metaphase chromosomes. The chromosomes are separated from cellular debris by centrifugation, and then are mixed with recipient cells. The host cells ingest the donor chromosomes, and subsequently, they degrade them in lysosomal vesicles. Presumably at a low frequency, small pieces of the donor chromosomes escape inactivation, and these "transgenomes" express at least some of their genes in a normal manner.

The size of the transgenome can be estimated roughly by somatic cell genetic procedures. It is known from hybrid mapping studies, for example, that structural genes for phosphoglycerate kinase (PGK) and glucose-6-phosphate dehydrogenase (G6PD) lie to either side of the HPRT locus in man. In experiments where HPRT is transferred into mouse cells, analysis shows that neither donor PGK nor G6PD enzymes are expressed in transformants which express donor HPRT (Willecke and Ruddle, 1975). The distance between PGK and G6PD can be approximated as being equivalent to 0.01 of the haploid genome, 3×10^7 nbp. In other experiments using Chinese hamster transformants (Ruddle and McBride, 1977), and human transformants (Willecke, et al, 1976), it was shown that the structural gene for galactose kinase is cotransferred in about 20% of transformants. The distance between TK and GK is not precisely known. However, hybrid mapping studies indicate the distance between these two markers is no more than .001 of the haploid genome, corresponding to approximately 1×10^6 nbp. We can, therefore, use this figure to estimate the maximum size of the transgenome. It could, of course, be considerably smaller.

Very little is known about the structural nature of the transgenome or its location in host cells. Several reports have provided convincing evidence for two types of transformed cells, namely unstable and stable transformants. These are defined with regard to the retention or loss of the selected marker under selection conditions.

A possible correlate of stability might be the integration of the donor transgenome into the host chromosome set. In order to test this hypothesis, Fournier and Ruddle have carried out experiments which seek to demonstrate a correlation between the transgenote genetic markers and a specific host chromosome in stably transformed cells. If integration has occurred, then we predict that a particular donor mouse chromosome will show a positive correlation with the human form of HPRT. Using the microcell method of chromosome transfer, we have obtained such a result which thus provides support for the integration hypothesis. It should be emphasized that this kind of experiment merely establishes an association between the transgenome and a specific chromosome; it does not prove the covalent integration of the transgenome, but it is consistent with this hypothesis.

CONCLUSIONS. In this brief report, I have sketched three modes by which genetic materials may be exchanged between mammalian cells. The three methods can be used to transfer whole chromosomes, one or several chromosomes, or relatively small parts of chromosomes. It has previously been amply shown that donor chromosomes may exchange segments with host chromosomes, and in this report, we provide first evidence that very small functional pieces of donor chromosomes may be functionally integrated into the host chromosome set. These procedures naturally lend themselves to gene mapping in mammalian species. Already, considerable information has been obtained for man in a relatively short period using these approaches. Furthermore, the ability to work with quite small subfractions of the genome now opens up the possibility for gene mapping in mammalian species at a high level of resolution. To date, our increasing capacity to physically manipulate the mammalian genome has been used primarily to map genes. We can look forward with some confidence to the not too distant time when these capabilities will begin to yield information regarding the regulation of gene expression.

ACKNOWLEDGEMENT

This work was supported by grants from the National Foundation - March of Dimes (CRBS-307) and from the NIH (2 R01 GM 09966).

REFERENCES

Ege, T. and Ringertz, N. R. (1974) Experimental Cell Research 87, 378-382.

Fournier, R.E.K. and Ruddle, F. H. (1977) Proceedings of the National Academy of Sciences USA 74, 319-323.

Johnson, R. T., Mullinger, A. M. and Skaer, R. J. (1975) Proceedings for the Royal Society of London B 18, 591-602.

Littlefield, J. W. (1966) Experimental Cell Research 41, 190-196.

McBride, O. W. and Ozer, H. L. (1973) Proceedings of the National Academy of Sciences USA 70, 1258-1262.

McKusick, V. A. and Ruddle, F. H. (1977) Science, In press.

Ruddle, F. H. and Creagan, R. P. (1975) Annual Review of Genetics 9, 407-486.

Ruddle, F. H. and McBride, O. W. (1977) In: The Molecular Biology of the Mammalian Genetic Apparatus, vol. II (Editor, P.O.P. Ts'o) ASP Biological and Medical Press, Amsterdam, in press.

Willecke, K. and Ruddle, F. H. (1975) Proceedings of the National Academy of Sciences USA 72, 1792-1796.

Willecke, K., Lange, R., Kruger, E. and Teber, T. (1976) Proceedings of the National Academy of Sciences USA 73, 1274-1278.

WORKSHOP SUMMARY: Mammalian Genetic Systems by Immo Scheffler, University of California at San Diego, La Jolla, California

A great deal of progress has been made in recent years in the mapping of genes on mammalian chromosomes. This has been possible in the mouse because of the possibility of doing large scale breeding experiments; and in the case of human cells because of the segregation of human chromosomes in somatic cell hybrids made from human and rodent cells. In the latter case advantage is taken of the differences in physical properties of many enzymes (proteins) between different species, such that in most cases there is no need for specific mutations.

However, in order to take full advantage of the genetic approach to the solution of problems in mammalian cell biology it becomes essential to have mutations affecting a large variety of functions. The program of the workshop was arranged to illustrate the selection and characterization of a variety of mammalian cell mutants, and there was some discussion of the usefulness of such mutants in the study of differentiation, energy metabolism, regulation of enzyme activities and other aspects of cellular physiology.

The selection of mutants resistant to drugs is relatively straightforward, if the target and action of the drug is well understood. Several groups presented the results of their studies of α-amanitin. Diploid and heteroploid cells contain both wild type and altered enzyme, and when such cells are grown in the presence of α-amanitin, only the altered enzyme is found, but at elevated levels to compensate for the inactivation of the wild type enzyme. In somatic cell hybrids α-amanitin resistance behaves as a co-dominant phenotype. J. Ingles then reported that among a collection of α-aminitin CHO cells selected at 34°C one could find some that were temperature sensitive at 39°C, and evidence was presented, supporting the interpretation, that in such mutants one or more alterations in the primary structure of the enzyme lead to the formation of a temperature-sensitive, α-amanitin resistant RNA polymerase II. M. Pearson presented data which indicate that some α-amanitin resistant mutants of the rat myoblast line (L6) also have lost the ability to fuse and form myotubes. This might be interpreted to mean that in such mutants an altered RNA polymerase II no longer is capable of recognising certain signals (promoters) which are essential for the expression of differentiated functions (e.g. creatine kinase).

Another set of drug resistant mutants of the S49 lymphoma cells were reported by P. Coffino. Wild type cells respond to cAMP by inducing cAMP phosphodiesterase, halting growth in the G_1 phase of the cell cycle, and subsequently dying (cytolysis). A new class of "death-less" mutants was isolated which respond to cAMP as do wild type cells (induction of the phosphodiesterase, normal cAMP-dependent protein kinase, arrest in G_1), but they do not undergo cytolysis. It is hoped that such mutants will be useful in studying the pathway distal to the protein kinase which leads to cytolysis and cell death.

I. Scheffler reported on the selection and characterization of a novel class of mammalian mutants (Chinese hamster) which are defective in oxidative energy metabolism (respiration). Six complementation groups, defined by somatic cell hybridization, are available so far, of which three have been characterized biochemically. One such mutant has a defective NADH-coenzyme Q reductase (complex I of the electron transport chain), another is almost totally lacking in succinate dehydrogenase, and a third mutant is defective in mitochondrial protein synthesis, leading to severe deficiencies in cytochrome c oxidase, oligomycin-sensitive ATPase, and cytochrome b. Such mutants are of interest in the study of energy metabolism in mammalian cells in culture and they may also help in the study of the biogenesis of mammalian mitochondria.

L. Thompson summarized the development of selection procedures for the isolation of conditionally lethal mutations affecting protein synthesis in Chinese hamster CHO cells. In all of the mutants found so far (6 complementation groups) the defect could be shown to involve a specific aminoacyl - tRNA synthetase. A number of the mutants were simply temperature sensitive, in some of these a temperature-sensitive synthetase activity could be demonstrated _in vitro_. Most others had an increased requirement at 34°C and/or 39°C for one of the following amino acids: arg, asn, glu, his, met, or leu. In these mutants impaired aminoacylation was found _in vivo_ under restrictive conditions (normal amino acid concentration, 39°C. It is concluded that all of the mutants have point mutations in the structural genes of the aminoacyl-tRNA synthetases.

J. Irr presented results which are compatible with the hypothesis suggesting that arginine controls the rate of expression of the genes coding for arginino succinate synthetase. The synthetase is one of the enzymes required for the conversion of citrulline to arginine, a pathway operating in established human lymphoblasts which is repressed when arginine is present in the medium. Two mutants were found by selection, and in both of these the synthetase activity appeared to be constitutive. In particular, in one of the mutants in the presence of arginine the activity was 200-fold greater than in the parental cells under the same conditions. The amount of activity was correlated with the amount of specific synthetase protein present, and it was suggested that the mutants were regulatory gene mutants. This would be the first such mutation described in a mammalian cell.

GENETIC CONTROL OF ARGININOSUCCINATE SYNTHETASE IN HUMAN LYMPHOBLASTS

Joseph D. Irr and Lee B. Jacoby

Genetics Unit, Massachusetts General Hospital
and Department of Pediatrics, Harvard Medical School,
Boston, MA 02114

ABSTRACT. The activity of argininosuccinate synthetase, one of the arginine biosynthetic enzymes, showed marked derepression when established lines of human lymphoblasts were transferred to arginine deficient medium. This derepression occurred both in normal lymphoblasts and in those derived from an individual with a deficiency in argininosuccinate lyase, the next enzyme of the pathway. The fully repressed level of argininosuccinate synthetase was obtained by the re-addition of arginine. Repression was due to normal turnover of pre-existing enzyme in the absence of new synthesis rather than to preferential degradation of the enzyme. Two variant cell lines, isolated from normal cells by two different selection techniques, had argininosuccinate synthetase activities that were not as extensively repressed by arginine. When grown in the presence of arginine, one line produced about 200 times as much argininosuccinate synthetase as its progenitor. Argininosuccinate synthetase activities were correlated with the amounts of synthetase protein in the variant and normal cells. The rates of turnover of argininosuccinate synthetase were also compared. The results suggest that arginine controls the rate of expression of the gene coding for argininosuccinate synthetase and that the variants may harbor mutations in regions of the genome that regulate argininosuccinate synthetase gene expression.

INTRODUCTION

The pathway for arginine metabolism in mammals is shown in Figure 1. The activity of argininosuccinate synthetase, the first of the two enzymes needed to convert citrulline to arginine, is repressed in normal human lymphoblasts grown in the presence of arginine and is partially derepressed in cells grown in medium containing citrulline in place of arginine (1, 2, 3). The activity of argininosuccinate lyase is the same in lymphoblasts under the two growth conditions (1).

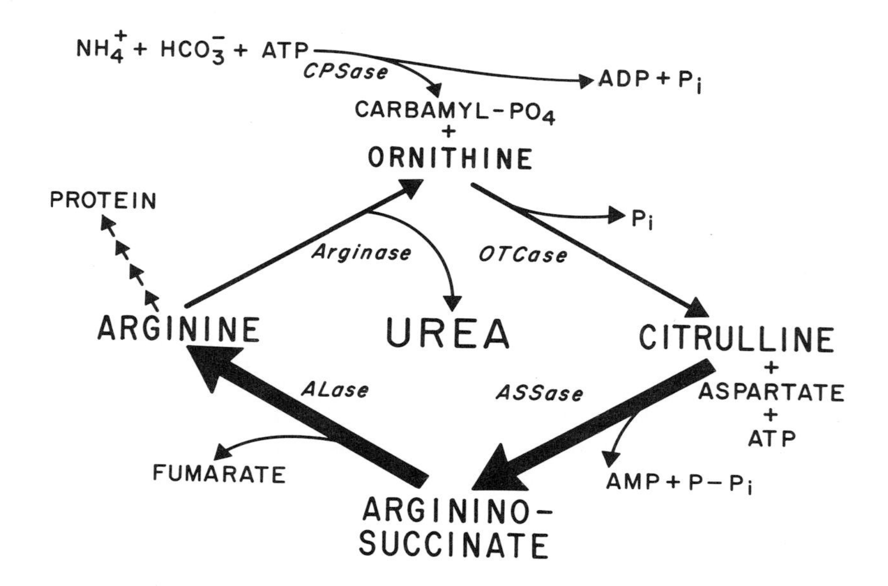

Figure 1. The Urea Cycle. Enzymes: carbamyl-phosphate synthetase (CPSase), ornithine transcarbamylase (OTCase), argininosuccinate synthetase (ASSase), argininosuccinate lyase (ALase) and arginase. Broad arrows indicate activities in lymphoblasts.

Schimke, however, found that both of these activities were partially derepressed in human HeLa and KB cells and in mouse L cells (8). Schimke postulated but did not prove that arginine repressed synthesis of the two enzymes. He then went on to show that arginine caused de novo synthesis, i.e. induction, of arginase in HeLa cells (9).

We shall provide evidence that arginine controls the rate of synthesis of argininosuccinate synthetase in normal lymphoblast lines. We also shall describe some properties of two types of variant cell lines that appear to bear mutations in regions of the genome responsible for regulating synthesis of argininosuccinate synthetase.

ARGININOSUCCINATE SYNTHETASE DEREPRESSION

When the normal lymphoblast line MGL8B2 was transferred from arginine to citrulline supplemented medium, there was a growth lag of 48 hours during which argininosuccinate synthetase activity increased (Fig. 2). Cells maintained in arginine supplemented medium grew without a lag and showed no change in synthetase activity. Cycloheximide added at 1 µg/ml at the time of the medium change blocked the rise in synthetase activity.

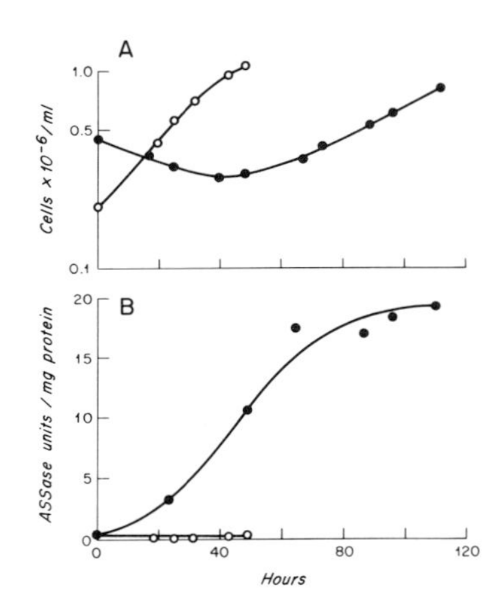

Figure 2. MGL8B2

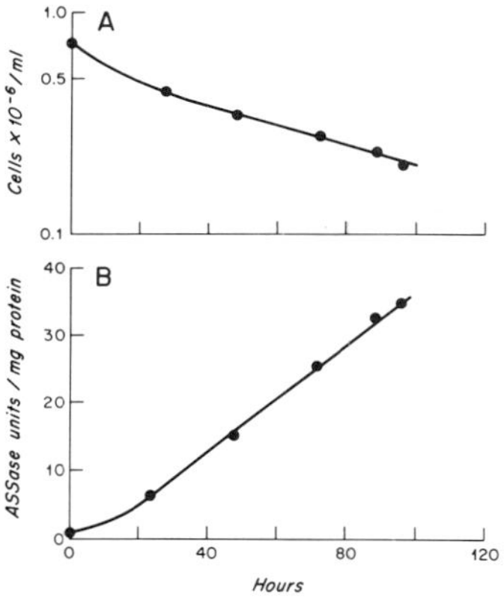

Figure 3. MGL14

Figures 2 and 3. Effect of Replacing Arginine with Citrulline on Growth and Argininosuccinate Synthetase Activity of Normal (MGL8B2) and Argininosuccinate Lyase Deficient (MGL14) Lymphoblasts. MGL8B2 cells growing in Eagle's MEM containing 0.6 mM arginine, 15% fetal calf serum and nonessential amino acids were transferred at 0 time to fresh medium containing 0.6 mM arginine (O) or 0.6 mM citrulline (●). MGL14 cells were transferred from arginine to citrulline medium. Cell counts (A) and synthetase activities (B) were determined on samples taken as indicated. Synthetase activity was determined by a modification of the method of Schimke (8). One unit of activity equals one nanomole of product formed per hour at 37°C.

When the argininosuccinate lyase deficient line MGL14 was transferred from arginine to citrulline supplemented medium, there was a similar but more pronounced increase in argininosuccinate synthetase activity without renewed growth (Fig. 3). Mixing of crude extracts of MGL8B2 grown in arginine or citrulline medium or gel filtration of these crude extracts produced no changes in enzyme activity. Thus, either arginine or citrulline affects synthesis or degradation of argininosuccinate synthetase in these lymphoblast lines.

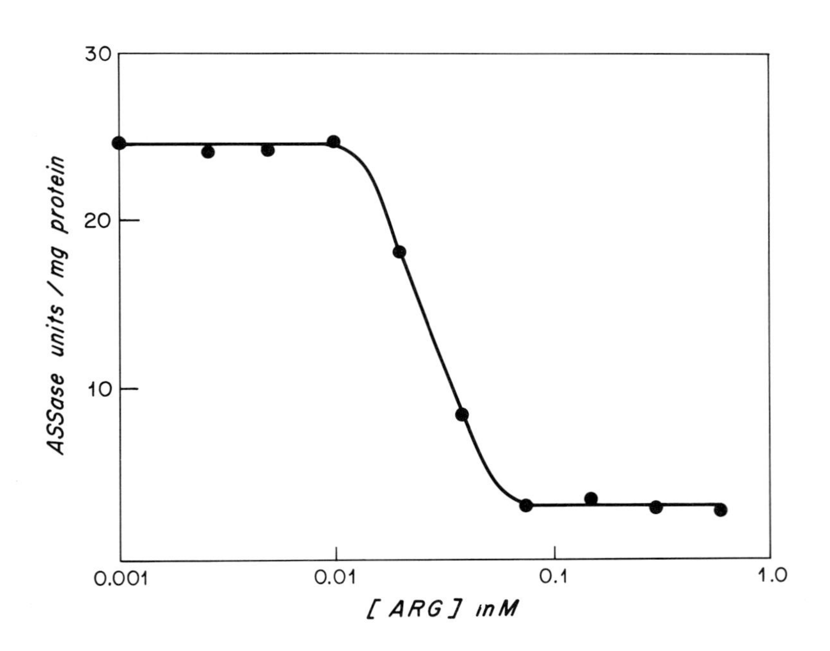

Figure 4. Effect of Arginine on Argininosuccinate Synthetase Activity in Normal Lymphoblasts. MGL8B2 cells growing in citrulline medium were transferred to medium containing 0.6 mM citrulline and the indicated amounts of arginine. Cells were harvested for synthetase assays after 2.3 generations of additional growth.

IDENTITY OF THE EFFECTOR

Whether citrulline was a positive effector or arginine a negative effector of argininosuccinate synthetase was tested in cultures of MGL8B2. Citrulline added at a range of 0.2 to 8 mM to cells growing in arginine medium had no effect upon synthetase activity. However, when arginine at concentrations greater than 0.01 mM was added to cells growing in citrulline medium, the synthetase activity was reduced (Fig. 4). While argininosuccinate synthetase activity seems to be independent of citrulline, there is a reciprocal relationship between activity and exogenous arginine that strongly suggests arginine or a metabolite acts as a negative effector to control the intracellular concentration of this enzyme.

ARGININOSUCCINATE SYNTHETASE IN VARIANTS

Table 1 illustrates the differences in argininosuccinate synthetase activity common to normal cell lines grown in arginine or citrulline medium. Also shown are the synthetase activities found in two different types of variants derived from the two normal lines. MGL8D1 is a variant selected from MGL8B2 that is resistant

TABLE I

ARGININOSUCCINATE SYNTHETASE ACTIVITIES IN NORMAL AND VARIANT LYMPHOBLASTS GROWN IN ARGININE OR CITRULLINE SUPPLEMENTED MEDIUM

	Argininosuccinate Synthetase[1]	
Cell Line	Arginine	Citrulline
Normals		
MGL8B2	0.5	26
MGL33	0.3	19
Variants		
MGL8D1	213	227
MGL33D57	20	40

Cells were grown for 10 generations in medium containing 0.6 mM arginine or 0.6 mM citrulline.

[1] Nanomoles of product formed per hour per mg. protein.

to canavanine, a toxic arginine analogue (4,5). The synthetase activity in this line was not affected by arginine. MGL33D57 is a variant that has regained the ability to utilize citrulline for growth, a trait lacking in its immediate parent but present in its progenitor MGL33 (6). The synthetase activity in MGL33D57 was repressed only 50% by growth in arginine medium.

The argininosuccinate synthetase activity measurements suggested that differences in enzyme concentration might exist in these normal and variant cells. We determined the amount of synthetase protein produced by partially purifying the enzyme from cells labeled with ^{35}S-methionine. We believe the enzyme is in one of six major protein bands detected by 0.1% SDS-10% polyacrylamide gel electrophoresis of a preparation enriched 300-fold for the synthetase present in MGL8D1. This protein band is called #4 since we have not yet proven that it is the

TABLE 2

INCORPORATION OF ^{35}S-METHIONINE INTO PROTEINS

Cell Line	Growth Conditions	Protein #4 cpm	Protein #2 cpm
MGL8B2	Arginine	283	2195
MGL8D1	Arginine	2371	3682
MGL33D57	Arginine	630	3052
MGL8B2	Citrulline	615	3568
MGL8D1	Citrulline	4820	6298
MGL33D57	Citrulline	1025	2038

Cells were grown for 72 hours in arginine or citrulline medium containing ^{35}S-methionine (2 μci/0.03 μmoles/ml). The synthetase was partially purified by ammonium sulfate fractionation and DEAE-cellulose column chromatography and lyophylized. Samples were reduced in the presence of 2-mercaptoethanol by the method of Weber and Osborn (10) and characterized further by electrophoresis in vertical slab gels containing 0.1% SDS and 10% bis-acrylamide. Protein bands identified by staining with Coomassie blue were extracted with perchloric acid and hydrogen peroxide (7) and quantified for radioactivity by scintillation counting.

synthetase nor that it is a homogeneous species.

In arginine grown cells, MGL8B2 contained only 12% as much protein #4 as MGL8D1 and only 45% as much protein #4 as MGL33D57 (Table 2). Although these differences between cell lines are not as striking as the enzyme activity measurements, they indicate a similar trend. The differences in the amounts of the other proteins were not as marked, and only one representative protein band is shown in Table 2.

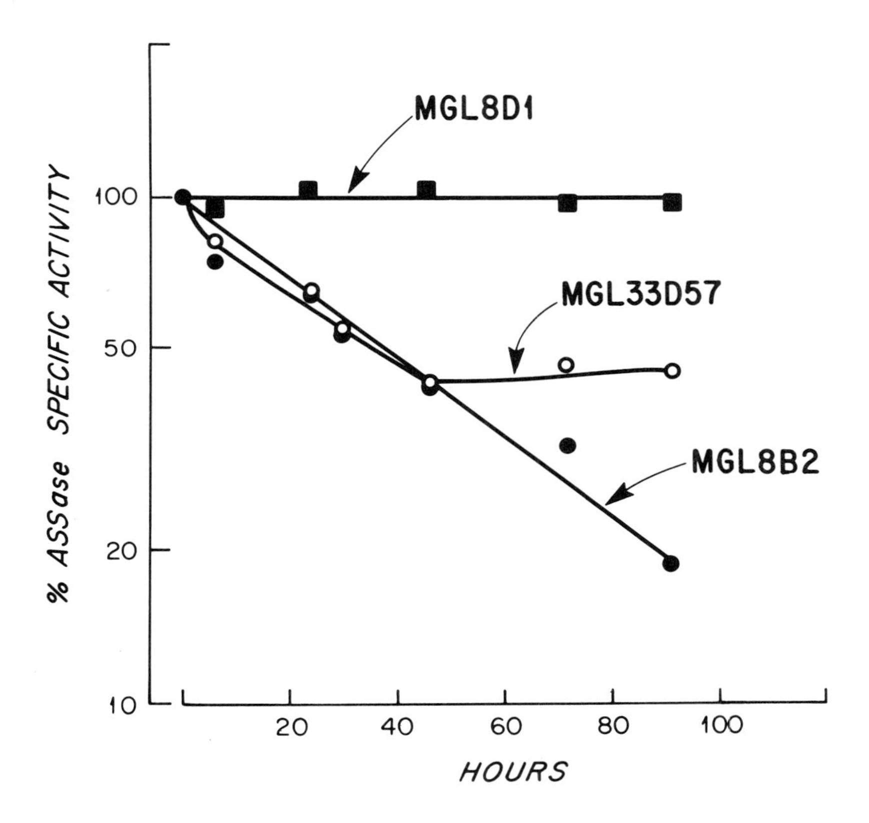

Fig. 5. Kinetics of Change in Argininosuccinate Synthetase Activity after Addition of Arginine. At 0 time, 0.6 mM arginine was added to MGL8B2 cells (●), MGL8D1 cells (■) and MGL33D57 cells (0) growing in citrulline medium. Samples were removed for synthetase assays at the indicated times.

The citrulline grown MGL8B2 cells contained about 14% as much protein #4 as MGL8D1, but the other 5 proteins from MGL8D1 had on the average 50% more radioactivity than those in MGL8B2 so the differences in protein #4 may be less than indicated. MGL33D57 produced 67% more protein #4 than MGL8B2 yet had similar amounts of the other proteins. These differences in amounts of radioactively labeled protein #4 parallel the differences in enzyme activities shown in Table 1.

TURNOVER STUDIES

Figure 5 shows the rate of decay of argininosuccinate synthetase activity after adding arginine to cells growing in citrulline medium. Exponential decay of enzyme activity occurred in normal MGL8B2 cells but did not occur in MGL8D1. The identical decay rates for synthetase in MGL8B2 and MGL33D57 suggest that synthesis rather than degradation of the enzyme is altered in MGL33D57.

Turnover studies of protein #4, the putative argininosuccinate synthetase protein, suggest that preformed protein #4 turned over at a rate comparable to many other cellular proteins in MGL8B2 (Table 3).

TABLE 3

TURNOVER OF ^{35}S-METHIONINE LABELED PROTEINS

Cell Line	Addition	Protein #4 cpm	Protein #4 %	Protein #3 cpm	Protein #3 %
MGL8B2	0 Time	374	100	3474	100
	Arg & CH	215	58	1428	41
	Arg	222	59	1886	54
	None	230	61	2312	67
MGL8D1	0 Time	976	100	1759	100
	Arg & CH	281	29	926	53

Cells grown 72 hours in citrulline medium containing ^{35}S-methionine (2 μci/0.03 μmoles/ml) were transferred at 0 time to and incubated 24 hours in citrulline medium containing unlabeled methionine. Where indicated the cultures also contained 0.6 mM arginine and 1 μg/ml cycloheximide (CH). Samples were analyzed as in Table 2.

Furthermore, the rate of turnover was about the same when protein synthesis was blocked with cycloheximide or when the cells were in citrulline medium either with or without 0.6 mM arginine. In MGL8D1, synthetase turnover occurred at a somewhat faster rate than the other proteins and more rapidly than in MGL8B2. If the mutation giving rise to MGL8D1 had affected the rate of synthetase breakdown, a considerably decreased rate of turnover would have been seen in this test. Therefore, MGL8D1 probably has elevated synthetase activity because these cells synthesize the enzyme at a greatly accelerated rate.

DISCUSSION

Arginine appears to control the rate of synthesis of argininosuccinate synthetase in human lymphoblastoid cell lines. In addition, the two classes of mutants described here may prove to be highly informative tools for further studies on the mechanism controlling the expression of the gene or genes coding for the synthetase. Mutants represented by MGL8D1 are potential regulator gene mutants that no longer recognize arginine as an effector and thus allow maximum expression of the structural genes under its control. MGL33D57 could be an operator type mutant that has an altered site for regulator gene product action. Complementation tests between normal and mutant lymphoblasts may provide further clues towards resolving the mechanisms altered by the mutations.

ACKNOWLEDGEMENTS

Supported by research grant CA14534 from the NCI and by fellowships from NIH and the National Foundation to JDI. Thanks go to Carol Milbury and Barbara Fleming for their excellent assistance.

REFERENCES

1. Irr, J. and Jacoby, L. (1977) Submitted for publication.
2. Irr, J., Jacoby, L. and Erbe, R. (1975) Am. J. Hum. Genet. 27, 50a.
3. Jacoby, L.B. (1974) Exp. Cell Res. 84, 167.
4. Jacoby, L.B. (1976) J. Cell Biol. 70, 134a.
5. Jacoby, L.B. (1977) Submitted for publication.
6. Jacoby, L.B. and Irr, J. (1975) J. Cell Biol. 67, 189a.
7. Mahin, D. and Logberg, R. (1966) Anal. Biochem. 16, 500.

8. Schimke, R. (1964) J. Biol. Chem 239, 136.
9. Schimke, R. (1964) Metabolic Control Mechanisms Nat. Cancer Inst. Monograph 13, 197.
10. Weber, K and Osborn, M. (1969) J. Biol. Chem. 244, 4406.

SMALL STABLE RNA MOLECULES IN THE NUCLEUS: POSSIBLE MEDIATORS IN GENE EXPRESSION

David Apirion, Imre Berek*, Bikram S. Gill, and Uma S. Podder

Department of Microbiology and Immunology, Division of Biology and Biomedical Sciences, Washington University School of Medicine, St. Louis, Missouri 63110

ABSTRACT. The distribution of small stable nuclear (ssn)RNAs in mouse L cells was investigated. Some of these molecules were found in chromatin. We compared the ssRNAs of a number of cell lines from a number of mammalian species. While species specificity was detected, most tested lines from the same species have qualitatively similar molecules, but quantitative differences exist between lines. The ssRNA of mouse organs show great similarities with clear quantitative differences. The patterns of the organs are radically different from those observed in the cell lines. In all three tumors tested (one myeloma, and two hepatomas), some of the ssRNA molecules were greatly reduced or completely missing.

In order to achieve differential gene expression mediated by ssnRNA it is only necessary to assume different distributions of these molecules along the genome (association with middle repetitive DNA). Thus, neither qualitative nor quantitative differences in ssnRNA patterns among different cell lines from the same organism are required to achieve specificity. The fact that qualtitative and quantitative differences in the patterns of ssRNA between cells from the same organism were observed does not establish the function of ssRNA molecules, but it does not support the concept that all these molecules have a non-specific general function, such as being part of a nuclear backbone, or participating in transport of RNA within the nucleus, or from the nucleus to the cytoplasm.

INTRODUCTION

Among the known RNA molecules in the cell, there is a class which consists of small, relatively stable RNA molecules, most of which are localized in the nucleus, the function of which is unknown (1). Since in many models to explain gene expression in complex organisms small RNA molecules play a major role (2-4) and since in our own way of thinking it is conceivable that ssnRNA could participate in the determination and maintenance of the differentiated state, we decided to investigate this class of molecules.

*Permanent address: Department of Genetics, Attila Jozsef University, Szeged, Hungary.

The observations presented here, while they do not establish the function of these molecules, suggest that not all these molecules are involved in general cellular functions.

RESULTS

Distribution of ssRNA in mouse L-cells. While extensive studies of these molecules were carried out in HeLa cells (5), or in Novikoff rat hepatoma cells (6), mouse L cells have not been studied extensively. Since we wanted to use different mouse strains and mouse cell lines we decided to first study mouse L cells. In order to minimize possible artefacts, and to label mainly the more stable RNA species, monolayers of mouse L-cells were labeled with ^{32}Pi for about 24 hours in MEM medium devoid of phosphate and supplemented with 10% fetal calf serum (FCS), and chased for another day with regular MEM medium. After a number of washes, and while the cells were still attached to the plastic petri dishes, a cracking buffer containing sodium dodecyl sulphate and diethyl pyrocarbonate was added (7) and some of the suspension was applied directly to a slab polyacrylamide tandem gel, consisting of 5% at the top, while the rest of the gel is 10 or 12% (8). Also, similarly labeled cells were separated to nuclei and cytoplasm (9) and portions of each resuspended in the cracking buffer and loaded on the gel. The results of one such experiment are shown in Fig. 1. The ssRNA molecules are numbered 1 to 17. Most of the bands which appear in the nucleus and cytoplasm can be clearly seen in the unfractionated cells. Since the pattern observed in the total cells was obtained under minimum handling of the samples, it is rather unlikely that the patterns observed are due to RNase action after the cells were opened. Moreover, labeled, phenol extracted RNA added to unlabeled cells, did not change their pattern. The large RNA molecules and the DNA (see below) are retained in the 5% gel. (Since the vast majority of the ^{32}P label is retained in this part of the gel, this part of the gel was oftened cut off before autoradiography.) One of the bands 12, 13, clearly contains two molecules, 5S rRNA no. 13, which is cytoplasmic and gains three different conformations (10) during the treatment to separate the cytoplasm from nucleus (see Fig. 1 3rd lane), and a nuclear species no. 12 which appears only in one conformation. Species 17 (Fig. 1 right) is the 5.8S rRNA (11), which can be observed after certain treatments of the ribosomes. Species 5 is also cytoplasmic. Bands 5 and 12 are easily distinguishable and can be used as landmarks to recognize the other species. Species 16 contains the tRNAs and is clearly cytoplasmic. Below 16 a large amount of ^{32}Pi labeled material appears which is not nucleic acids (insensitivity to RNase and DNase, and does not label with thymidine), and is most likely the cellular phospholipids. The bands 1 to 16 are

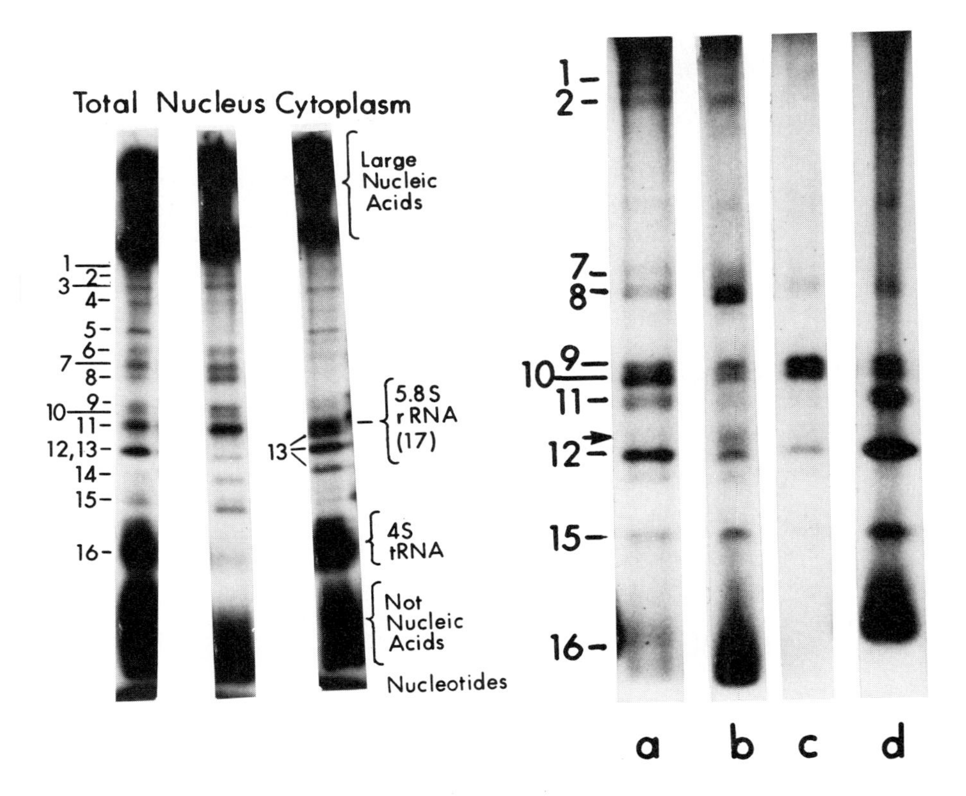

Fig. 1. (Left). Comparison of the distribution of ssRNA molecules in nuclear and cytoplasmic fractions. Exponentially growing mouse L cells were labeled with 15 μCi/ml ^{32}Pi for 21 hours chased for 24 hours and harvested. Nuclei were prepared by the citric acid procedure (9). Equal counts of ^{32}Pi labeled material were applied to each slot of the gel.

Fig. 2. (Right). Comparison of ssRNA distribution in subnuclear fractions prepared by the NaCl procedure (14) and displayed on a 5-12% SDS-polyacrylamide tandem gel. a. nucleus, b. nucleolus, c. chromatin, d. material eluted from the nuclei by a wash with a buffer containing 0.14 M NaCl. Lane d was on the extremity of the gel and therefore the migration of band 16 is somewhat slower than the same band in lane b.

sensitive to RNase and can be labeled with uridine, but not with thymidine, and therefore they are RNA.

The size of these molecules was estimated from their mobility on gels. Such determinations suggest that molecule no. 1 contains 348 bases and no. 15 contains 85 bases. In order to quantitate the no. of the molecules per cell we made use of the fact that on gels consisting of 1.5% polyacrylamide and 0.5% agarose, the DNA is either retained in the origin or migrates as a single band (12). Thus, for the purpose of quantitation, gels were dried, autoradiographed, cut and radioactivity counted. The precentage of 5S rRNA (band 13), from the total nucleic acids, was determined and compared to the DNA. (Independently the no. of 28S and 18S molecules were also determined from a 1.5% gel.) By assuming a molecular weight for mouse DNA and for 5S rRNA,the no. of molecules was readily computed. In order to determine the no. of the other ssRNAs, they were compared to 5S rRNA. We found that for this purpose cutting the gel and counting is less reliable than comparing densitometer tracings. This is due to the fact that the backgrounds are relatively high. Using this technique we estimated that the no. of ssRNA molecules per cell varies from 2.1×10^4 for no. 1 to 8.7×10^5 for no. 11. These figures are in agreement with calculations carried out by other techniques for HeLa cells (13).

Since we are interested to know if the ssRNA participates in gene action we wanted to find out how are they distributed within the nucleus. In order to do so we chose a technique (9) which allows separation of a nucleus to nucleolus, chromatin, HnRNP particles, and post HnRNP (nucleoplasm). Most of the ssRNA species were found associated with HnRNP particles, which did not contain appreciable amounts of DNA. However some of the species were found in the chromatin fraction, especially species 9 and 10. Other species such as 1,4,7,11, 12,14 and 15 were also found in the chromatin. The post HnRNP particles supernatant did not contain any ssRNA.

In order to ascertain the reliability of these findings chromatin was prepared by two other techniques. One of the preparations (14) is shown in Fig. 2 where nuclei, nucleoli, and chromatin are displayed. We can see that the nucleoli are very rich in species 8 while the chromatin is clearly enriched for species 9 and 10. We also can see from Fig. 2d that the different ssnRNA species are bound in the nuclei with different strengths, since material eluted with 0.14N NaCl from the nuclei (10) contains differential amounts of the different species (compare 2a and 2d.)

Comparison of ssRNA in different cell lines. Rein and Penman (15) showed that the ssRNA from different cell lines from the same mammal show similar patterns but that there was a species specificity in the ssRNA patterns. Since these experiments were carried out by labeling RNA with tritium, and cutting and counting gel slices after electrophoresis, we decided to repeat some of these experiments and to quantitate in them the no. of ssnRNA molecules per nucleus. We analyzed four human cell lines, HeLa, Chang liver, FL and KB; three mouse lines, L929, neuroblastoma, and BALB/c 3T3, as well as rat liver line and BHK. Some of the patterns observed are shown in Fig. 3 where it can be clearly seen that the two human lines HeLa and Chang liver have very similar or identical patterns and that these are very different from those observed in mouse L cells. The patterns observed in the mouse and rat lines were very similar but not identical. However the pattern of ssRNA in L cells and BALB/c 3T3 are very different below the 5S rRNA (see Fig. 6). (To the best of our knowledge this is the first report of such a difference within a species.)

Quantitative determinations suggest, that the species in the human lines which migrate in the region corresponding to bands 6-11 in the L cells, exist in different amounts per nucleus in two different cell lines such as HeLa and Chang liver.

Comparison of ssRNA in mouse organs. The observation of clear differences in the pattern of ssRNA molecules, between BALB/c 3T3 cells and mouse L cells, encouraged us to investigate the ssRNA molecules in the mouse itself. In these experiments BALB/c mice (6-8 weeks old) were injected intraperitoneously with $^{32}P_i$ and 48 hours later the animals were sacrificed and various organs dissected, the RNA isolated by cold phenol extraction (17) and the material analyzed on gels. The results of one such experiment are shown in Fig. 4; for comparison, material from labeled L cells is also shown. The patterns of the ssRNA in various organs are very similar, but there are clear quantitative differences. For instance in Fig. 4 on the right hand side an arrow points to a band which is very prominent in the heart. Other less dramatic differences can also be observed. These results are extremely reproducible.

The most startling differences are observed when cell lines and organs are compared. For instance, below the 5S rRNA a very prominant band appears in all the organs which is very insignificant in the cell lines (see Figs. 4,5). At the top of the gel of the organs a band appears (above band no. 5) which is not observed in the cell lines. Also band 11 in the cell lines does not seem to have a counter part in the organs. (For band numbers consult Fig. 1). Finally between band 11 and 12 a very prominent band appears in the organs which is missing in cell lines (right arrow in Fig. 5). All these bands are RNA since the material was phenol extracted four to eight times, and it is sensitive to ribonuclease.

In order to find out if these differences could be due to the fact that cells in tissue culture grow faster than cells in 8 week old mice, hepatectomized mice were labeled at

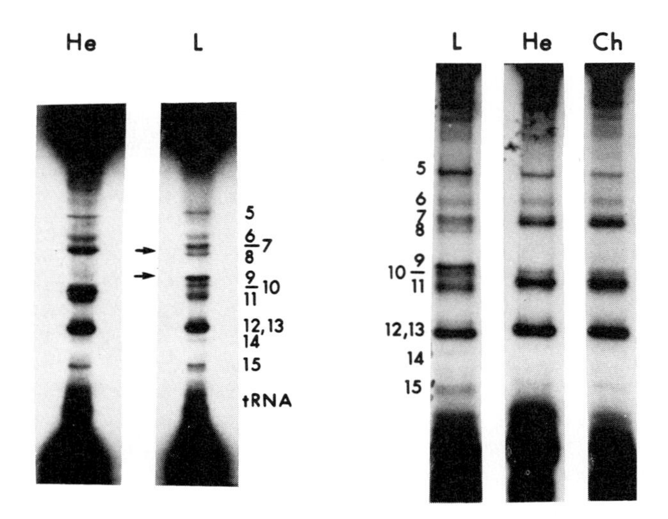

Fig. 3. Comparison of ssRNA distribution in different cell lines. Exponentially growing mouse L, human HeLa, and Chang liver cells were labeled with 15 μCi/ml ^{32}Pi for 18-20 hrs. chased for 20 hrs. and harvested. Equal counts of ^{32}Pi labeled material were applied to each slot of a tandem (5-10%) polyacrylamide gel.

Fig. 4. Comparison of ssRNA molecules in mouse L cells and different mouse organs. Exponentially growing mouse L cells were labeled with 15 μCi/ml ^{32}Pi for 20 hrs. and chased for 20 hrs. and harvested. The L cells were lysed in cracking buffer. RNA was extracted by phenol from organs of male BALB/c mice (6-8 wk.old) labeled with 2 mCi of ^{32}Pi per mouse for 48 hrs. (intraperitoneal injection). Phenol extracted RNA was suspended in cracking buffer. Samples were applied to the slots of a tandem (5-10%) polyacrylamide gel.

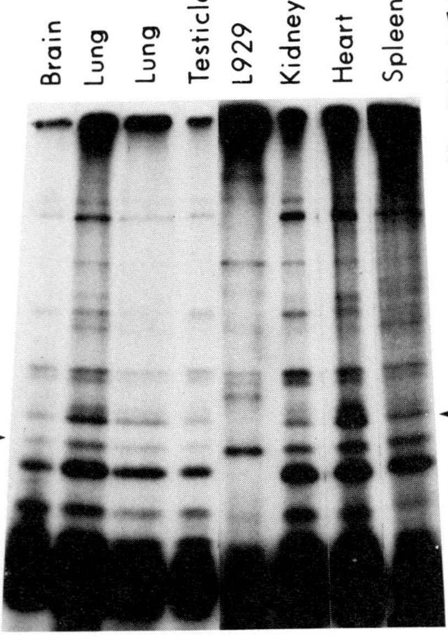

various intervals after surgery, and their small RNAs were displayed in gels. With the exception of one band which appeared inconsistently at the top of the gels, the patterns from hepatectomized livers, and livers from control animals were identical. Animals were labeled for 48 hr at various intervals after the hepatectomy (0 up to 16 days). These experiments also show that the RNA molecules observed in the cell lines but not in the organs could not exist in the organs undetected as could have been the case if they were excessively stable, and therefore would not have picked up sufficient label.

ssRNA in tumors. Since gene expression is altered dramatically in tumors, we decided to analyze ssRNA in tumors. Thus far only three tumors, a myeloma and two hepatomas were analyzed. As can be seen in Fig. 5 the patterns of ssRNA observed in tumors is very interesting indeed, since two of the bands corresponding to bands 10 and 11 are clearly missing or are greatly reduced, (recall that band 10 is clearly in chromatin, see Fig. 2) while the equivalent of band 9 is more intense. Moreover a band which is present in very low amounts in normal mice becomes very intense in animals carrying tumors; this is observed very clearly in the kidney (see Fig. 5 middle arrow on the left). In Fig. 5 myeloma cells are presented; the patterns observed in the hepatomas were very similar.

DISCUSSION

The experiments presented here show rather clearly that there are large consistant variations between small stable RNAs in cells under different growth conditions.

The differences observed, could not be artefactual, since in all cases when mixing experiments were carried out prior to the analysis of unlabeled cells of one type, with labeled cells of another type, the pattern of the labeled cells was retained and was not perturbed by the other type of cells. To the best of our knowledge these are the first experiments to report clear differences, quantitative and qualitative, in these molecules among different cells.

Most of the ssRNA molecules reside in the nucleus, and are distributed in the various nuclear components. Not all of these molecules have to participate in similar functions. The experiments presented here suggest rather strongly, that all these molecules are unlikely to participate in a general cellular function, since thus far molecules, such as rRNA are similar in tissue culture cells and in cells from organs. We would like to think that at least some of these molecules participate in gene expression. In order to do so they have to be in the nucleus, but they do not have to be necessarily associated tightly with chromatin. Since they exist in large quantities, they could interact with certain regions of the

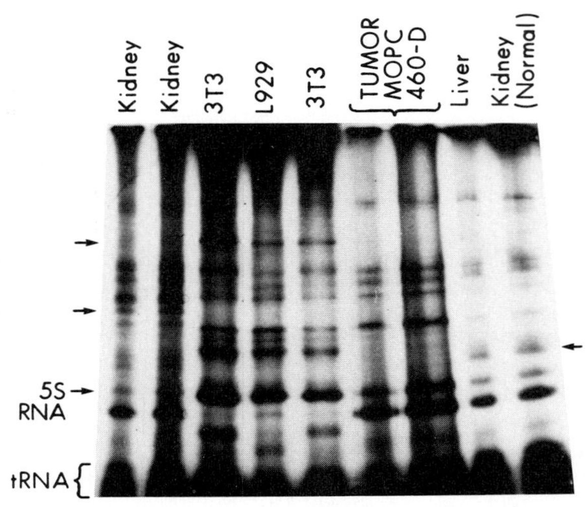

Fig. 5. Comparison of ssRNA in different mouse cell lines, mouse organs, and a tumor. Exponentially growing mouse L929 and BALB/c 3T3 cells were labeled with 15 μCi/ml ^{32}Pi for 20 hrs. and chased for 20 hrs. and harvested. The cells were lysed in cracking buffer. RNA was extracted by phenol from liver and kidney of BALB/c mice, from the tumor, liver and kidney of BALB/c mice carrying MOPC-460-D, a solid plasmacytoma tumor. Phenol extracted samples were resuspended in cracking buffer.

chromatin, even though their association with it is not very firm. Of special interest is ssRNA no. 10 which is tightly bound to chromatin and is greatly reduced or completely missing in tumors.

There are a number of possible models to explain the involvement of ssRNA in gene expression. Moreover, it is also possible that not all of these molecules have the same function in gene expression. At present it is tempting to suggest that molecules 9 and 10 have different functions from those of the others.

One particular model which is currently useful to consider is the following. In this model the ability of a gene to respond to environmental stimuli and to be transcribed depends on the association (via base pairing) of ssRNA with the control DNA of this gene, i.e. with middle repetitive DNA preceding the unique DNA to be expressed. Since there are only about 150 types of cells in a mammal, it is necessary to have 150 different programs for gene expression. Each one could be specified by sequences in the middle repetitive DNA which needs to have at least two signals. The different combinations of these signals could provide for 150 different arrangements. In each cell the association of ssRNA with this middle repetitive DNA allows it to be functional. This association is determined during differentiation and is maintained during the life of the individual.

This type of model does not even require quantitative differences between different cells in order for the ssRNA to participate in the maintenance of the differentiated functions, but only for the ssRNA to be distributed differentially along the middle repetitive DNAs of different cell types. The virtue of this model is that one can test some of its specific predictions.

ACKNOWLEDGEMENTS

Aided by a grant from the National Foundation-March of Dimes 1-326, and an NIH grant CA-15389.

REFERENCES

1. Ro-Choi, T.S. and Busch, H. (1974) in "The Cell Nucleolus" vol. 3, Busch,H. Ed. New York,N.Y., Academic Press p.151.
2. Britten, R.J. and Davidson, E.H. (1969) Science 165, 349.
3. Monahan, J.J. and Hall, R.H. (1974) CRC Crit.Rev.Biochem. 2, 67.
4. Goldstein, L. (1976) Nature 261, 519.
5. Zieve, G. and Penman, S. (1976) Cell, 8, 19.
6. Rai, N.B.K., Ro-Choi, T.S. and Busch, H. (1975) Biochem. 14, 4380.
7. Gegenheimer, P. and Apirion, D. (1975) J.Biol.Chem.250, 2407.
8. Kaplan, R. and Apirion, D. (1975) J. Biol. Chem. 250,1854.
9. Bhorjee, J.S. and Pederson, T. (1973) Biochem. 12, 2766.
10. Weinberg, R.A. and Penman, S. (1969) Biochim. Biophys. Acta 190, 10.
11. Pene, J.J., Knight, E., Jr. and Darnell, J.E. Jr. (1968) J. Mol. Biol. 33, 609.
12. Caras, M., Bailey, S.C. and Apirion, D. (1977) FEBS Lett. in press.
13. Weinberg, R.A. and Penman, S. (1968) J. Mol. Biol. 38,289.
14. Monahan, J.J. and Hall, R.H. (1974) Analyt. Biochem. 62, 217.
15. Rein, A. and Penman, S. (1969) Biochim. Biophys. Acta 190, 1.
16. Girard, M. (1967) Methods in Enzymology 12A, Grossman, L. and Moldave, K., Eds., Academic Press, p. 581.

Author Index

(Article numbers are shown following the names of contributors. Affiliations are listed on the title page of each article)

A

Abelson, J., 23
Apirion, D., 45

B

Barea, J. L., 28
Barnitz, J. T., 24
Beckmann, J. S., 23
Berek, I., 45
Berg, P., 15
Bigelis, R., 19
Botstein, D., 25
Boyer, H., 4

C

Cappecchi, M. R., 40
Capecchi, N. E., 40
Carr, S., 36
Case, M. E., 28
Chovnick, A., 35
Cockburn, A. F., 33
Cramer, J. H., 24

D

Dahlberg, J., 9
Davis, R. W., 8
Donoghue, D. J., 6

E

Epstein, H. F., 39

F

Fareed, G. C., 13, 14, 16
Farrelly, F. W., 24
Fiddes, J. C., 9
Fink, G. R., 19
Firtel, R., 30, 32, 33
Frankel, G. A., 33
Fuhrman, S. A., 23

G

Garcea, R. L., 39
Geiduschek, E. P., 20
Gelbart, W., 35
Gilbert, W., 2
Giles, N. H., 28
Gill, B. S., 45
Goff, S. P., 15
Gorenstein, C., 22
Greenfield, L., 11

H

Harris, H. E., 39
Hautala, J. A., 28
Hearst, J. E., 17
Hecht, R., 36
Hereford, L. M., 25
Herskowitz, I., 21
Hicks, J. B., 21
Hilmen, M., 3
Hirsh, D., 36
Hogness, D. S., 34

I

Irr, J. D., 44

J

Jacobson, A., 31
Jacobson, J. W., 28
Jacoby, L. B., 44
Johnson, C. D., 38
Johnson, P. F., 23
Jordan, J. M., 18

K

Kaplan, D. A., 11, 12
Keesey, J., 19
Kindle, K. L., 32

L

LaPointe, J. W., 39
Lee, C. S., 8
Loomis, W., 30
Lucas, M. C., 28

M

Mabie, C. T., 31
Maizels, N., 26
Maxam, A. M., 2
McCarron, M., 35
Metzenberg, R. L., 27

N

Nelson, R. E., 27
Newkirk, M. J., 33

O

Ohsumi, M., 7

P

Palatnik, C. M., 31
Patel, V. B., 28
Petes, T. D., 25
Philippsen, P., 8
Podder, U. S., 45
Puck, T. T., 41

R

Rand, J. B., 38
Ray, D. S., 5
Reinert, W. R., 28
Rine, J., 21
Rownd, R. H., 24
Ruddle, F. H., 42
Russell, R. L., 38

S

Schachat, F. H., 39
Scheffler, I., 43
Scherer, S., 38
Shapiro, L., 10
Sharp, P. A., 6
Shen, C.-K. J., 17
Silverman, M., 3
Simon, M., 3
Skalka, A. M., 10
Skolnik, H., 16
Skryabin, K. G., 2
Strathern, J. N., 21
Stroman, P., 28
Sveda, M. M., 40

T

Taylor, W. C., 33
Tizard, R., 2

U

Upcroft, J. A., 16
Upcroft, P., 16

V

Vanderslice, R., 36
Vonder Haar, R. A., 40
Vovis, G. F., 7

W

Warner, J. R., 22
Weiss, R., 29
Wilcox, G., 11, 12
Wilkins, C., 31
Wood, W. B., 36, 37

Y

Young, M., 33

Z

Zieg, J., 3
Zinder, N. D., 7
Zwass, M. S., 38

Subject Index

(Citations are to article number. The article numbers appear in the articles on top of the left hand pages in front of the names of the contributors.)

A

Acetylcholine metabolism
  and behavioral mutants, 38
  in *Caenorhabditis elegans,* 38
  enzymes of, 38
  mutants in, 38
Actin genes
  Dictyostelium discoideum, 32
  in recombinant plasmids, 32
Antigen detection, 10
  in situ immunodiffusion assay, 10
Arginine
  in human lymphoblasts, 44
  in regulatory mutants, 44
  repression of argininosuccinate synthetase by, 44
Argininosuccinate synthetase
  mutants derepressed for synthesis of, 44
  repression by arginine, 44
Autoradiography
  double-label, 27
  of *Neurospora crassa,* 27

B

Behavioral mutants
  and acetylcholine metabolism, 38
  in *Caenorhabditis elegans,* 38
  in nematodes, 38

C

Caenorhabditis elegans, 36, 38, 39
  behavioral mutants of, 38
  enzymes of, 38
  maternal influence, 36
  muscle defective mutants, 39
  mutants in acetylcholine metabolism, 38
  myosin, 39
  temperature sensitive mutants, 36
cAMP and reverse transformation, 41
Carrier state, 18
  SV40 DNA infection, 18
Cassette model of interconversion, 21
Cell lines, 45
  small stable RNA, 45
Cell fractions, 45
  small stable RNA, 45
Cell free protein synthesis, 22, 28
  Neurospora crassa, 28
  of ribosomal proteins, 22
  qa gene cluster, 28
  quinic acid metabolism, 28
Chimera, 7
  bacteriophage fl, 7
  plasmids, 7
  pSC101, 7
CIS acting control element in *Drosophila,*
  35
  Rosy Locus, 35

Cloning, 3, 8, 11, 23
double digestion, 11
E. coli ara B gene, 8
flagellar antigen genes, 3
H1 and H2 genes, 3
λh80dara, 11
noncomplementary ends, 11
ribosomal DNA, 23, 24
Salmonella typhimurium, 3
satellite DNA, 8
tRNA genes, 23
use of λgt-ara B, 8
yeast, 23
Cloning efficiency noncomplementary ends, 11
Colcemid and microtubules, 41
Cytochalasin B and microfilaments, 41

D

Denaturation mapping of ribosomal DNA, 24
Dictyostelium discoideum 26, 31–33
actin genes, 32
classes of poly(A) containing RNA in, 31
developmental changes in, 31
gene sequences, 32
mRNA stability in, 31
poly(A) metabolism in, 31
restriction enzyme digestion, 33
ribosomal DNA, 26, 33
shortening of poly(A) with age, 31
Distamycin A
ecoRI, 26
restriction enzyme digestion, 26
ribosomal DNA, 26
DNA restriction fragments, 12
electrophoresis of, 12
of satellite DNA, 8
DNA sequencing
of 5s rDNA, 2
of φx 174 DNA, 1
overlapping genes, 1
plus and minus technique, 1
of promoters, 1, 2
purine/pyrimidine bias in, 2
of ribosome binding sites, 1
transcription unit, 2
Double-digestion
cloning, 11
cloning efficiency, 11
noncomplementary ends, 11
restriction enzymes, 11
Drosophila nasutoides satellite II, 8
Drosophila melanogaster, 34, 35
cis acting control element in, 35
gene mapping of, 34
mRNA, 34
recombinant DNA, 34
repeated genes, 34
restriction enzyme digestion, 34
Rosy locus, 35
Rosy locus structural element, 35
structural genes, 34
Xanthine dehydrogenase, 35

E

EcoRI
distamycin A, 26
restriction enzyme digestion with, 26
EK2 vector, phage, 6
Electrophoresis, 12, 17, 45
agarose gel electrophoresis, 17
of DNA restriction fragments, 12
small stable RNA, 45
Enzyme replacement liposomes, 27
Episomal states, SV40 Su+ III recombinants, 16
Escherichia coli, 7
bacteriophage f1, 7
phage vector, 7
plasmid, 7
pSC101, 7

F

5s rDNA
DNA sequence of, 2
promoter structure of, 2
restriction mapping of, 2
transcription unit, 2
in yeast, 2
4,5,8 trimethylpsoralen
photochemical crosslinkage by, 17
trioxsalen, 17
Flagellar antigens
cloning of genes, 3
expression of genes H1 and H2, 3
Salmonella typhimurium, 3

G

Gene mapping, 25, 34, 42
Drosophila, 34
repeated genes, 34
restriction enzyme digestion, 34

ribosomal DNA, 25
somatic cell hybrids, 42
structural genes, 34
yeast, 25
Gene sequences, *Dictyostelium discoideum,* 32
Gene transfer, 42
microcells, 42
model systems, 6
phagocytosis, 42
somatic cell hybrids, 42

H

Hairpins
DNA sequences with twofold symmetry, 17
electron microscopic mapping of, 17
map of SV40 DNA, 17
SV40, 17
His gene cluster
multifunctional proteins in, 19
and proteases, 19
in yeast, 19
H1 gene, 3
cloning, 3
flagellar antigen expression, 3
Salmonella typhimurium, 3
H2 gene, 3
cloning, 3
flagellar antigen expression, 3
Salmonella typhimurium, 3
Human lymphoblasts
arginine in, 44
control of argininosuccinate synthetase in, 44
regulatory mutants, 44

I

Initiation of transcription SV40, 17
In situ immunodiffusion assay, 10
antigen detection, 10
lambda lysogen, 10
Intercistronic sequences in ϕX174 DNA, 1
Immunofluorescence
of microfilaments, 41
of microtubules, 41

K

Kanamycin resistance in phage hybrids, 6

L

λh80dara cloning, 11
λgt-ara B as a vector, 8
Liposomes enzyme replacement, 27
Lysogen induction
antigen detection, 10
in situ immunodiffusion assay, 10
lambda lysogen, 10

M

Mammalian cell genetics
nonsense mutants, 40
supressor tRNA, 40
Maternal influence, 36
Caenorhabditis elegans, 36
nematodes, 36
temperature sensitive mutants, 36
Mating type interconversion, 21
Microcells, 42
gene transfer, 42
Microinjection into mammalian cells of supressor tRNAs, 40
Microfilaments
and colcemid, 41
and cytochalasin B, 41
immunofluorescence of, 41
in transformed cells, 41
Mini-Col El
colicin E1 factor, 6
hybrids with phage, 6
Monkey cells, 18
SV40 infection, 18
m RNA, 31, 34
Drosophila, 34
repeated genes, 34
structural genes, 34
m RNA stability, poly(A) metabolism, 31
Myosin, 39
Caenorhabditis elegans, 39
muscle-defective mutants, 39
nematodes, 39
subunit composition, 39
Multifunctional proteins
His gene cluster, 19
and proteases, 19
in yeast, 19

N

Nematodes, 36, 38, 39
behavioral mutants of, 38

Nematodes (*cont.*):
Caenorhabditis elegans, 36, 38, 39
enzymes of, 38
maternal influence, 36
muscle defective mutants, 39
mutants in acetylcholine metabolism, 38
myosin, 39
temperature sensitive mutants, 36
Neurospora crassa, 27, 28
autoradiography of, 27
cell free protein synthesis, 28
phosphatase in, 27
regulation of phosphorous metabolism in, 28
qa gene cluster, 28
quinic acid metabolism, 28
Noncomplementary ends
cloning, 11
cloning efficiency, 11
double digestion, 11
restriction enzymes, 11
Nonsense mutants
in mammalian cells, 40
mammalian cell genetics, 40
somatic cell genetics, 40

O

Overlapping genes
DNA sequence of, 1
in ϕX174 DNA, 1

P

Phage, 6
deletion mutants, 6
EK 2 vector, 6
hybrids with Kanamycin resistance, 6
hybrids with mini-Co1E1, 6
recombinants with Colicin E1 factor, 6
as a vector, 6
Phage, SV40, hybrids, 14
Phage lambda lysogen, 10
antigen detection, 10
in situ immunodiffusion assay, 10
lysogen induction, 10
Phage vector, 7
E. coli, 7
bacteriophage f1, 7
plasmid, 7
recombinant DNA, 7
pSC101, 7
ϕX174 DNA
intercistronic sequences in, 1
overlapping genes in, 1
plus and minus DNA sequencing technique, 1
promoters in, 1
ribosome binding sites in, 1
sequence of, 1
Phagocytosis, 42
gene transfer, 42
Phase variation
cloning of H1 and H2 genes, 3
expression of flagellar antigens, 3
Salmonella typhimurium, 3
Phosphatase
in *Neurospora crassa,* 27
regulation of, 27
repression of, 27
Phosphorous metabolism
in *Neurospora crassa,* 27
regulation of, 27
repression of, 27
Photochemical crosslinkage
4,5′,8 trimethylpsoralen, 17
stabilization of double helical regions by, 17
Plasmids 6, 23, 24, 32
bacteriophage f1, 7
containing yeast ribosomal DNA, 24
mini-ColE1, 6
pSC101, 7
recombinant DNA, 7
recombinants containing actin sequences, 32
ribosomal DNA, 24
tRNA genes, 23
Poly(A)-containing RNA
classes of, 31
in *Dictyostelium discoideum,* 31
shortening of poly(A) with age, 31
Poly(A) metabolism
classes of poly(A) containing RNA, 31
developmental changes, 31
in *Dictyostelium discoideum,* 31
mRNA stability, 31
shortening of poly(A) with age, 31
Poly dT-poly dA
ribosomal DNA, 2
yeast, 2
Polynucleotide kinase, ribosomal DNA, 26
Plus and minus DNA sequencing technique, ϕX174 sequence, 1
Processing of RNA precursors, SV40, 14
Promoters
DNA sequence of, 1, 2

5s rDNA, 2
in ϕX174 DNA, 1
RNA polymerase III, 2
structure of, 2
yeast, 2
pSC101, bacteriophage f1, 7
Psoralen, 17
Proteases
and His gene cluster, 19
and multifunctional proteins, 19
in yeast, 19
Purine/pyrimidine bias, DNA sequencing, 2

Q

qa gene cluster, 28
cell free protein synthesis, 28
Neurospora crassa, 28
quinic acid metabolism, 28
Quinic acid metabolism, 28
Neurospora crassa, 28
qa gene cluster, 28

R

Recombinant DNA, 3, 6, 7, 8, 15, 16, 23, 24, 34
bacteriophage f1, 7
cloning of satellite DNA, 8
Drosophila, 34
flagellar antigen genes, 3
H1 and H2 genes, 3
λgt-ara B as vector, 8
model systems, 6
mRNA, 34
phage-mini ColE1, 6
phage vector, 7
plasmid, 7
pSC101, 7
repeated genes, 34
restriction enzyme digestion, 15
ribosomal DNA, 23, 24
Salmonella typhimurium, 3
screening, 23
structural genes, 34
SV40, 15
SV40-su^{+} *III* recombinants, 18
transducing virus, 15
tRNA genes, 23
yeast, 23
Regulatory mutants, 44
arginine, 44
human lymphoblasts, 44
Repeated genes, 33, 34
Dictyostelium discoideum, 33
Drosophila, 34
recombinant DNA, 34
ribosomal DNA, 33
structural genes, 34
Replication origin in SV40, 17
Restriction enzyme digestion, 3, 11, 15, 23, 26, 33, 34
Dictyostelium, 33
distamycin A, 26
double digestion, 11
Drosophila, 34
flagellar antigen genes, 3
H1 and H2 genes, 3
ribosomal DNA, 26, 33
Salmonella typhimurium, 3
SV40, 15
transducing virus insertion, 15
Restriction enzymes, 8, 11, 12, 26
cleavage of satellite DNA, 8
double digestion, 11
eco RI, 26
noncomplementary ends, 11
Restriction mapping, 2, 24
in yeast, 2
of 5s rDNA, 2
ribosomal DNA, 2, 24
Reverse transformation, 41
role of cAMP in, 41
of transformed cells, 41
Ribosomal DNA, 2, 23–26, 33
cloning of, 23, 24
denaturation mapping of, 24
Dictyostelium, 26, 33
distamycin A, 26
5s, 2
in plasmids, 24
organization of, 25
poly dT-poly dA in, 2
polynucleotide kinase, 26
recombinant DNA, 23, 24
restriction mapping of, 2, 24, 26, 33
sequence of, 2
yeast, 2, 23–25
Ribosomal proteins, 22
cell-free synthesis of, 22
in yeast, 22
Ribosomal RNA, 2, 22, 24
cloning of genes for, 24
in yeast, 2, 22, 24
Ribosome binding sites, 1
DNA sequence of, 1
in ϕX174 DNA, 1

RNA polymerase III, 2
promoter, 3
yeast, 2
Rosy locus control element, 35
cis acting, 35
in *Drosophila,* 35
Xanthine dehydrogenase, 35
Rosy locus structural element, 35
in *Drosophila,* 35
Xanthine dehydrogenase, 35

S

Saccharomyces cerevisiae, 22, 25
ribosomal proteins in, 22
Salmonella typhimurium, 3
cloning, 3
flagellar antigens, 3
genes H1 and H2, 3
phase variation, 3
recombinant DNA, 3
restriction enzyme digestion, 3
Satellite DNA, 8
cleavage by restriction enzymes, 8
cloning of, 8
D. nasutoides satellite II, 8
Shortening of poly(A) with age, 31
Dictyostelium discoideum, 31
poly(A) containing RNA, 31
poly(A) metabolism, 31
Small stable RNA, 45
in cell fractions, 45
in cell lines, 45
electrophoresis of, 45
nuclear, 45
in organs, 45
in tumors, 45
Somatic cell genetics, 40, 42
gene mapping, 42
gene transfer, 42
mammalian cell hybrids, 42
microcells, 42
nonsense mutants, 40
phagocytosis, 42
suppressor tRNA, 40
Structural genes, 34
Drosophila, 34
mRNA, 34
recombinant DNA, 34
repeated genes, 34
Supercoiled DNA, 17
Suppressor tRNA, 14, 16, 40
mammalian cell genetics, 40
microinjection into mammalian cells, 40
somatic cell genetics, 40
SV40, 14
SV40-su^+ *III* recombinants, 16
SV40, 14–18
carrier state, 18
hairpins, 17
hybrids, 14
initiation of transcription, 17
processing of RNA precursors, 17
regulation of translation, 17
replication origin, 17
suppressor tRNA, 14
su^+ *III* recombinants, 16
termination of transcription, 17
twofold symmetry, 17
SV40 DNA, hairpin map of, 17
SV40-su^+ *III* recombinants, 16
episomal states, 16
recombinant DNA, 16
suppressor tRNA, 16
SV40 vector, suppressor tRNA, 14

T

Temperature sensitive mutants, 36
Caenorhabditis elegans, 36
maternal influence, 36
nematodes, 36
Termination of transcription in SV40, 17
Transcription unit, 2
DNA sequence of, 2
5s rDNA, 2
yeast, 2
Transducing virus, 15
Transformed cells, 41
microfilaments in, 41
microtubules in, 41
reverse transformation of, 41
Translational regulation in SV40, 17
Trioxsalen, 4,5,8-trimethylpsoralen, 17
tRNA genes, 23
cloning, 23
recombinant DNA, 23
yeast, 23
Tumors, small stable RNA in, 45
Twofold symmetry, 17
hairpin sequences in DNA, 17
SV40, 17

X

Xanthine dehydrogenase, 35
in *Drosophila,* 35

Rosy locus, 35
Rosy locus structural element, 35

Y

Yeast, 2, 21–25
cassette model, 21
cloning, 23
5s rDNA in, 2
His gene cluster in, 19
location of ribosomal genes, 25
mating types, 21
multifunctional proteins in, 19
poly dT-poly dA, 2
promoter structure in, 2
proteases in, 19
recombinant DNA, 23
restriction mapping of, 2
ribosomal DNA, 2, 23–25
ribosomal proteins in, 22
ribosomal RNA, 2, 22, 24
RNA polymerase III, 2
Saccharomyces cervisiae, 22, 25
transcription unit, 2
tRNA genes, 23

A
B 7
C 8
D 9
E 0
F 1
G 2
H 3
I 4
J 5

Date Due
MAY 04 1987